Cambridge Studies in Biological and Evolutionary Anthropology 25

Human growth in the past: studies from bones and teeth

Until now, studies of dental and skeletal growth and development have often been treated as independent disciplines within the literature. *Human growth in the past* takes a fresh perspective by bringing together these two related fields of enquiry in a single volume whose purpose is to place methodological issues of growth and development in past populations within a strong theoretical framework. Contributions examine a variety of aspects of human growth in the past, drawing from both palaeoanthropological and bioarchaeological data. The book covers a wide spectrum of topics, from patterns of growth in humans and their close relatives, innovative methods and applications of techniques and models for the study of growth, to estimation of age at death in subadults and infant mortality in archaeological samples. *Human growth in the past* will be of interest to biological anthropologists and those in the related fields of dental anatomy, evolutionary biology and developmental biology.

ROBERT D. HOPPA is an Assistant Professor in the Department of Anthropology at the University of Manitoba, Canada. His research interests include skeletal biology, growth and development and demography of past populations.

CHARLES M. FITZGERALD is a Postdoctoral Research Fellow and Adjunct Assistant Professor in the Department of Anthropology at McMaster University, Canada. In addition to growth and development, his research interests include skeletal biology and palaeoanthropology.

Cambridge Studies in Biological and Evolutionary Anthropology

Selected titles also in the series

11 *Genetic Variation and Human Disease* Kenneth M. Weiss 0 521 33660 0 (paperback)
12 *Primate Behaviour* Duane Quiatt & Vernon Reynolds 0 521 49832 5 (paperback)
13 *Research Strategies in Biological Anthropology* Gabriel W. Lasker & C. G. N. Mascie-Taylor (eds.) 0 521 43188 3
14 *Anthropometry* Stanley J. Ulijaszek & C. G. N. Mascie-Taylor (eds.) 0 521 41798 8
15 *Human Variability and Plasticity* C. G. N. Mascie-Taylor & Barry Bogin (eds.) 0 521 45399 2
16 *Human Energetics in Biological Anthropology* Stanley J. Ulijaszek 0 521 43295 2
17 *Health Consequences of 'Modernisation'* Roy J. Shephard & Anders Rode 0 521 47401 9
18 *The Evolution of Modern Human Diversity* Marta M. Lahr 0 521 47393 4
19 *Variability in Human Fertility* Lyliane Rosetta & C. G. N. Mascie-Taylor (eds.) 0 521 49569 5
20 *Anthropology of Modern Human Teeth* G. Richard Scott & Christy G. Turner II 0 521 45508 1
21 *Bioarchaeology* Clark S. Larsen 0 521 49641 (hardback), 0 521 65834 9 (paperback)
22 *Comparative Primate Socioecology* P. C. Lee (ed.) 0 521 59336 0
23 *Patterns of Human Growth*, second edition Barry Bogin 0 521 56438 7 (paperback)
24 *Migration and Colonisation in Human Microevolution* Alan Fix 0 521 59206 2

Human growth in the past

studies from bones and teeth

EDITED BY

ROBERT D. HOPPA
Department of Anthropology
University of Manitoba
Winnipeg, Manitoba, Canada

CHARLES M. FITZGERALD
Department of Anthropology
McMaster University
Hamilton, Ontario, Canada

CAMBRIDGE UNIVERSITY PRESS
Cambridge, New York, Melbourne, Madrid, Cape Town, Singapore, São Paulo

Cambridge University Press
The Edinburgh Building, Cambridge CB2 2RU, UK

Published in the United States of America by Cambridge University Press, New York

www.cambridge.org
Information on this title: www.cambridge.org/9780521631532

First published 1999
This digitally printed first paperback version 2005

A catalogue record for this publication is available from the British Library

Library of Congress Cataloguing in Publication data

Human growth in the past: studies from bones and teeth / edited by
Robert D. Hoppa, Charles M. FitzGerald.
p. cm. – (Cambridge studies in biological and evolutionary
anthropology)
ISBN 0 521 63153 X
1. Human growth. 2. Teeth — Growth. 3. Human skeleton –
Growth. 4. Dental anthropology. 5. Physical anthropology.
I. Hoppa, Robert D., 1967– . II. FitzGerald, Charles M., 1947– .
III. Series.
GN62.9.H85 1999
599.9 – dc21 99–11536 CIP

ISBN-13 978-0-521-63153-2 hardback
ISBN-10 0-521-63153-X hardback

ISBN-13 978-0-521-02122-7 paperback
ISBN-10 0-521-02122-7 paperback

Both editors are friends who have shared much of their experience in higher education. We would like to mutually dedicate this book to two groups of people special to us, many of whom we hold in common: to our teachers and mentors, who through their encouragement or example have had the greatest impact on our developing careers in this field; and to our families who have given us their support and loyalty throughout:

Shelley Saunders, Chris Dean, Ann Herring, Robert Foley, Jerry Melbye, Jerry Rose, Larry Sawchuk and Charlotte Roberts

and

Trich, Katie and Emily

Anita, Gerald and Ron, who are sorely missed, and Dorian, Eugénie, Mary, and Susan

Contents

	List of contributors	xi
	Preface	xiii
	Acknowledgements	xviii
1	From head to toe: integrating studies from bones and teeth in biological anthropology ROBERT D. HOPPA AND CHARLES M. FITZGERALD	1
2	Heterochrony: somatic, skeletal and dental development in *Gorilla*, *Homo* and *Pan* MIKE DAINTON AND GABRIELE A. MACHO	32
3	Relative mandibular growth in humans, gorillas and chimpanzees LOUISE T. HUMPHREY	65
4	Growth and development in Neandertals and other fossil hominids: implications for the evolution of hominid ontogeny ANDREW J. NELSON AND JENNIFER L. THOMPSON	88
5	Hominoid tooth growth: using incremental lines in dentine as markers of growth in modern human and fossil primate teeth CHRISTOPHER DEAN	111
6	New approaches to the quantitative analysis of craniofacial growth and variation PAUL O'HIGGINS AND UNA STRAND VIDARSDOTTIR	128
7	Invisible insults during growth and development: contemporary theories and past populations SARAH E. KING AND STANLEY J. ULIJASZEK	161
8	What can be done about the infant category in skeletal samples? SHELLEY R. SAUNDERS AND LISA BARRANS	183
9	Sources of variation in estimated ages at formation of linear enamel hypoplasias ALAN H. GOODMAN AND RHAN-JU SONG	210
10	Reconstructing patterns of growth disruption from enamel microstructure SCOTT W. SIMPSON	241

11 Estimation of age at death from dental emergence and implications for studies of prehistoric somatic growth 264
LYLE KONIGSBERG AND DARRYL HOLMAN

12 Linear and appositional long bone growth in earlier human populations: a case study from Mediaeval England 290
SIMON A. MAYS

Index 313

Contributors

Lisa Barrans
Department of Anthropology, McMaster University, 1280 Main Street West, Hamilton, Ontario, Canada L8S 4L9

*Mike Dainton
Hominid Palaeontology Research Group, Department of Human Anatomy and Cell Biology, University of Liverpool, Liverpool L69 3GE, UK

Christopher Dean
Evolutionary Anatomy Unit, Department of Anatomy and Developmental Biology, University College London, Gower Street, London WC1E 6BT, UK

Charles M. FitzGerald
Department of Anthropology, McMaster University, 1280 Main Street West, Hamilton, Ontario, Canada L8S 4L9

*Alan H. Goodman
Natural Science, Hampshire College, Amherst, MA 01002, USA

Darryl Holman
Pennsylvania State University, University Park, 0419 Carpenter Building, University Park, PA 16802, USA

*Robert D. Hoppa
Department of Anthropology, University of Manitoba, Winnipeg, Manitoba, Canada R3T 2N2

*Louise T. Humphrey
Human Origins Group, Department of Palaeontology, Natural History Museum, Cromwell Road, London SW7 5BD, UK

*Sarah E. King
Department of Biological Anthropology, University of Cambridge, Downing Street, Cambridge CB2 3D2, UK

*Lyle Konigsberg
Department of Anthropology, University of Tennessee, Knoxville, TN 37996, USA

*Gabriele A. Macho

Hominid Palaeontology Research Group, Department of Human Anatomy and Cell Biology, University of Liverpool, Liverpool L69 3GE, UK

*Simon A. Mays
Ancient Monuments Laboratory, English Heritage, 23 Savile Row, London W1X 1AB, UK

*Andrew J. Nelson
Department of Anthropology, University of Western Ontario, Ontario, Canada N6A 5C2

*Paul O'Higgins
Evolutionary Anatomy Unit, Department of Anatomy and Developmental Biology, University College London, Rockefeller Building, University Street, London WC1 6JJ, UK

*Shelley R. Saunders
Department of Anthropology, McMaster University, 1280 Main Street West, Hamilton, Ontario, Canada L8S 4L9

*Scott W. Simpson
Department of Anatomy, School of Medicine, Case Western Reserve University, 10900 Euclid Avenue, Cleveland, OH 44106, USA

Rhan-Ju Song
Department of Anthropology, University of Massachusetts, Amherst, MA 01003, USA

Jennifer L. Thompson
Department of Anthropology, University of Nevada, Las Vegas, 4505 Maryland Parkway, Box 45501, Las Vegas, NV 89154–5002, USA

Stanley J. Ulijaszek
Institute of Biological Anthropology, University of Oxford, 58 Banbury Road, Oxford OX2 6QS, UK

Una Strand Vidarsdottir
Evolutionary Anatomy Unit, Department of Anatomy and Developmental Biology, University College London, Rockefeller Building, University Street, London WC1 6JJ, UK

*Asterisks denote corresponding authors.

Preface

In the past, studies of dental and linear growth and development have too often been treated as independent entities within the literature. Frequently such publications have also been prone to a manual or catalogue approach. This book takes a fresh perspective by bringing together these two related fields of enquiry in a single coherent volume that places methodological issues of growth and development within a strong theoretical framework. Individual contributions illustrate the potential benefits of growth and development studies for understanding and interpreting past populations, as well as some of their problems and pitfalls. The contributors are researchers in biological anthropology, and it was an explicit goal of the editors to ensure that there was a mix of well-recognised authors and new scholars in the field. A total of 20 scholars from Canada, the USA and the United Kingdom have contributed to the volume.

We hope that this monograph will serve as an exemplar, demonstrating the advantages of bringing together researchers in the field of growth and development whose focus is on humans who lived in the past. Specifically, this book is intended to integrate areas of study that for a number of reasons (explored in Chapter 1) have tended to remain separate. One obvious way to have structured a volume of this nature would have been to group all of the dental papers together and all of the skeletal papers together. However, such an approach would of course have flown in the face of the purpose of this book. While we fully acknowledge the importance of individual studies in each of these fields, we think that there are significant benefits to be had by integrating studies from bones *and* teeth. Indeed, in no other way can a complete picture of the evolutionary processes of human growth and development be painted. We have therefore opted to organise the volume in a fashion that follows a more chronological approach, although in so doing we have also tried to avoid creating another obvious (and related) schism: dividing studies into palaeoanthropological and bioarchaeological categories. The theme of this book is the gathering together of all researchers interested in growth and development of past populations under one umbrella. In fact, in Chapter 1 the editors propose the use of the term 'palaeoauxology' for such studies. In this first chapter we explain the book's motif and establish its overall tone, with an overview of the state of studies of growth from bones and teeth. A brief summary of the development of the two fields within biological anthropology provides a basis on which to build an argument for the natural integration of dental and skeletal studies.

Chapters 2 to 6 broadly examine various aspects of the evolution of human ontogeny. Dainton and Macho (Chapter 2) explore methodological issues related to heterochrony in humans, chimpanzees and gorillas. They examine patterns of growth in body mass and carpal and dental development, observing that heterochronic processes can be determined from analysis of patterns alone. They also caution researchers about substituting other variables as estimates of body mass and the inclusion of adults in studies of ontogenetic allometry, since both of these practices can lead to serious biases and misinterpretations of the data.

Humphrey examines mandibular growth, tooth emergence and chronological age in gorillas, chimpanzees and modern humans in Chapter 3. In her analysis, all three species show variation in the pattern of growth of the different parts of the mandible, and while the sequence of growth is similar among these species, clear differences between humans and the apes are noted. She concludes from this study that it is not possible to adequately predict skeletal growth attainment on the basis of chronological age or dental emergence, and cautions other researchers about making inferences regarding the rate of development of extinct hominids based solely on dental evidence.

Nelson and Thompson, in Chapter 4, investigate Neandertal growth and development. They examine two juvenile males (Teshik-Tash 1 and Le Moustier 1) in detail to try to assess whether Neandertals experienced an extended period of slow childhood growth and a rapid adolescent growth spurt. They focus on estimates of chronological age from dental development, and estimate stature as a basis for comparison with other cold-adapted populations. These authors observe that, in contrast to studies of *Homo erectus* juveniles, the Neandertal juveniles have a dental age that exceeds their proportional height age and therefore do not demonstrate a primitive ape-like pattern of growth and development. However, because the Neandertal age estimates differ, on the basis of developmental standards, it is argued that they also contrast with modern humans. Two alternative hypotheses are explored to explain the observed results: delayed skeletal growth or advanced dental development.

Dean, in Chapter 5, reviews the use of dental microstructures in dentine as a method of assessing development of both extinct and extant hominoid species. Most workers in the field are familiar with Dean's pioneering work on dental enamel incremental microstructures, which are now accepted as incremental markers of growth. These structures provide endogenous developmental information and many studies in the last 18 years or so in anthropology, a significant number of which have involved Dean, have utilised them to explore various aspects of hominid (and hominoid) growth. Once again Dean is a trailblazer, this time initiating the use of microstructural growth markers in dentine, the bulk of the hard tissue of teeth and therefore a potentially even more useful tool for exploring development than enamel. However, dentine presents a greater technical challenge to researchers, and in this chapter Dean provides an excellent primer on its use. He presents a solid basic introduction to dentine's structure, function and the nature of the regular chronological periodicity of its microstructures, and he also offers practical techniques for histological analysis of dentine and illustrates the advantages of utilising it with examples from recent studies.

O'Higgins and Strand Vidarsdottir present some recent approaches to assessing morphological variation in Chapter 6. They illustrate the use of geometric morphological techniques through a case study of comparative craniofacial growth in Aleutian and Alaskan population samples. The results of their case study suggest that the influences on postnatal facial development are identical between these closely related populations, although a fundamental difference in facial morphology is established during infancy. Of particular interest is the observation of differences in facial morphology between the two study populations that are independent of age. Their review of the methods available to researchers demonstrate the potential for understanding the ontogenetic development of morphological variation and its relationship to evolutionary adaptation and divergence.

In Chapter 7, King and Ulijaszek examine the basis for growth reduction as a measure of reduced health and well-being from archaeological studies. They argue that, despite some problems inherent to archaeological material, comparisons of skeletal growth profiles between different archaeological groups and/or modern populations can provide information on the timing and extent of growth faltering. This chapter provides an overview of bioarchaeological studies of skeletal growth, and discusses previous interpretations of growth faltering in the light of current literature on environmental factors that can influence growth. In particular, they note that many previous interpretations of growth faltering have been influenced by the types of archaeological evidence available and the prevailing interpretations of palaeopathological indicators of stress. From their review of the growth literature, they suggest that interpretations of archaeological studies may need some reconsideration.

Saunders and Barrans (Chapter 8) review the potentials and the problems of examining the infant cohort in skeletal samples. They begin by defining the infant category demographically (i.e. those who have died prior to their first birthday) and call on all researchers in the field to universally apply this definition. Placing infancy in the context of studies of growth and development, they address issues of variation in infant mortality, determination of sex from infant skeletal remains, identification of feeding practices, causes of death, and seasonality in infant mortality. The authors argue that research into the skeletal biology of infant morbidity, mortality and growth offers great hope for the future, and that the most promising and exciting areas of research will be studies of 'cause of death' and dietary reconstruction.

Chapters 9 and 10 both examine issues of enamel hypoplasia and interpreting episodic periods of stress during the growing years. Traditional approaches to estimating the timing of these developmental insults have come under some criticism recently and the authors of these two chapters have both confronted these concerns in different and innovative ways. In Chapter 9, Goodman and Song provide a brief historical survey of studies utilising linear enamel hypoplasias (LEH) and the methods that have been employed to time their formation, the accuracy of which is key to their utility as stress indicators. The authors then deconstruct the possible sources of variation between estimated and true ages of formation of LEH, taking account of some of the most recent studies in the timing of dental development. Having assessed the impact of each of the factors identified, they make a number of important recommendations to future researchers on how to correct or minimise variances when estimating timing of formation of LEH.

Simpson in Chapter 10 looks at two types of structure defects in teeth, LEH and pathological striae (or so-called Wilson bands). He employs a case study approach to illustrate the methodological modifications that he advocates that will improve the analysis of both types of defect and the accurate timing of their formation. Utilising a large sample of Native American mandibular canines from individuals who lived between AD 1 and AD 1704, he records the frequency and location of enamel defects. The sample is divided into four distinct periods corresponding to major changes in culture and subsistence through this period and Simpson verifies the hypothesis that the frequency of microdefects can be related to expected outcomes in each of these periods. Most interestingly, rather than taking a traditional approach using exogenous standards developed on the assumption of linear growth throughout crown formation, he establishes the chronology of defects by utilising microstructural growth markers of enamel in unworn crowns in the sample to generate development standards specific to this archaeological population. Simpson also discusses the nature and possible aetiology of these two defect types and concludes that they should not be considered different responses to a similar underlying cause, but are products of physiological disruptions with different courses, timings and durations.

In Chapter 11, Konigsberg and Holman present a rigorous statistical approach to estimating age in juveniles from dental eruption data. They discuss issues related to the use of standards of ageing to generate skeletal growth profiles in bioarchaeological studies. In particular, these authors argue that most studies of growth in the past do not adequately address the error in age estimation, and as such, many observed differences in skeletal growth profiles are probably methodological biases. Presenting a technique that accounts for the uncertainty in age estimates, they conclude that, while it is possible to get relatively unbiased estimates of growth from prehistoric populations, larger skeletal samples are necessary to make meaningful statistical statements regarding growth in the past.

Mays (Chapter 12) provides a case study of skeletal growth in the Mediaeval sample from Wharram Percy, England. Mays asserts that while assessments of linear growth are common, studies of appositional growth in bioarchaeology have been less frequent. The author notes that many archaeological assemblages from Roman times to the early modern era show very similar patterns of longitudinal bone growth that do not necessarily reflect similarities in childhood nutrition and disease experience. However, differences in the present study are highlighted when appositional growth is taken into account. The author advocates measurements of cortical bone in conjunction with longitudinal bone growth when investigating skeletal growth in earlier populations.

We think that no better case for the advantages of an integrated approach can be made than the consummate volume of chapters presented by our contributors. The book offers readers an opportunity to learn about some of the most current and interesting methodological and theoretical research activities occurring at the moment in the field of palaeoauxology. We hope that they get as much enjoyment from their reading as we the editors had in assembling this volume and working with its authors.

Rob Hoppa
Charles FitzGerald

Acknowledgements

We would like to extend our thanks to Robert Foley for his support in the early stages of developing the volume and to Tracey Sanderson, who has been invaluable through the whole duration of this project – we are deeply indebted to her for her words of wisdom and continued support at various stages of production. Mrs Sandi Irvine also deserves plaudits for her rigorous second-sighted editing of the volume. And of course, the greatest debt of gratitude is owed to all of our contributors who have taken the time and effort to provide us with original and exciting studies of growth and development in the past.

1 *From head to toe*:

integrating studies from bones and teeth in biological anthropology

ROBERT D. HOPPA AND CHARLES M. FITZGERALD

Introduction

From its inception physical anthropology has been preoccupied with human variation (Hrdlička 1927). Since growth is the process that produces variation, it is therefore not surprising to find that biological anthropologists have a long history of studying human variation and growth in numerous populations and within many temporal frameworks (cf. Goldstein 1940; Garn 1980; Beall 1982). Understanding variation in growth patterns among populations permits a better understanding of observed morphological differences (Johnston 1969; Eveleth and Tanner 1976; Johnston and Zimmer 1989). Moreover, this curiosity has not only been about population-wide growth, but also about growth as an ontogenetic process.

With the expansion of physical anthropology and the inevitable concomitant specialisation of interest, two broad streams of growth studies developed (this division, in fact, occurred early in the evolution of the discipline). In one stream, the application of auxological studies to populations from the distant past represented a natural extension for physical anthropologists interested in unravelling the mystery of human evolution. In the other, researchers interested in prehistoric skeletal populations were quick to apply techniques borrowed from anthropometric studies to their samples. Both clearly recognised that juvenile specimens contained a potential wealth of information on the evolution of human ontogeny. However, for the most part these two areas of enquiry, palaeoanthropological studies and bioarchaeological studies, have remained separated, the former tending toward increased scrutiny in dental development and the latter focusing more on linear growth in the appendicular skeleton. While the topical separation was clear – palaeoanthropological studies focused on very detailed analyses of a few fragments (many consisting of dental elements) and bioarchaeological studies aimed to assess population level indices of many individuals – it is unfortunate that the practitioners of each have remained so isolated from one another.

Our specific objective (and the leitmotif of this whole volume) is to reduce the barriers between those two subdisciplines and to bring closer together all of those interested in human growth and development of past populations, whatever their specialty. We believe that there are obvious significant benefits in such an integration and that it might be more easily achieved with the use of more inclusive vocabulary. We propose the term 'palaeoauxology' to classify or group together growth and development studies of past populations. First used by Tillier (1995), this word aptly describes the generality of such studies, and it carries with it no divisive historical connections.

Of course, an interest in human growth is not confined to biological anthropologists, there are other disciplines also enmeshed in the subject. In its broadest sense it impinges on oral biology, dental morphology, forensic odontology, developmental biology, and a variety of clinical medicine specialties, to name just a few. However, in order to stay within reasonable bounds of space, the focus here is narrow. It is restricted to a discussion of growth and development derived from bones and teeth within the two streams of biological anthropology just identified, and even more particularly to studies that can be related directly to *Homo sapiens* or its evolution. This topic is too expansive to cover comprehensively and we have therefore been very selective in presenting an overview of the relevant research.

The first topic to be addressed then, is an important one faced by all investigators confronted with unknown human skeletal or dental material: estimating age at death and determining sex. Since our perspective is growth and development, interest will of course be focused on methods of estimating sex and age at death in non-adult remains.

The basics: estimation of age and sex in non-adults

Determination of sex from the skeleton has, to the regret of many researchers, been restricted to those who have survived past adolescence and who then manifest changes in the skeleton reflective of sex. While a variety of studies have investigated traits that might be sexually dimorphic in infants and juveniles (Thompson 1899; Reynolds 1945; Boucher 1955, 1957; Sundick 1977; Weaver 1980; Schutkowski 1986, 1987, 1989, 1993; DeVito and Saunders 1990; Hunt 1990; Mittler and Sheridan 1992; Introna *et al.* 1993; Majo *et al.* 1993) only a few have had sufficient levels of accuracy to warrant their application in osteological analyses. More promising, but still restricted by cost and time, is the possibility of determining sex by extracting ancient DNA from bones or teeth of individuals (e.g. Fattorini *et al.* 1989; Faerman *et al.* 1995; Stone *et al.* 1996; Lassen 1997). As a result of the limited reliability of morphometric techniques for sexing pre-pubescent individuals, non-adults have remained, for most investigations, lumped within a single group representing both males and females.

Age estimation of children is based, in order of precision, on dental development, epiphyseal closure, and diaphyseal length. Estimation of age in the non-adult is much easier and more accurate by far than in the adult. While there may be fewer techniques than are available for adult age estimation, each has a smaller range of error, as the processes being measured are finite in the sense that there is a beginning and an end to each phase.

In general, most investigators agree that dental development shows a slight degree of sexual dimorphism, but far less than other osseous traits (Gleiser and Hunt 1955; Hunt and Gleiser 1955; Lewis and Garn 1960; Lauterstein 1961, for a recent review of sex determination using teeth see also Teschler-Nicolar and Prossinger 1998). Further, while variation occurs in almost all forms of maturation within the body, tooth formation has proved no more variable than other factors (Lewis and Garn 1960). Of all of the methods of assessing the age of subadults that rely on exogenous references, dental development standards, particularly those assessing tooth formation (e.g. Schour and Massler 1940a,b; Moorrees *et al.* 1963a,b; Demirjian *et al.* 1973; Gustafson and Koch 1974; Anderson *et al.* 1976; Demirjian and Goldstein 1976; Staaf *et al.* 1991) have traditionally been seen as the most accurate, under the strongest genetic control, and least subject to external pressures and population differences. Most studies of dental development have been carried out on the permanent dentition, despite the fact that most skeletal samples contain large proportions of infants and young children; there is still a distinct shortage of detailed standards for the early formation of deciduous crowns that can be applied to foetal and neonatal skeletons (Skinner and Goodman 1992). In addition to tooth formation, dental eruption standards that assess emergence of teeth through the alveolar bone or gums have also been widely utilised, but are a poor surrogate for tooth formation. Recent investigations of dental metrics (Liversidge *et al.* 1993) and more rigorous statistical approaches to estimating age (Jungers *et al.* 1988; cf. Konigsberg and Holman, Chapter 11, this volume) hold promise for improving the reliability of estimates of developmental age. Perhaps most promising of all are techniques based on analysis of dental microstructures, since these obviate the need to apply exogenous standards. These are discussed in a later section of this chapter.

While tests of age prediction using children of known age have been reported from studies of living children (for a review of this literature, see Smith 1991) only a few researchers (Bowman *et al.* 1992; Saunders *et al.* 1993a; Liversidge 1994) have examined the accuracy of dental age estimates in archaeological samples. Crossner and Mansfield (1983) observed that 70% of tooth formation estimates from permanent mandibular and anterior maxillary teeth in 23 childrens' teeth fell within $\pm$ 3 months of true age and discrepancies of no more than 6 months were found for age estimates when tested against the standards of Liliequist and Lundberg (1971) and Gustafson and Koch (1974). Haag and Matsson (1985) compared several dental formation standards using permanent teeth and found standard deviations of the difference between dental and chronological age to be approximately 10% of age. Demirjian and colleagues (1973) estimated a subject's chronological age within 15–25 months with 95% confidence. Tests of these standards on other population samples produced mixed results. Proy and colleagues (1981) observed a mean advancement of 9 months in the dental development of French children, while Kataja and co-workers (1989) found that the models predicted dental age reliability (Kataja *et al.* 1989). More studies exploring interpopulation efficacy of standards need to be conducted.

Bowman and colleagues (1992) recently examined the skeletal remains of 26 juveniles from the crypts of St Bride's Church, Fleet Street, 16 of which were fully documented as having died between the years of 1794 and 1841. These investigators examined the relationship between the skeletal and dental age, and documented chronological age at death, observing that, while long bone growth progressively underestimated age, dental calcification and eruption were similar to modern populations. Saunders and co-workers (1993a) also examined the accuracy of dental formation standards, on a 19th century archaeological sample (St Thomas' Church, Belleville, Ontario) with a small subsample of children of documented age. Dental age estimates in the St Thomas' sample based on a single tooth have a standard deviation of ± 0.94, while the average standard deviation when all possible teeth are used is ± 0.38 years (Saunders *et al.* 1993a).

Smith (1991) has stated that none of the tested systems are particularly suited to age prediction, and has provided a series of recommended formation values for age prediction based on her determination of the appropriate method for constructing chronologies of growth stages. She argued that, for age prediction, it is more appropriate to assign an age that is the midpoint between the mean age of attainment of a subject's current stage of formation and the subsequent one since, at the time of observation, the subject is in between the attainment of one stage and the next. Smith's own test of age prediction accuracy utilised four Canadian children of British origin who were used to test dental age standards in the study by Anderson and co-workers (1976). The chronological ages of these children were compared to her calculated values for predicting age from the stages of permanent mandibular tooth formation derived from the data of Moorrees and colleagues (1963a). The results were 'remarkably accurate' (Smith, 1991: 162) differing by a maximum of 0.2. Within-individual inaccuracy based on a single tooth yielded a standard deviation of ± 0.56 years, while mean values for five or more teeth decreased the standard deviation to ± 0.09 years, suggesting that dental age can be estimated to within 2 months for young children. Saunders *et al.* (1993a) calculated within-individual coefficients of variation for mean dental age for all individuals aged by more than one tooth. In contrast to the lower mean coefficient of variation (CV) (10) reported by Smith (1991), within-individual CVs in the St Thomas's sample had an average value of 20. However, most of the individuals from their sample had ages estimated from five teeth or fewer. Saunders and colleagues (1993a) found little difference in the average level of accuracy of age estimation when using age prediction versus age-of-attainment tables.

The accuracy of several methods of age estimation based on developing teeth was tested on an archaeological population of children by Liversidge (1994). The sample consisted of the dental remains of 63 individuals of known age at death, between 0 and 5.4 years, from Christ Church, Spitalfields, London, interred between 1729 and 1856. Liversidge tested the atlas method of Schour and Massler (1941), Gustafson and Koch's (1974) diagram method, Moorrees *et al.* (1963a,b) mineralisation standards and Smith's (1991) modification of them, and her own quantification standards for length, incorporating regression equations for weight from Deutsch *et al.* (1985). Her results show that the atlas and diagram methods were considerably more accurate for this population and age group than the other methods tested, although the quantification method performed well for the youngest age children. Liversidge agreed with Smith (1991) that prediction methods gave better results than age-of-attainment methods of estimating age using dental development.

However, analysing any results obtained from testing exogenous standards against individuals of known age is not easy. Interpretations are confounded by the difficulties of comparison because of methodological inconsistencies and differing approaches in data collection and assessment of tooth maturity from study to study. For instance, it is well known that tooth formation as traditionally measured by the appearance of growing teeth on radiographs is fraught with methodological problems (Risnes

1986; Aiello and Dean 1990; Beynon *et al.* 1991; Beynon *et al.* 1998a). Also, some studies have used as few as three fractional stages to gauge tooth development, others as many as 20. Not only do these and other difficulties make cross-study comparisons difficult, but they raise a more fundamental question: is the high ontogenetic intra- and interpopulation variability commonly seen in dental standards really a feature of modern human dental development, or is some or all of it an artefact of the methodology used to determine it (FitzGerald *et al.* 1999)?

From head to toe: assessing growth in skeletal samples

Both palaeoanthropological and bioarchaeological studies have, for the most part, focused on specific areas of the skeleton with regard to growth and development. In the case of palaeoanthropology this has largely been the result of many studies having to rely on the limited number of skeletal elements associated with a particular site. In the case of bioarchaeology it is often because of the need to examine growth variables, such as long bone length, that can be easily placed within a comparative framework with other archaeological samples. Relatively few studies have examined growth and development within skeletal samples from a broader perspective, although research by Steyn and Henneberg (1996a,b) on Iron Age remains or Smith (1993) on the Turkana Boy provide good examples against which future research strategies should be modelled.

While there is a wealth of information on skeletal growth and development, physical anthropologists have been relatively selective in the kinds of data that have been explored for past populations. Many studies, particularly bioarchaeological studies, have been primarily descriptive in nature, with theoretical and methodological issues forming a secondary role in the literature. Yet, there has been in the human biology and medico-legal disciplines a significant amount of literature dealing with specific methodologies for examining subadults that can be exploited by current researchers in the field of palaeoauxology.

Assessments of growth in length (or width) of the long bones are the most commonly employed assessment of statural growth in skeletal remains. Studies of allometric growth, often utilised in palaeoanthropology studies, are seen less frequently in bioarchaeological research. More often, simple measures of diaphyseal length, clavicle length or iliac breadth are made, and the distributions analysed in the context of some independent assessment of age – usually dental development. In doing so, a cross-sectional 'growth curve' or skeletal growth profile (SGP) can be constructed to

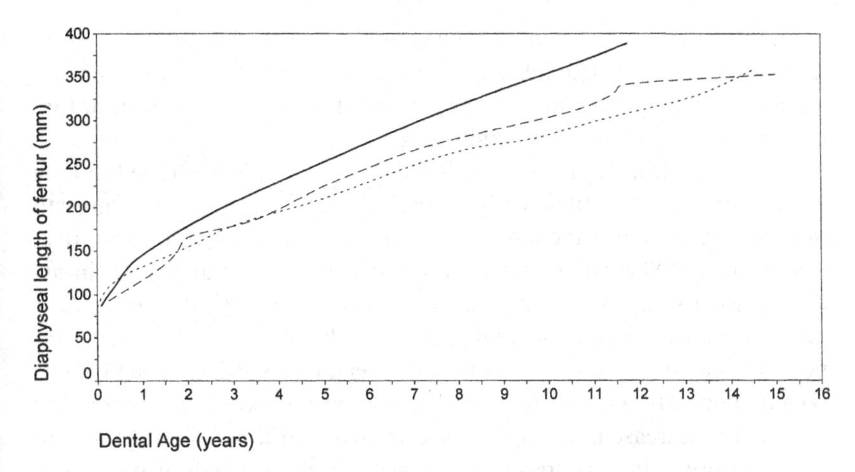

Fig. 1.1. Typical skeletal growth profile (SGP) plotting diaphyseal length of femur against chronological age estimated from dental development standards. Here comparisons of two archaeological samples, 9th century Slavic (dotted line; Stloukal and Hanáková 1978) and Altenerding, a 6th–7th century collection from Munich (dashed line; Sundick 1978), are made with a modern reference sample (continuous line, Maresh 1970).

exhibit the age-progressive trend in long bone length up to the time of epiphyseal fusion (Fig. 1.1). In contrast, studies of appositional bone growth in earlier populations have been less frequent (Armelagos *et al.* 1972; Huss-Ashmore *et al.* 1982; Hummert 1983; Mays 1985, 1995; Van Gerven *et al.* 1985; Saunders and Melbye 1990; cf. Mays, Chapter 12, this volume).

The major limitation of cross-sectional data is that it does not allow one to observe individual variability in the rate or velocity of growth or in the timing of the adolescent growth spurt, although attempts to examine the rate of growth have been undertaken (e.g. Lovejoy *et al.* 1990). In comparative analyses between populations, however, the means and variations of the 'population' rather than the patterns unique to the individual are often more important (Eveleth and Tanner 1976). As others have noted though, the nature of the data used to construct the SGPs often prevents adequate statistical comparison between population samples, and more powerful techniques are likely to reveal little more than is observable from simple examination of the graphs (Merchant 1973). Further, differences between SGPs may be, in part, a result of differences in adult stature related to longer-term adaptations (Eveleth 1975). In order to control for this, some researchers have constructed skeletal growth profiles as a percentage of mean adult long bone length or mean adult stature for individual populations (Lovejoy *et al.* 1990; Wall 1991; Hoppa 1992; Goode *et al.* 1993;

Saunders *et al.* 1993b). Utilising mean adult long bone lengths is recommended as this will take into account variation in limb proportions and differences in the regression equations utilised to reconstruct mean stature (Hoppa and Saunders 1994; Sciulli 1994).

Studies of infrancranial growth, however, need not be restricted to long bones. For example, Miles and Bulman (1994, 1995) have examined growth in shoulder, hip, hand and foot bones from a 16th to 19th century Scottish island sample. Several studies have investigated vertebral growth in archaeological samples (Clark *et al.* 1986; Porter and Pavitt 1987; Clark 1988; Grimm 1990; Kneissel *et al.* 1997). Clark (1988) examined vertebral neuronal canal size and vertebral body height in the Dickson Mounds skeletal population in an attempt to determine why previous research has observed a decrease in diaphyseal growth over time, but no difference in adult stature. Clark proposes that since vertebral canals usually stop growing by early childhood, but vertebral body height continues to grow through young adulthood, if canal size is reduced but body height is not, this implies an early disruption in growth followed by catch-up growth. In contrast, if both areas are reduced in size, growth disruption was probably chronic. From his analysis of the Dickson Mounds sample, Clark (1988) concluded that early interruptions in growth were followed by catch-up growth.

The appearance and eventual union of primary and secondary centres of ossification in bone are also of interest. As Stewart (1954) has pointed out, the age of *appearance* of various primary centres has limited application in the analysis of archaeological skeletal remains, as these are rarely recovered during excavation. However, the development and union of such remains are more easily assessed in older individuals at which time the various epiphyses are beginning to unite. Nevertheless, some studies of epiphyseal fusion in past populations have been explored (e.g. Hoppa 1992; Albert 1995).

Palaeoanthropological studies

The pattern of anatomically modern human growth is characterised by an extended period of infant dependency, prolonged childhood and a rapid increase in adolescent growth at the time of sexual maturation. Most agree with Bogin (1988: 61) that 'the prolonged delay in human growth due to the evolution of a childhood period of growth is one feature of the human growth curve that distinguishes it from all others' (although see Leigh 1996). Therefore, understanding the evolutionary development of delayed maturation and prolonged infant dependency among modern humans will reveal important information about the way our ancestors lived and how they were related to each other. Clearly, increased brain size associated with the complex behaviours connected with human culture necessitated the extension of the childhood stage. More time was required for learning. However, such an adaptation could not be made without other socio-biological consequences. As a result, the processes and mechanisms that have allowed early hominids to survive and evolve are of considerable interests to physical anthropologists (cf. Bogin 1997; Leigh and Park 1998).

The question remains whether these uniquely human life history variables evolved in concert or in a mosaic fashion, and, if the latter, which appeared when in our evolutionary history. Of course, life cycles cannot be studied directly from skeletal or dental material, particularly material as fragmentary as that with which palaeoanthropologists have to contend most of the time. To a large extent inferences have been made by reconstructing patterns based on assessments of chronological or biological growth derived from the skeletal material, although other approaches have also been used to investigate ontogeny. For instance, enamel thickness, traditionally studied as a phylogenetic indicator, has also been recruited to help reconstruct life history (Macho 1995). The focus here will be on growth assessment. However, that raises a fundamental issue: just how does one go about assessing the development of long extinct life forms?

Since until recently no reliable ways existed by which to endogenously determine development (see discussion below), in practical terms the preoccupation has been with deciding what reference standards to use to assess patterns of development in early hominids. The problematic nature and the applicability of exogenous standards to non-referent populations has already been intimated in an earlier section in the context of anatomically modern humans. In the case of extinct hominids, these difficulties are compounded. The choice of whether to apply modern human or modern ape growth standards has significant consequences for the estimate of a specimen's chronological age (Smith 1986, 1994; Mann *et al.* 1987, 1990; Lampl *et al.* 1993; cf. Nelson and Thompson, Chapter 4, this volume). To address this issue, comparative studies of human and ape growth have become the focus of many researchers (e.g. Dean and Wood 1981; Smith 1989, 1994; Smith *et al.* 1994; Simpson *et al.* 1996; Braga 1998; Reid *et al.* 1998; Dainton and Macho 1999; cf. Dainton and Macho, Chapter 2, and Humphrey, Chapter 3, this volume).

Some earlier studies, predicated on what must be said to be circular reasoning (arising from the use of modern human dental standards), inferred from dental development that australopithecines had a delayed maturation similar to that of modern humans (Mann 1975). However, Bromage and Dean (1985) demonstrated from their analysis of dental enamel microstructures that australopithecines and early *Homo* may have developed at rates that were more similar to living apes, than to modern humans. These investigators applied a histological approach that relies on the interpretation of certain microscopic features of enamel and dentine as incremental growth markers (see Fig. 1.2). These microstructures are formed during the regular appositional pattern of growth of both of these dental hard tissues. Although initially challenged, the time dependency of dental microstructures is no longer seriously disputed (cf. Dean 1987a; FitzGerald 1998; Shellis 1998) and most concede that they record normal growth in a way that permits the developmental and chronological history of a tooth to be accurately reconstructed. Estimates of age and development from dental microstructures overcome one of the principal handicaps associated with most other age estimation techniques – the precise and accurate correlation between biological age and chronological age. More importantly in the context of palaeoanthropology, using standards derived from the fossils themselves overcomes the problems associated with relying on those based on maturation studies of modern humans or apes (some of which were identified earlier).

The premise of different development patterns between early and late hominid groups first identified by Bromage and Dean has been supported by a number of subsequent studies of enamel microstructure (e.g. Dean 1985a,b, 1987a,b, 1989; Beynon and Wood 1987; Beynon and Dean 1988; Ramirez-Rozzi 1991, 1993, 1995) and dentine microstructure (Dean 1995, 1998; Dean and Scandrett 1995, 1996; see also Dean, Chapter 5, this volume). Independent corroboration has also come from investigators using non-histological approaches. For instance, Smith's studies (e.g. 1986, 1987, 1989, 1992, 1994) and Smith and Tompkins' (1995) study of dental development concluded that early hominids had a developmental pattern that was unlike that of any living primate, many specimens more closely resembling pongid patterns. Also, Tompkins' (1996) analysis of dental development among recent human populations, Upper Palaeolithic specimens, and Neandertals, identified dental development differences between the two early groups and the modern one. Upper Palaeolithic and Neandertal specimens seem to share a more similar pattern of dental development, which differed somewhat from that of modern populations. What has clearly emerged from these and other studies is that attempts to characterise any hominid taxon's growth as being either 'ape-like' or 'human-like' is a gross oversimplification. Not surprisingly perhaps, it appears that each hominid group probably has a unique suite of dental and

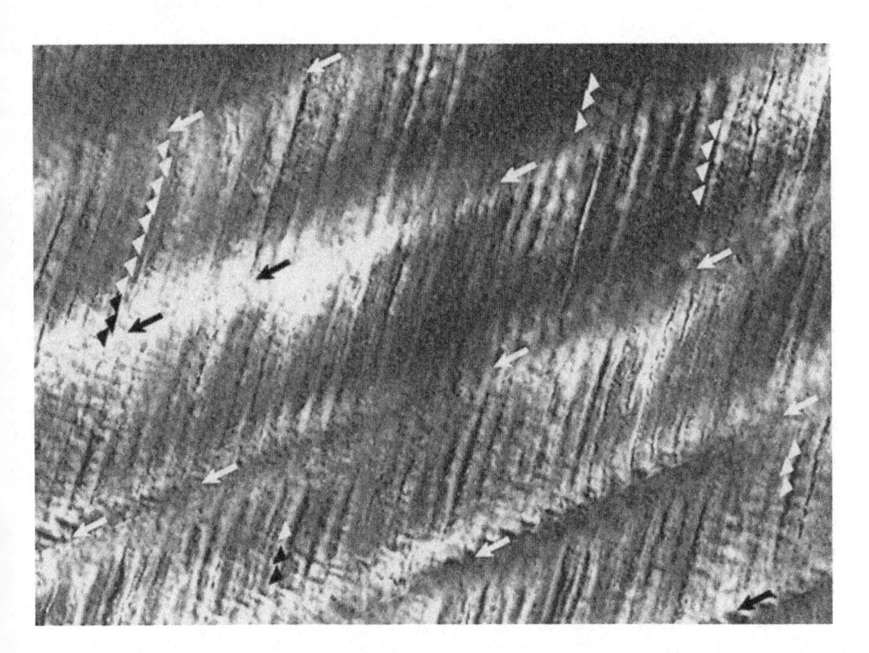

Fig. 1.2. This is a photomicrograph at 400 × magnification of a longitudinal section of enamel near the tooth surface taken in polarised light. It illustrates the two types of incremental growth marker used in enamel histological ageing techniques. Enamel prisms can be seen, inclined at a slight angle, running from the top of the photo, which is toward the enamel–dentine junction, to the bottom of the photo and the tooth surface, just out of frame. The triangles point to some examples of cross-striations, the circadian markers that appear as successive dark and light bands occurring along the full length of enamel prisms. Cross-striations therefore represent the amount of appositional enamel produced in one day, and carefully measuring them yields the daily rate of enamel production. The arrows point to brown striae of Retzius, long period markers that cross-cut prisms at regular circaseptan intervals (usually 7–10 days, although in this photo a sequent run of 11 cross-striations is indicated toward the top left). Imbricational (non-cuspal) striae emerge at the tooth surface as perikymata, visible as 'wrinkles' on the exterior of tooth crowns. The number of cross-striations between adjacent striae, called the circaseptan interval, differs from individual to individual, but is uniform within and among all of the teeth in one dentition. These markers can be utilised in several different ways to arrive at crown formation times; for instance, in one approach the number of striae of Retzius are counted and then multiplied by the circaseptan interval, which need be established only once within a tooth, to yield formation time in days. For a full and recent discussion of these and other techniques and examples of their application, see the special issue, of the *Journal of Human Evolution* (1998) **35**(4/5).

developmental characteristics that reflect their distinctive evolutionary trajectories.

Despite this debate, which throws into question the appropriateness of any modern development standards to assess non-modern populations, studies of hominid ontogeny using traditional comparative approaches have proceeded. Antón (1997) recently re-evaluated the Mojokerto child remains from Java, whose developmental and taxonomic status have been controversial. On the basis of her comparisons of the partial calvaria with a series of *H. sapiens*, Neandertal and *H. erectus* juveniles, she concluded that the Mojokerto child represents a *H. erectus* child of between 4 and 6 years of age. Smith's (1993) study of the well-preserved juvenile *H. erectus* skeleton, KNM-WT 15 000, observed a pattern of growth and development that was more modern when compared to other earlier hominids. However, comparisons of dental and somatic indicators of growth implied an advanced skeletal maturation relative to the dentition, making it less similar to the modern rate and pattern of growth. A variety of studies of archaic *H. sapiens* and Neandertal juvenile specimens have also been undertaken over the past decades (e.g. Heim 1982; Tillier 1983, 1988; Dean *et al.* 1986; Hublin and Tillier 1988; Zollikofer *et al.* 1995; Tompkins 1996), although it is still not clear how the Neandertal pattern of growth fits between the more primitive *H. erectus* and modern human patterns (cf. Nelson and Thompson, Chapter 4, this volume).

Bioarchaeological studies

Given the abundance of remains from archaeological populations relative to those of earlier hominids, it is hardly surprising that a larger corpus of growth studies has accumulated in bioarchaeology. The basic assumption of growth-related studies of past populations is that the growth of a child reflects his or her health and nutritional status better than any other single index (Johnston 1969; Eveleth and Tanner 1990). Anthropometric studies support this statement, with many researchers observing higher rates of morbidity and subsequent mortality associated with varying degrees of stunting and wasting in children. Like anthropometric studies of living populations, studies of skeletal growth from archaeological collections often use linear growth as a proxy for health and thus make interpretations regarding the overall health and well-being of a population from the apparent growth of children. Since many studies imply that long bone growth is differentially affected by the nutritional and health status of the individual, osteologists have attempted to utilise cross-sectional analyses of long bone growth as a non-specific indicator of nutritional status within subadult skeletal samples (cf. Johnston and Zimmer 1989; Saunders 1992). The premise of such studies stems from experimental and contemporary studies that demonstrate the permanent effects of a variety of stressors on skeletal growth and bone dimensions (Buikstra and Cook 1980; cf. King and Ulijaszek, Chapter 7, this volume). The demonstration of differential growth between samples is then employed as evidence for differential health status between entire populations, either geographically or temporally. It must be noted though, that growth-related measurements remain non-specific indicators of health, and are sensitive to many factors. As such, they can reveal that there is a problem, but say very little about its cause (Martorell and Ho 1984: 51).

The analysis of human skeletal growth in archaeological samples first became popular with the works of Stewart (1954) and Johnston (1962). The 20 years that followed saw a variety of publications dealing specifically with growth (Johnston 1968; Mahler 1968; Walker 1969; Armelagos *et al.* 1972; Sundick 1972, 1978; Merchant 1973; y'Edynak 1976; Merchant and Ubelaker 1977; Stloukal and Hanáková 1978). In the 1980s the issue of growth stunting for assessing health within past populations became a popular interpretative tool associated with palaeopathological studies (Hummert 1983; Hummert and Van Gerven 1983; Cohen and Armelagos 1984; Jantz and Owsley 1984a,b; Mensforth 1985; Owsley and Jantz 1985; Storey 1986). Beginning in the late 1980s and early 1990s there was a resurgence of interest in studies dealing specifically with subadult growth (Jungers *et al.* 1988; Hühne-Osterloh 1989, Hühne-Osterloh and Grupe 1989; Johnston and Zimmer 1989; Molleson 1989, 1990; Lovejoy *et al.* 1990; Wall 1991; Hoppa 1992; Saunders 1992; Saunders *et al.* 1993b; Jantz and Owsley 1994; Miles and Bulman 1994, 1995; Ribot and Roberts 1996; Steyn and Henneberg 1996a; Hutchins 1998).

Most of these osteological studies of growth and development have compared linear size measurements of long bones and skeletal maturity, although studies of appositional bone growth and cross-sectional geometry have also been undertaken. Interpretations of health from such studies are derived primarily from comparisons of skeletal and dental development with previously published studies or with modern standards (e.g. Maresh 1943, 1955, 1970; Anderson and Greene 1948; Gindhart 1973). However, interpretations of the resultant differences and their significance concerning conditions of health in past populations are often difficult given (1) the lack of a single consistent methodology when constructing skeletal growth profiles from archaeological samples and (2) whether the reference samples utilised are in fact appropriate.

An issue that has had considerable attention for such studies is the fact that growth as reflected in the subadult cohort associated with a burial sample are essentially non-survivors. That is, they represent individuals who, for whatever reason, have not survived to complete maturation and therefore whose level or pattern of growth and development might not be representative of the true pattern of growth in the population. Long recognised as a theoretical obstacle by researchers, this issue re-emerged with the publication of the *Osteological Paradox* in which Wood and colleagues (1992) argued that skeletal samples are intrinsically biased because they are the products of selective mortality or non-random entry. Selective mortality refers to the fact that skeletal samples do not represent all susceptibles for a given age cohort, but only those individuals who have died at that age. For example, 5 year old individuals in the skeletal sample represent only those 5 year olds who died and not all of the 5 year olds who were alive in the population at risk; other susceptibles who survived, went on to contribute to older mortality cohorts. In a review of the child survival literature, Saunders and Hoppa (1993) examined this issue of mortality bias specifically with respect to growth in children. They concluded that while there did appear to be statistically significant differences between the growth of survivors and non-survivors in the clinical literature, the actual magnitude of this difference for cross-sectional studies of long bone growth would be minimal, and probably less important than the error introduced by methodological issues like ageing standards.

A serious problem that confronts both osteological and dental development studies of human growth is the general lack of a consistent methodological approach for the collection and analysis of data. Since comparative data for growth studies are sparse, the lack of comparable methodologies between various investigators has resulted in many studies becoming isolated analyses whose interpretation cannot be adequately evaluated against other data (Saunders *et al.* 1993b). Teeth in intermediate stages of growth must be characterised in arbitrary fractional stages of completion and there is a variety of idiosyncratic approaches that have been adopted to accomplish this, making cross-study comparisons difficult or impossible. Dental data may also be collected cross-sectionally, semi-longitudinally, and rarely, longitudinally, adding to comparison problems.

Skeletal data are often placed into age cohorts representative of a group of individuals who are developmentally similar. While such categorisation does accommodate the presentation of variation in length, it does not allow for the presentation or analysis of the equally important variance of the age distribution within individual cohorts (Hoppa 1992). Many studies provide standard deviations to illustrate the distribution of lengths around a cohort mean, but few, if any, calculate the confidence limits of individual cohort means. Naturally, the precision of such confidence limits will be dependent on both sample size and variance, with variance expected to increase with age as well as from the pooling of data for both sexes. In

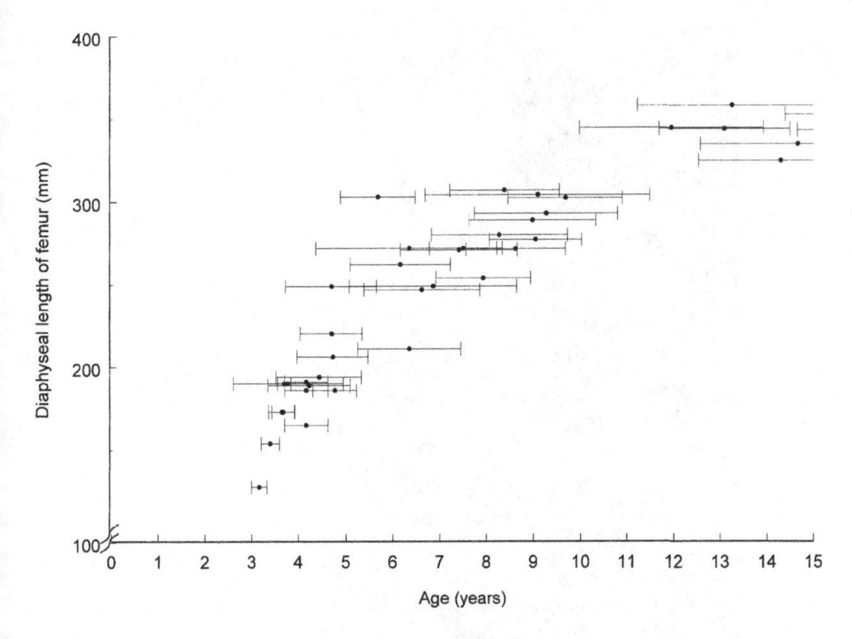

Fig. 1.3. A skeletal growth profile showing the variability in individual age estimates as 95% confidence intervals, based on the pooled variance in the original reference sample (adapted from Hoppa 1992).

addition, the proportions of older children, particularly adolescents, are often small in archaeological samples. As a result, many skeletal growth profiles will overlap with modern reference standards and subsequent interpretations of health based on any observed differences may be erroneous, and simply the product of inadequate sample sizes. Even when apparently clear differences in the mean growth profiles of two population samples are observed, interpretations based on these differences may be incorrect (cf. Konigsberg and Holman, Chapter 11, this volume). Similarly, the individual variation in dental development has seldom been addressed in bioarchaeological studies of growth (Lampl and Johnston 1996). Hoppa (1992) calculated the pooled variances for dental calcification standards from the original reference data represented in Anderson *et al.* (1976). The resultant variance for final individual estimates of mean dental age was then used to calculate confidence intervals on individuals when plotted on a skeletal growth curve to illustrate the potential effects on assessing growth (Fig. 1.3).

Other areas of interest for bioarchaeological studies of growth and development have been in examining specific developmental cohorts of juveniles such as infants and young children (e.g. Brothwell 1986; Storey

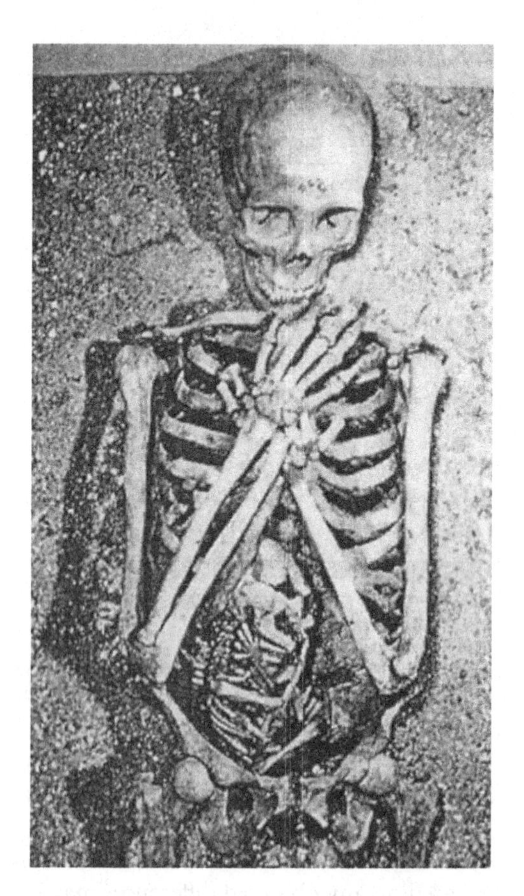

Fig. 1.4. Burial of a female skeleton with foetus *in situ* from Aebelholt cemetery, Denmark (Museum of Medical History, Copenhagen).

1986, 1992; Herring *et al.* 1998; Hutchins 1998; cf. Saunders and Barrans, Chapter 8, this volume). Identification of obstetric deaths (see e.g. Fig. 1.4) are uncommon in the archaeological record and often the focus of special attention when found (e.g. Högberg *et al.* 1987; Sjøvold *et al.* 1974; Hawkes and Wells 1975; Wells 1978), although it is unlikely that specific causes can be postulated, with the exception perhaps of obstructed labour. Similarly, many neonatal deaths resulting from congenital abnormalities even if recovered archaeologically, will have died as a result of malformations or damage to specific organs or bodily systems.

Since infant mortality has such a profound effect on the crude death rate of a population, it is considered to be a sensitive indicator of overall population fitness. However, poor preservation of small, fragile infant bones or differential burial is commonly cited as a reason for under-representation of infants in skeletal samples (e.g. Guy *et al.* 1997; although see Saunders 1992). Given these constraints, some have tried to make use of infant mortality data more thoroughly. Higgins' (1989) proposed an interpretative model whereby the physiological manifestations of various health relationships (e.g. maternal–infant, infant–child, child–adolescent etc.) can be correlated with potential skeletal manifestations. For example, she suggested that poor maternal health within a population will result in an increase in young female mortality, an increase in infant mortality and a reduction of pathological lesions of the skeleton as a result of reduced immune responses and greater incidence of mortality from acute rather than chronic infections (Higgins 1989).

A variety of studies have also attempted to detect evidence of the weaning process in infants and young children from demographic (e.g. Herring *et al.* 1998), auxological (Wall 1991) and biochemical analyses (Blakeley 1989; Hühne-Osterloh and Grupe 1989; Katzenberg and Pfeiffer 1995; Katzenberg *et al.* 1996; Schurr 1997; Wright and Schwarcz 1998) and by evaluation of enamel hypoplasias (e.g. Goodman *et al.* 1984, 1987; Corruccini *et al.* 1985; Lanphear 1990; Moggi-Cecchi *et al.* 1994; although see Hillson and Bond 1997; Saunders and Keenleyside 1999). Developmental defects of tooth enamel have also been more broadly studied as non-specific indicators of stress throughout the growing years (Goodman and Rose 1990, 1991; Skinner and Goodman 1992; see Goodman and Song, Chapter 9 and Simpson, Chapter 10, this volume).

Palaeoauxology: integrating studies from bones and teeth

Studies of craniofacial growth (excluding dental development), are well represented in palaeoanthropological studies (Tillier 1982; Dean and Wood 1984; Bromage 1986; Minugh-Purvis 1988, 1993; Antón 1997; Antón and Franzen 1997; Mann and Vandermeersch 1997; Moggi-Cecchi *et al.* 1998), but with few exceptions they have been, until relatively recently, scarce among bioarchaeological studies (e.g. Markowitz 1995; Sejrsen *et al.* 1997; Steyn and Henneberg 1996b; Humphrey 1998). This is due primarily to the evolutionary importance of comparative studies of cranial morphology in hominid specimens, particularly in relation to increased cranial capacity among *H. erectus* and *H. sapiens*. In contrast, bioarchaeological studies of infracranial growth, particularly linear growth, have been numerous. However, studies of growth and development in archaeological samples need not be restricted to simple linear growth. As outlined above, a variety of other comparative data that might be useful in bioarchaeological reconstructions is also available and should not be overlooked by palaeo-auxological studies. Quantitative assessments of cortical bone growth (e.g. Armelagos *et al.* 1972; Hummert 1983; Saunders and Melbye 1990), growth in the vertebrae (Clark 1988), finger and foot bones, shoulder and hip (Miles and Bulman 1994, 1995) and trabecular changes (from primary to secondary) offer some further areas of exploration. Biochemical studies for dietary reconstruction are also a valuable source of information to be integrated into growth studies. Incorporating qualitative assessments of stress indicators in studies of growth provides further support for interpretation of health and well being among past populations. Evaluation of growth-related studies in conjunction with other non-specific indicators of stress is another venue for continued research (Cook 1981; Mays 1985, 1995; Porter and Pavitt 1987; Hühne-Osterloh 1989; Hühne-Osterloh and Grupe 1989; Grimm 1990; Elliott 1992; Ribot and Roberts 1996). Porter and Pavitt (1987) examined vertebral growth in two archaeological populations, and found other indicators of stress such as dental hypoplasias and Harris lines to be correlated with reduced vertebral growth. In contrast, Ribot and Roberts (1996) found a less clear association between stress indicators and reduced diaphyseal growth.

A recurring theme throughout this whole chapter has centred on the use of appropriate exogenous reference standards to assess growth and development of past populations (indeed, a concern has been raised about the applicability of any modern reference standards to past populations). This represents one of the central methodological problems facing palaeoauxology. It may be a question of whether a modern North American growth curve is appropriate to assess growth deficit in 10th century Anglo-Saxon children, or whether dental developmental patterns derived from 20th century North American children are suitable to estimate chronological age in prehistoric children, or, more broadly, whether human or ape patterns are most suitable to assess early hominid dental development. The issue that future research must address is how to compensate for, or at least minimise, the impact of potentially inappropriate reference standards. The solution here will not be simple and alternative approaches may offer the best hope. Two that show great promise are the use of comparative developmental curves (e.g. heterochrony), and endogenous developmental ageing using dental microstructure analysis. An excellent example of the kind and quality of information that can be obtained using the latter approach is illustrated in a study by Beynon *et al.* (1998b) on the dental development of the non-hominid *Proconsul heseloni*. A variety of histological techniques were used to analyse the microstructural features of both enamel and dentine, and the authors were able to establish that *Proconsul* dental development presented a hominoid mosaic, most similar to *Pongo*, but with some features shared with *Pan* and *Homo*.

While finding their greatest application so far in evolutionary ontogeny and studies of pre modern hominid growth, dental microstructural analyses of enamel and more recently dentine are beginning to be exploited in bioarchaeological studies (e.g. Huda and Bowman 1995; Reid *et al.* 1998b; FitzGerald *et al.* 1999). While the time and labour required to make assessments on large samples of subadult individuals have so far precluded widespread use of this technique for bioarchaeological studies, its potential benefits are clear. Improvement and standardisation of methods and incorporation of computer-based digital imaging procedures will make dental microstructural analysis more accessible to non-specialists and make assessment of large samples feasible (FitzGerald *et al.* 1999). Using such techniques, developing teeth may be aged and used to form endogenous or population-specific rates of dental development that can be used to assess other somatic indicators of growth.

Ultimately, all of those who study past human growth and development pursue a common research goal – to understand the developmental patterns of growth and make inferences about the evolutionary forces, either macro or micro, that have shaped those patterns. Despite this shared objective, the particular demands of palaeoanthropological and bioarchaeological studies have resulted in a continuing dichotomisation of these two fields of enquiry. In this very brief review we have tried to establish the theme for this book: the promotion of a more explicit integration of studies of human growth in the past from bones and teeth, which we suggest should be grouped together under the umbrella of palaeoauxological studies. It is our hope that this monograph, which has integrated a variety of studies into a single, coherent volume, will help to remind investigators of the commonality of their goals, and the consequent benefits to be gained through sharing information about growth and development in past populations.

References

Aiello L and Dean C (1990) *An Introduction to Human Evolutionary Anatomy*. New York: Academic Press.

Albert AM (1995) Assessment of the variability in the timing and pattern of epiphyseal union associated with stress in teenage and young adult skeletons from Medieval Kulubnarti, Sudanese Nubia. PhD dissertation, University of

Colorado at Boulder.

Anderson M and Greene WT (1948) Lengths of the femur and the tibia. *American Journal of Diseases of Childhood* **75**, 279–290.

Anderson DL, Thompson GW and Popovich F (1976) Age of attainment of mineralization stages of the permanent dentition. *Journal of Forensic Science* **21**, 191–200.

Antón SC (1997) Developmental age and taxonomic affinity of the Mojokerto child, Java, Indonesia. *American Journal of Physical Anthropology* **102**, 497–514.

Antón SC and Frazen JL (1997) The occipital torus and developmental age of Sangiran-3. *Journal of Human Evolution* **33**, 599–610.

Armelagos G, Mielke J, Owen K, Gerven DV, Dewey J and Mahler P (1972) Bone growth and development in prehistoric populations from Sudanese Nubia. *Journal of Human Evolution* **1**, 89–119.

Beall CM (1982) An historical perspective on studies of human growth and development in extreme environments. In *A History of American Physical Anthropology*, ed. F. Spencer, pp. 447–465. New York: Academic Press.

Beynon AD and Dean MC (1988) Distinct dental development pattern in early fossil hominids. *Nature* **335**, 509–514.

Beynon AD and Wood BA (1987) Patterns and rates of enamel growth in the molar teeth of early hominids. *Nature* **326**, 493–496.

Beynon AD, Dean MC and Reid DJ (1991) Histological study of the chronology of the developing dentition in gorilla and orang-utan. *American Journal of Physical Anthropology* **86**, 189–203.

Beynon AD, Clayton CB, Ramirez Rozzi FV and Reid DJ (1998a) Radiographic and histological methodologies in estimating the chronology of crown development in modern humans and great apes: a review, with some applications on studies for juvenile hominids. *Journal of Human Evolution* **35**, 351–370.

Beynon AD, Dean MC, Leakey MG, Reid DJ and Walker A (1998b) Comparative dental development and microstructure of *proconsul* teeth from Rusinga Island, Kenya. *Journal of Human Evolution* **35**, 163–209.

Blakely RL (1989) Bone strontium in pregnant and lactating females from archaeological samples. *American Journal of Physical Anthropology* **80**, 173–185.

Bogin B (1988) Rural-to-urban migration. In *Biological Aspects of Human Migration*, ed. C.G.N. Mascie-Taylor and G.W. Lasker, pp. 90–129. Cambridge: Cambridge University Press.

Bogin B (1997) Evolutionary hypotheses for human childhood. *Yearbook of Physical Anthropology* **40**, 63–89.

Boucher BJ (1955) Sex differences in the foetal sciatic notch. *Journal of Forensic Medicine* **2**, 51–54.

Boucher BJ (1957) Sex differences in the foetal pelvis. *American Journal of Physical Anthropology* **15**, 581–600.

Bowman JE, MacLaughlin SM and Scheuer JL (1992) The relationship between biological and chronological age in the juvenile remains from St. Bride's Church, Fleet Street. *Annals of Human Biology* **19**, 216.

Braga J (1998). Chimpanzee variation facilitates the interpretation of the incisive suture closure in South African plio-pleistocene hominids. *American Journal of Physical Anthropology* **105**, 121–135.

Bromage TG (1986) A comparative scanning electron microscope study of facial growth and remodelling in early hominids. PhD thesis, University of Toronto.

Bromage TG and Dean MC (1985) Re-evaluation of the age at death of immature fossil hominids. *Nature* **317**, 525–527.

Brothwell D (1986) The problem of the interpretation of child mortality in earlier populations. *Antropologia Portuguesa* **4/5**, 135–143.

Buikstra JE and Cook DC (1980) Palaeopathology: an American account. *Annual Reviews of Anthropology* **9**, 433–470.

Clark GA (1988) New method for assessing changes in growth and sexual dimorphism in paleoepidemiology. *American Journal of Physical Anthropology* **77**, 105–116.

Clark GA, Hall NR, Armelagos GJ, Borkan GA, Panjabi MM and Wetzel FT (1986) Poor growth prior to early childhood: decreased health and life-span in the adult. *American Journal of Physical Anthropology* **70**, 145–160.

Cohen MN and Armelagos GJ (eds.) (1984) *Paleopathology at the Origins of Agriculture.* San Diego: Academic Press.

Cook DC (1981) Mortality, age structure and status in the interpretation of stress indicators in prehistoric skeletons: a dental example from the Lower Illinois Valley. In *The Archaeology of Death*, ed. R. Chapman, I. Kinnes and K. Randsborg, pp. 133–144. Cambridge: Cambridge University Press.

Corruccini RS, Handler JS and Jacobi KP (1985) Chronological distribution of enamel hypoplasias and weaning in a Caribbean slave population. *Human Biology* **57**, 699–711.

Crossner CG and Mansfield L (1983) Determination of dental age in adopted non-European children. *Swedish Dental Journal* **7**, 1–10.

Dainton M and Macho GA (1999) Did Knuckle walking evolve twice? *Journal of Human Evolution* **36**, 171–194.

Dean MC (1985a) The eruption pattern of the permanent incisors and first permanent molars in *Australopithecus* (*Paranthropus*) *robustus. American Journal of Physical Anthropology* **67**, 251–257.

Dean MC (1985b) Variation in the developing root cone angle of the permanent teeth of modern man and certain fossil hominids. *American Journal of Physical Anthropology* **68**, 233–238.

Dean MC (1987a) Growth layers and incremental markings in hard tissues; a review of the literature and some preliminary observations about enamel structure in *Paranthropus boisei. Journal of Human Evolution* **16**, 157–172.

Dean MC (1987b) The dental developmental status of six East African juvenile fossil hominids. *Journal of Human Evolution* **16**, 197–213.

Dean MC (1989) The developing dentition and tooth structure in hominoids. *Folia Primatologica* **53**, 160–176.

Dean MC (1995) The nature and periodicity of incremental lines in primate dentine and their relationship to periradicular bands in OH 16 (*Homo habilis*). In *Aspects of Dental Biology: Paleontology, Anthropology and Evolution*, ed. J. Moggi-Cecchi, pp. 239–265. Florence: International Institute for the Study of

Man.

Dean MC (1998) Comparative observations on the spacing of short-period (von Ebner's) lines in dentine. *Archives of Oral Biology* **43**, 1009–1021.

Dean MC and Scandrett AE (1995) Rates of dentine mineralization in permanent human teeth. *International Journal of Osteoarchaeology* **5**, 349–358.

Dean MC and Scandrett AE (1996) The relation between enamel cross stations and long-period incremental markings in dentine in human teeth. *Archives of Oral Biology* **41**, 233–241.

Dean MC and Wood BA (1981) Developing pongid dentition and its use for ageing individual crania in comparative cross-sectional growth studies. *Folia Primatologica* **36**, 111–127.

Dean MC and Wood BA (1984) Phylogeny, neoteny and growth of the cranial base in hominoids. *Folia Primatologica* **43**, 157–180.

Dean MC, Stringer CB and Bromage TG (1986) Age at death of the Neandertal child from Devil's Tower, Gibraltar and the implications for studies of general growth and development in Neandertals. *American Journal of Physical Anthropology* **70**, 301–309.

Demirjian A and Goldstein H (1976) New systems for dental maturity based on seven and four teeth. *Annals of Human Biology* **3**, 411–421.

Demirjian A, Goldstein H and Tanner JM (1973) A new system of dental age assessment. *Human Biology* **45**, 211–227.

Deutsch D, Tam O and Stack MV (1985) Postnatal changes in size, morphology, and weight of developing postnatal deciduous anterior teeth. *Growth* **49** 202–217.

DeVito C and Saunders SR (1990) A discriminant function analysis of deciduous teeth to determine sex. *Journal of Forensic Sciences* **35**, 845–858.

Elliott BF (1992) The ontogeny of pre-pubescent skeletal growth, as seen in a prehistoric central California Indian collection. MA thesis, San Jose State University.

Eveleth PB (1975) Differences between ethnic groups in sex dimorphism of adult height. *Annals of Human Biology* **2**, 35–39.

Eveleth PB and Tanner JM (1976) *Worldwide Variation in Human Growth.* Cambridge: Cambridge University Press.

Eveleth PB and Tanner JM (1990) *Worldwide Variation in Human Growth, 2nd edn. Cambridge: Cambridge University Press.*

Fattorini F, Caccio S, Gustincich S, Altamura B and Graziosi G (1989) Sex determination from skeleton. A new method using a DNA probe. *Acta Medicinae Legalis et Socialis* **39**, 201–205.

Faerman M, Filon D, Kahila G, Greenblatt CL, Smith P and Oppenheim A (1995) Sex identification of archaeological human remains based on amplification of the X and Y amelogenin alleles. *Gene* **167**, 327–332.

FitzGerald CM (1998) Do enamel microstructures have regular time dependency? Conclusions from the literature and a large-scale study. *Journal of Human Evolution* **35**, 371–386.

FitzGerald CM, Saunders SR, Macchiarelli R and Bondioli L (1999) Large scale histological assessment of deciduous crown formation. In *Proceedings of the*

11th Symposium of Dental Morphology, Oulu, Finland, ed. J.T. Mayhall and T. Heikkinen, in press.

Garn SM (1980) Human growth. *Annual Review of Anthropology* **9**, 275–292.

Gindhart PS (1973) Growth standards for the tibia and radius in children aged one month through eighteen years. *American Journal of Physical Anthropology* **39**, 41–48.

Gleiser I and Hunt EE (1955) The permanent mandibular first molar: its calcification, eruption and decay. *American Journal of Physical Anthropology* **13**, 253–283.

Goldstein MS (1940) Recent trends in physical anthropology. *American Journal of Physical Anthropology* **26**, 191–209.

Goode H. Waldron T and Rogers J (1993) Bone growth in juveniles: a methodological note. *International Journal of Osteoarchaeology* **3**, 321–323.

Goodman AH and Rose JC (1990) Assessment of systemic physiological perturbations from dental enamel hypoplasias and associated histological structures. *Yearbook of Physical Anthropology* **33**, 59–110.

Goodman AH and Rose JC (1991) Dental enamel hypoplasias as indicators of nutritional status. In *Advances in Dental Anthropology*, ed. C. Larsen and M. Kelley, pp. 279–293. New York: Alan R. Liss.

Goodman AH, Armelagos GJ and Rose JC (1984) The chronological distribution of enamel hypoplasias from prehistoric Dickson Mounds populations. *American Journal of Physical Anthropology* **65**, 259–266.

Goodman AH, Allen LH, Hernandez GP, Amador A, Arriola LV, Chavez A and Pelto GH (1987) Prevalence and age at development of enamel hypoplasias in Mexican children. *American Journal of Physical Anthropology* **72**, 7–19.

Grimm H (1990) [Secular trend of body height and 'acceleration' or fluctuation in growth in height in child development – on questions of methodology in relation to pre-historical skeletal remains]. *Ärztliche Jugendkunde* **81**, 437–440.

Gustafson G and Koch G (1974) Age estimation up to 16 years of age based on dental development. *Odontologisk Revy* **25**, 297–306.

Guy H, Masset C and Baud C (1997) Infant taphonomy. *International Journal of Osteoarchaeology* **7**, 221–229.

Haag U and Matsson L (1985) Dental maturity as an indicator of chronological age. The accuracy and precision of three methods. *European Journal of Orthodontics* **7**, 25–34.

Hawkes SC and Wells C (1975) An Anglo-Saxon obstetric calamity from Kingsworthy, Hampshire. *Medical and Biological Illustration* **25**, 47–51.

Heim J-L (1982) *Les enfants de La Farrassie*. Paris: Masson.

Herring DA, Saunders SR and Katzenberg MA (1998) Investigating the weaning process in past populations. *American Journal of Physical Anthropology* **105**, 425–439.

Higgins V (1989) A model for assessing health patterns from skeletal remains. In *Burial Archaeology: Current Research, Methods and Developments*, ed. C.A. Roberts, F. Lee and J. Bintliff, BAR British Series 211, pp. 175–204.

Hillson S and Bond S (1997) Relationship of enamel hypoplasia to the pattern of tooth crown growth: a discussion. *American Journal of Physical Anthropology* **104**, 89–104.

Högberg U, Iregren E, Siven C and Diener L (1987) Maternal deaths in medieval Sweden: an osteological and life table analysis. *Journal of Biosocial Science* **19**, 495–503.

Hoppa RD (1992) Evaluating human skeletal growth: an Anglo-Saxon example. *International Journal of Osteoarchaeology* **2**, 275–288.

Hoppa RD and Saunders SR (1994) The $\delta 1$ method for examining bone growth in juveniles: a reply. *International Journal of Osteoarchaeology* **4**, 261–263.

Hrdlička A (1927) Anthropology and medicine. *American Journal of Physical Anthropology* **10**, 1–10.

Hublin JJ and Tillier AM (1988) Mousterian children of Jebel Irhoud (Morocco): comparison with juvenile European Neandertals. *Bulletins et Mémoires de la Société d'Anthropologie de Paris*, 237–246.

Huda TF and Bowman JE (1995) Age determination from dental microstructure in juveniles. *American Journal of Physical Anthropology* **92**, 135–150.

Hühne-Osterloh G (1989) [Causes of paediatric mortality in a medieval skeletal series.] *Antropologischer Anzeiger* **47**, 11–25.

Hühne-Osterloh G and Grupe G (1989) Causes of infant mortality in the Middle Ages revealed by chemical and palaeopathological analyses of skeletal remains. *Zeitschrift für Morphologie und Antropologie*, **77**, 247–58.

Hummert JR (1983) Childhood growth and morbidity in a medieval population from Kulubnarti in the 'Batn El Hajar' of Sudanese Nubia. PhD dissertation, University of Colorado at Boulder.

Hummert JR and Van Gerven DP (1983) Skeletal growth in a medieval population from Sudanese Nubia. *American Journal of Physical Anthropology* **60**, 471–478.

Humphrey LT (1998) Patterns of growth in the modern human skeleton. *American Journal of Physical Anthropology* **105**, 57–72.

Hunt DR (1990) Sex determination in the subadult ilia: an indirect test of Weaver's nonmetric sexing method. *Journal of Forensic Sciences* **35**, 881–885.

Hunt EE and Gleiser I (1955) The estimation of age and sex of pre-adolescent children from bones and teeth. *American Journal of Physical Anthropology* **13**, 479–487.

Huss-Ashmore R, Goodman AH and Armelagos GJ (1982) Nutritional inference from palaeopathology. *Advances in Archaeological Method and Theory* **5**, 395–474.

Hutchins LA (1998) Standards of infant long bone diaphyseal growth from a late nineteenth century and early twentieth century almshouse cemetery. MA thesis, University of Wisconsin-Milwaukee.

Introna F, Cantatore F, Dragone M and Colonna M (1993) [Sexual dimorphism of deciduous teeth in medico-legal identification.] *Bollettino – Societa Italiana Biologia Sperimentale* **69**, 223–230.

Jantz RL and Owsley WD (1984a) Long bone growth variation among Arikara skeletal populations. *American Journal of Physical Anthropology* **63**, 13–20.

Jantz RL and Owsley WD (1984b) Temporal changes in limb proportionality among skeletal samples of Arikara Indians. *Annals of Human Biology* **11**, 157–163.

Jantz RL and Owsley DW (1994) Growth and dental development in Arikara children. In *Skeletal Biology in the Great Plains: Migration, Warfare, Health, and Subsistence*, ed. D.W. Owsley and R.L. Jantz, pp. 247–258, Washington, DC: Smithsonian Institution Press.

Johnston FE (1962) Growth of the long bones of infants and young children at Indian Knoll. *American Journal of Physical Anthropology* **20**, 249–254.

Johnston FE (1968) Growth of the skeleton in earlier peoples. In *The Skeletal Biology of Earlier Human Populations*, ed. D.R. Brothwell, pp. 57–66. London: Pergamon Press.

Johnston FE (1969) Approaches to the study of developmental variability in human skeletal populations. *American Journal of Physical Anthropology* **31**, 335–341.

Johnston FE and Zimmer LO (1989) Assessment of growth and age in the immature skeleton. In *Reconstruction of Life from the Skeleton*, ed. M.Y. Iscan and K.A.R. Kennedy, pp. 11–21. New York: Alan R. Liss.

Jungers WL, Cole TM and Owsley DW (1988) Multivariate analysis of relative growth in the limb bones of Arikara Indians [published erratum appears in *Growth, Development and Ageing* 1989; **53**(4), 140]. *Growth, Development, and Ageing* **52**, 103–107.

Kataja M, Nystrom M and Aine L (1989) Dental maturity standards in southern Finland. *Proceeding of the Finnish Dental Society* **85**, 187–197.

Katzenberg MA and Pfeiffer S (1995) Nitrogen isotope evidence for weaning age in a nineteenth century Canadian skeletal sample. In *Bodies of Evidence: Reconstructing History through Skeletal Analysis*, ed. A.L. Grauer, pp. 139–160. New York: John Wiley and Sons, Inc.

Katzenberg MA, Herring DA and Saunders SR (1996) Weaning and infant mortality: evaluating the skeletal evidence. *Yearbook of Physical Anthropology* **39**, 177–199.

Kneissel M, Roschger P, Steiner W, Schamall D, Kalchhauser G, Boyde A and Teschler-Nicola M (1997) Cancellous bone structure in the growing and ageing lumbar spine in a historic Nubian population. *Calcified Tissue International* **61**, 95–100.

Lampl M and Johnston FE (1996) Problems in the ageing of skeletal juveniles: perspectives from maturation assessments of living children. *American Journal of Physical Anthropology* **101**, 345–355.

Lampl M, Monge JM and Mann AE (1993) Further observations on a method for estimating hominoid dental development patterns. *American Journal of Physical Anthropology* **90**, 113–127.

Lanphear KM (1990) Frequency and distribution of enamel hypoplasias in a historic skeletal sample. *American Journal of Physical Anthropology* **81**, 35–43.

Lassen C, Hummel S and Herrmann B (1997) [Molecular sex determination in skeletal remains of premature and newborn infants of the Aegerten, Switzerland, burial field.] *Anthropologischer Anzeiger* **55**, 183–191.

Lauterstein AM (1961) A cross-sectional study in dental development and skeletal age. *Journal of the American Dental Association* **62**, 161–167.

Leigh SR (1996) Evolution of human growth spurts. *American Journal of Physical Anthropology* **101**, 455–474.

Leigh SR and Park PB (1998) Evolution of human growth prolongation. *American Journal of Physical Anthropology* **107**, 331–350.

Lewis AB and Garn SM (1960) The relationship between tooth formation and other maturational factors. *The Angle Orthodontist* **39**, 70–77.

Liliequist B and Lundberg M (1971) Skeletal and tooth development. *Acta Radiologica* **11**, 97–112.

Liversidge HM (1994) Accuracy of age estimation from developing teeth of a population of known age (0–5.4 years). *International Journal of Osteoarchaeology* **4**, 37–46.

Liversidge HM, Dean MC and Molleson TI (1993) Increasing human tooth length between birth and 5.4 years. *American Journal of Physical Anthropology* **90**, 307–313.

Lovejoy CO, Russell KF and Harrison ML (1990) Long bone growth velocity in the Libben population. *American Journal of Human Biology* **2**, 533–541.

Macho GA (1995) The significance of hominid enamel thickness for phylogenetic and life-history reconstruction. In *Aspects of Dental Biology; Paleontology, Anthropology and Evolution*, ed. J. Moggi-Cecchi, pp. 51–68. Florence: International Institute for the Study of Man.

Mahler PE (1968) Growth of the long bones in a prehistoric population from Sudanese Nubia. MA thesis, University of Utah.

Majo T. Tillier AM and Bruzek J (1993) [Test of Schutkowskis discriminant function analysis for the determination of sex when applied to the ilium in an immature sample of known age and sex.] *Bulletins et Mémoires de la Société d'Anthropologie de Paris* **5**, 61–68.

Mann A (1975) *Paleodemographic Aspects of the South African Australopithecines.* Pennsylvania: University of Pennsylvania Publications in Anthropology.

Mann A and Vandermeersch B (1997) An adolescent female Neandertal mandible from Montgaudier Cave, Charente, France. *American Journal of Physical Anthropology* **103**, 507–527.

Mann A, Lampl M and Monge JM (1987) Maturational patterns in early hominids. *Nature* **328**, 673–675.

Mann A, Lampl M and Monge JM (1990) Patterns of ontogeny in human evolution: evidence from dental development. *Yearbook of Physical Anthropology* **33**, 111–150.

Maresh M (1943) Growth of major long bones in healthy children. *American Journal of Diseases in Children* **66**, 227–254.

Maresh M (1955) Linear growth of long bones of extremities from infancy through adolescence. *American Journal of Diseases in Children* **89**, 725–742.

Maresh M (1970) Measurements from roentgenograms. In *Human Growth and Development*, ed. R.W. McCammon, pp. 157–200. Illinois: Charles C. Thomas.

Markowitz DL (1995) Arikara subadult craniofacial development: a cross-sectional growth study in three dimensions. PhD dissertation, University of Pennsylvania.

Martorell R and Ho TJ (1984) Malnutrition, morbidity and mortality. In *Child*

Survival: Strategies for Research, ed. W.H. Mosley and L.C. Chen, pp. 49–68. Population Development Review, Supplement to vol. 10.

Mays SA (1985) The relationship between Harris line formation and bone growth and development. *Journal of Archaeological Science* **12**, 207–220.

Mays S (1995) The relationship between Harris lines and other aspects of skeletal development in adults and juveniles. *Journal of Archaeological Science* **22**, 511–520.

Mensforth RP (1985) Relative tibia long bone growth in the Libben and Bt-5 prehistoric skeletal populations. *American Journal of Physical Anthropology* **68**, 247–262.

Merchant VL (1973) A cross-sectional growth study of the protohistoric Arikara from skeletal material associated with the Mobridge site (39WWI), South Dakota. MA thesis, The American University.

Merchant VL and Ubelaker DH (1977) Skeletal growth of the protohistoric Arikara. *American Journal of Physical Anthropology* **46**, 61–72.

Miles AW and Bulman JS (1994) Growth curves of immature bones from a Scottish Island population of sixteenth to mid-nineteenth century: limb bone diaphyses and some bones of the hands and feet. *International Journal of Osteoarchaeology* **4**, 121–136.

Miles AW and Bulman JS (1995) Growth curves of immature bones from a Scottish Island population of sixteenth to mid-nineteenth century: shoulder girdle, ilium, pubis and ischium. *International Journal of Osteoarchaeology* **5**, 15–28.

Minugh-Purvis N (1988) Patterns of craniofacial growth and development in upper pleistocene hominids. PhD dissertation, University of Pennsylvania.

Minugh-Purvis N (1993) Reexamination of the immature hominid maxilla from Tangier, Morocco. *American Journal of Physical Anthropology* **92**, 449–461.

Mittler DM and Sheridan SG (1992) Sex determination in subadults using auricular surface morphology: a forensic science perspective. *Journal of Forensic Science* **37**, 1068–1075.

Moggi-Cecchi J, Pacciani E and Pinto-Cisternas J (1994) Enamel hypoplasia and age at weaning in 19th-century Florence, Italy. *American Journal of Physical Anthropology* **93**, 299–306.

Moggi-Cecchi J, Tobias PV and Beynon AD (1998) The mixed dentition and associated skull fragments of a juvenile fossil hominid from Sterkfontein, South Africa. *American Journal of Physical Anthropology* **106**, 425–465.

Molleson TI (1989) Social implications of mortality patterns of juveniles from Poundbury Camp, Romano-British cemetery. *Anthropologischer Anzeiger* **47**, 27–38.

Molleson TI (1990) Retardation of growth and early weaning of children in prehistoric populations. *Acta Musei Nationalis Prague* **46b**, 182–188.

Moorrees CFA, Fanning EA and Hunt EE (1963a) Formation and resorption of three deciduous teeth in children. *American Journal of Physical Anthropology* **21**, 205–213.

Moorrees CFA, Fanning EA and Hunt EE (1963b) Age variation of formation stages for 10 permanent teeth. *Journal of Dental Research* **42**, 1490–1502.

Owsley WD and Jantz RL (1985) Long bone lengths and gestational age distribu-

tions of post-contact period Arikara Indian perinatal infant skeletons. *American Journal of Physical Anthropology* **68**, 321–328.
Porter RW and Pavitt D (1987) The vertebral canal I. Nutrition and development, an archaeological study. *Spine* **12**, 901–906.
Proy E, Sempe M and Ajacques JC (1981) etude comparée des maturations dentaire et squelettique chez des enfants et adolescents français. *Revue d'Orthopédie Dento-faciale* **15**, 3.
Ramirez-Rozzi FV (1991) Differences in molar teeth development between *Paranthropus aethiopicus* and *P. boisei*. *American Journal of Physical Anthropology Supplement* **12**, 148.
Ramirez-Rozzi FV (1993) Tooth development in East African *Paranthropus*. *Journal of Human Evolution* **24**, 429–454.
Ramirez-Rozzi FV (1995) Enamel microstructure as a tool for taxonomic attribution of Plio-Pleistocene hominids. In *Aspects of Dental Biology: Palaeontology, Anthropology and Evolution*, ed. J. Moggi-Cecchi, pp. 217–238. Florence: International Institute for the Study of Man.
Reid DJ, Schwartz GT, Dean MC and Chandrasekera (1998a) A histological reconstruction of dental development in the common chimpanzee, *Pan troglodytes*. *Journal of Human Evolution* **35**, 427–448.
Reid DJ, Beynon AD and Rozzi FVR (1998b) Histological reconstruction of dental development in four individuals from a medieval site in Picardie, France. *Journal of Human Evolution* **35**, 364–477.
Reynolds EL (1945) The bony pelvic girdle in early infancy. *American Journal of Physical Anthropology* **3**, 321.
Reynolds EL (1947) The bony pelvis in prepuberal childhood. *American Journal of Physical Anthropology* **5**, 165–200.
Ribot I and Roberts C (1996) A study of non-specific stress indicators and skeletal growth in two medieval subadult populations. *Journal of Archaeological Science* **23**, 67–79.
Risnes S (1986) Enamel apposition rate and the prism periodicity in human teeth. *Scandinavian Journal of Dental Research* **94**, 394–404.
Saunders SR (1992) Subadult skeletons and growth related studies. In *Skeletal Biology of Past Peoples: Research Methods*, ed. S.R. Saunders and M.A. Katzenberg, pp. 1–20. New York: Wiley-Liss.
Saunders SR and Hoppa RD (1993) Growth deficit in survivors and non-survivors: biological mortality bias in subadult skeletal samples. *Yearbook of Physical Anthropology* **36**, 127–151.
Saunders SR and Keenleyside A (1999) Enamel hypoplasia in a Canadian historic sample. *American Journal of Human Biology*, in press.
Saunders SR and Melbye FJ (1990) Subadult mortality and skeletal indicators of health in Late Woodland Ontario Iroquois. *Canadian Journal of Archaeology* **14**, 61–74.
Saunders SR, DeVito CA, Herring DA, Southern RA and Hoppa RD (1993a) Accuracy tests of tooth formation age estimations for human skeletal remains. *American Journal of Physical Anthropology* **92**, 173–188.
Saunders SR, Hoppa RD and Southern RA (1993b) Diaphyseal growth in a 19th

century skeletal sample of subadults from St. Thomas' Church, Belleville, Ontario. *International Journal of Osteoarcheology* **3**, 265–281.

Schour I and Massler M (1940a) Studies in tooth development: the growth pattern of human teeth. *Journal of the American Dental Association* **27**, 1779–1793.

Schour I and Massler M (1940b) Studies in tooth development: the growth pattern of human teeth. *Journal of the American Dental Association* **27**, 1918–1931.

Schour I and Massler M (1941) The development of the human dentition. *Journal of the American Dental Association* **28**, 1153–1160.

Schurr MR (1997) Stable nitrogen isotopes as evidence for the age of weaning at the Angel Site: a comparison of isotopic and demographic measures of weaning age. *Journal of Archaeological Science* **24**, 919–927.

Schutkowski H (1986) [Sex differences in child skeletons. Current knowledge and diagnostic significance.] *Zeitschrift für Morphologie und Anthropologie* **76**, 149–168.

Schutkowski H (1987) Sex determination of fetal and neonate skeletons by means of discriminant analysis. *International Journal of Anthropology* **2**, 347–352.

Schutkowski H (1989) [Age and sex diagnosis of the skeleton of immature individuals.] *Anthropologischer Anzeiger* **47**, 1–9.

Schutkowski H (1993) Sex determination of infant and juvenile skeletons. I. Morphological features. *American Journal of Physical Anthropology* **90**, 199–205.

Sciulli P (1994) Standardization of long bone growth in children. *International Journal of Osteoarchaeology* **4**, 257–259.

Sejrsen B, Jakobsen J, Skovgaard LT and Kjaer I (1997) Growth in the external cranial base evaluated on human dry skulls, using nerve canal openings as references. *Acta Odontologica Scandinavica* **55**, 356–364.

Shellis RP (1998) Utilisation of periodic markings in enamel to obtain information on tooth growth. *Journal of Human Evolution* **35**, 387–400.

Simpson SW, Russel KF and Lovejoy CO (1996) Comparison of diaphyseal growth between the Libben population and the Hammann-Todd chimpanzee sample. *American Journal of Physical Anthropology* **99**, 67–78.

Sjøvold T, Swedborg I and Diener L (1974) A pregnant woman from the Middle Ages with exostosis multiplex. *Ossa* **1**, 3.

Skinner M and goodman AH (1992) Anthropological uses of developmental defects of enamel. In *Skeletal Biology of Past Peoples*, ed. S.R. Saunders and M.A. Katzenberg, pp. 153–174. New York: Wiley-Liss, Inc.

Smith BH (1986) Dental development in Australopithecus and early Homo. *Nature* **323**, 327–330.

Smith BH (1987) Maturational patterns in early hominids. *Nature* **328**, 674–675.

Smith BH (1989) Dental development as a measure of life history in primates. *Evolution* **43**, 683–688.

Smith BH (1991) Standards of human tooth formation and dental age assessment. In *Advances in Dental Anthropology*, ed. M.A. Kelley and C.S. Larsen, pp. 143–168. New York: Wiley-Liss, Inc.

Smith BH (1992) Life history and the evolution of human maturation. *Evolutionary Anthropology* **1**, 134–142.

Smith BH (1993) The physiological age of KNM-WT 15000. In *The Nariokotome*

Homo erectus *Skeleton*, ed. A. Walker and R. Leakey, pp. 195–220. Cambridge MA: Harvard University Press.

Smith BH (1994) Patterns of dental development in *Homo*, australopithecines, *Pan* and *Gorilla*. *American Journal of Physical Anthropology* **94**, 307–325.

Smith BH and Tompkins RL (1995) Toward a life history of the Hominidae. *Annual Reviews of Anthropology* **24**, 257–279.

Smith BH, Crummett TL and Brandt KL (1994) Ages of eruption of primate teeth: a compendium for ageing individuals and comparing life histories. *Yearbook of Physical Anthropology* **37**, 177–231.

Staaf V, Mornstad H and Welander U (1991) Age estimation based on tooth development: a test of reliability and validity. *Scandinavian Journal of Dental Research* **99**, 281–286.

Stewart TD (1954) Evaluation of evidence from the skeleton. In *Legal Medicine*, ed. R.B.H. Gradwohl, pp. 407–450. St Louis: C. V. Mosby Co.

Steyn M and Henneberg M (1996a) Skeletal growth of children from the Iron site at K2 (South Africa). *American Journal of Physical Anthropology* **100**, 389–396.

Steyn M and Henneberg M (1996b) Cranial growth in the prehistoric sample from K2 at Mapungubwe (South Africa) is population specific. *Homo* **48**, 62–71.

Stone AC, Milner GR, Pääbo S and Stoneking M (1996) Sex determination of ancient human skeletons using DNA. *American Journal of Physical Anthropology* **99**, 231–238.

Storey R (1986) Perinatal mortality at pre-Columbian Teotihuacan. *American Journal of Physical Anthropology* **69**, 541–548.

Storey R (1992) The children of Copan: issues in paleopathology and paleodemography. *Ancient Mesoamerica* **3**, 161–167.

Stloukal V and Hanáková H (1978). Die Länge der Längsknochen altslawischer Bevölkerungen – Unter besonderer Berucksichtigung von Wachstumsfragen. *Homo* **29**, 53–69.

Sundick RI (1972) Human skeletal growth and dental development as observed in the Indian Knoll population. PhD dissertation, University of Toronto.

Sundick RI (1977) Age and sex determination of subadult skeletons. *Journal of Forensic Sciences* **22**, 141–144.

Sundick RI (1978) Human skeletal growth and age determination. *Homo* **29**, 228–249.

Teschler-Nicola M and Prossinger H (1998) Sex determination using tooth dimensions. In *Dental Anthropology: Fundamentals, Limits and Prospects*, ed. K.W. Alt, R.W. Rösing and M. Teschler-Nicola, pp. 479–500. Vienna: Springer-Verlag.

Thompson A (1899) The sexual differences of the foetal pelvis. *Journal of Anatomy, London* **33**, 359–380.

Tillier AM (1982) [The Neandertal children from Devil's Tower.] *Zeitschrift für Morphologie und Anthropologie* **73**, 125–148.

Tillier AM (1983) Le crâne d'enfant d'Engis 2: un exemple de distribution des caractères juvéniles, primitifs et néandertaliens. *Bulletin de la Société Belge d'Anthropologie et de Préhistoire* **94**, 51–75.

Tiller AM (1988) Devil's Tower remains (Gibraltar) and their place in Neandertal ontogeny. *Bulletins et Mémoires de la Société d'Anthropologie de Paris*, 257–266.

Tillier AM (1995) Biologie du squelette et populations anciennes: perspectives et limites de la paléoauxologie. Paper presented at the 22nd Congress of the Groupement des Anthropologistes de Langue Française, Brussels.

Tompkins RL (1996) Relative dental development of Upper Pleistocene hominids compared to human population variation. *American Journal of Physical Anthropology* **99**, 103–118.

Van Gerven DP, Hummert JR and Burr DB (1985) Cortical bone maintenance and geometry of the tibia in prehistoric children from Nubia's Batn el Hajar. *American Journal of Physical Anthropology* **66**, 275–280.

Walker PL (1969) The linear growth of long bones in Late Woodland Indian children. *Proceedings of the Indiana Academy of Science* **78**, 83–87.

Wall CE (1991) Evidence of weaning stress and catch-up growth in the long bones of a central California Amerindian sample. *Annals of Human Biology* **18**, 9–22.

Weaver DS (1980) Sex differences in the ilia of a known age and sex sample of fetal and infant skeletons. *American Journal of Physical Anthropology* **52**, 191–195.

Wells C (1978) A medieval burial of a pregnant woman. *Practitioner* **221**, 442–444.

Wood JW, Milner GR, Harpending HC and Weiss KM (1992) The osteological paradox: problems of inferring prehistoric health from skeletal samples. *Current Anthropology* **33**, 343–370.

Wright LE and Schwarcz HP (1998) Stable carbon and oxygen isotopes in human tooth enamel: identifying breast feeding and weaning in prehistory. *American Journal of Physical Anthropology* **106**, 1–18.

y'Edynak G (1976) Long bone growth in western Eskimo and Aleut skeletons. *American Journal of Physical Anthropology* **45**, 569–574.

Zollikofer CP, Ponce de León MS, Martin RD and Stucki P (1995) Neandertal computer skulls. *Nature* **375**, 283–285.

2 *Heterochrony:* *somatic, skeletal and dental development in* Gorilla, Homo *and* Pan

MIKE DAINTON AND GABRIELE A. MACHO

Introduction

Over the last decade much effort has been expended in elucidating the biological processes underlying the evolution of modern human morphology and development. Determination of ontogenetic trajectories between closely related taxa, using both comparative and experimental methods is a prerequisite for making inferences about heterochronic modifications of growth and development in extinct lineages (Hall 1992). Analyses of teeth, in particular, have proved useful in approximating the relative timing of developmental events in extinct taxa (Macho and Wood 1995). Notably, eruption of first permanent molars, which can be assessed in fossils, is highly correlated with other life history variables, including brain size, in primates (Smith 1989, 1991; Smith *et al.* 1995). Such clear-cut relationships are not the case, however, with regard to other dental developmental events, and the relationship between life history and homologous stages of skeletal maturation is even more obscure. For example, when determining age at death of an African *Homo erectus* youth, KNM-WT 15 000, on the basis of somatic maturity, stature and dental development, Smith (1993) obtained results conflicting by 2 to 5 years. The choice of an appropriate extant model for comparisons of somatic maturity is confounded, further, by the fact that the sequence and timing of epiphyseal fusion is different among hominoids (Schultz 1960), as are long bone growth trajectories (Jungers and Hartman 1988; Hartwig-Scherer 1992; Simpson *et al.* 1996); it must also be appreciated that extinct taxa may exhibit unique growth patterns not found in extant species.

Although developmental processes are generally determined by an interplay of genetic, epigenetic and environmental factors see (e.g. Atchley and Hall 1991), their respective influence on dental and somatic development differs substantially, thus resulting in some structures, like teeth, being better suited than others to reveal heterochronic modifications during evolution. Conversely, bone growth is generally regarded as an indicator of maturational status and hand/wrist bones are considered particularly useful in this regard (Greulich and Pyle 1959; Tanner *et al.* 1983; Smith 1993; Marzke *et al.* 1996; Winkler 1996). Surprisingly, however, a previous study (see Dainton and Macho 1999) indicated that carpal growth in *Pan* and *Gorilla* may stop before full adult body mass is attained, thus casting doubt on propositions that carpal development in extant hominoids can be employed to assess somatic maturation, although this may still be the case in earlier stages of development. Nonetheless, this finding raises fundamental questions about the sequence of development and maturation of different systems of the body, i.e. carpals, long bones and dentition. In fact, a recent study on modern humans confirms that skeletal elements stop growing at different times through ontogeny, the timing of which may be related to differing functional requirements (Humphrey 1998). It is the aim of this chapter to shed some light on these relationships, and to determine whether there are commonalities among modern humans and the African great apes that allow inferences to be made about the evolutionary processes underlying human and great ape growth patterns.

Methodological considerations

A number of analytical techniques have been employed to assess patterns of dental maturation (reviewed in Macho and Wood 1995), while skeletal growth patterns in extant and extinct hominoid species are commonly determined through methods of ontogenetic allometry (e.g. Shea 1981, 1986; Jungers and Hartman 1988; Inouye 1992, 1994; Jungers and Cole 1992; Taylor 1995). In such skeletal studies, body size (frequently described by mass), rather than age, is generally considered a useful variable for indicating homologous developmental stages during ontogeny, and against which to assess the development of other structures (Hall 1984; Jungers 1985). The relationship of a structure's growth to some size variable is typically described by the allometric power equation $Y = kX^b$ (Huxley 1932), which is linear when log-transformed, provided that their relative growth remains constant throughout the series (Lumer 1937). Owing to the limited records frequently associated with primate skeletal collections, and the incomplete preservation of fossil hominids, inferences about size or body mass are commonly made from intervening variables. This approach is, however, fraught with problems (Smith 1996), since intervening variables may not accurately reflect the true changes occurring in body mass, inter- or intraspecifically, especially where subadults are concerned, unless they meet strict criteria (Hartwig-Scherer and Martin 1992). A successful intervening variable should be highly correlated with body size (Jungers and German 1981) and maintain the same allometric exponent (Jungers 1985; Mosimann and James 1979) and coefficient (Hartwig-Scherer and Martin 1992) across all the species being compared. It is also important that the relationship between an intervening variable and body size is maintained throughout development. This is confounded by the fact that specific bones stop growing at different times during ontogeny (Humphrey 1998) through, for example, sequential fusion of epiphyseal plates. Furthermore, in primates body mass may continue to increase after the limbs have stopped growing (Sirianni and Swindler 1985). Unless body mass and long bone growth cessation coincide, the scaling relationship between the two variables will change and the dimensions of particular long bones will no longer reflect changing body mass. Even more important for comparative purposes is the fact that differences exist in the sequence and timing of epiphyseal union between species (Schultz 1960; Wintheiser *et al.* 1977). For example, in modern humans, the humerus fuses last, whereas in *Pan* and *gorilla* it is either the ulna or radius (Schultz 1960). Such differences in sequence between closely related species clearly caution against the use of a specific bone dimension as a substitute for body mass. Conversely, given that increase in long bone length will add to body mass, it stands to reason that the long bone, whose epiphyses fuse last, may most accurately mirror changes in body mass throughout ontogeny, and may thus be usefully employed for scaling purposes. If correct, this would greatly aid studies concerned with ontogenetic scaling.

Body mass and long bones

For the present study, body mass was assessed using the methods and equations given by Hartwig-Scherer and Martin (1992; see also Dainton and Macho 1999), for cross-sectional, mixed age, skeletal samples of wild-shot *Pan troglodytes* ($n = 31$) and *Gorilla gorilla* ($n = 30$) and a *Homo sapiens* sample of both African and European descent ($n = 60$). Least squares regressions are calculated for each species using linear, quadratic and cubic models on log-transformed data to determine the relationship between humerus length, radius length and body mass. While it is acknowledged that quadratic and cubic polynomials are unlikely to adequately describe the functional relationship between body mass and bone length, they do allow statistical assessment of whether there is a significant departure from linearity in a bivariate scatter plot. In order to appraise the relative goodness of fit of the three models, one must test the significance of any additional variance explained, beyond that of the linear equations, firstly by the quadratic and secondly by the cubic regressions (Fig. 2.1, Table 2.1). Linear regression seem to describe sufficiently the relationships between long bones and body mass in *Pan* and *Homo*, but quadratic equations give statistically better results in *Gorilla* (Table 2.1). To reiterate, log-transformation of measurements will only linearise the data scatter if the relative growth of the two dimensions remains constant (Lumer 1937; Sprent 1972). It follows therefore that there must be a relative change in the rate of growth of one or both dimensions whenever curvilinearity is observed. Examination of the *Gorilla* log–log plots (Fig. 2.1) and those of the residuals for the linear regressions reveals that this curvilinearity is brought about by the cessation of longitudinal growth of the radius and humerus prior to the attainment of full body mass in both sexes. Similar observations have been reported for other primates and it is likely that this situation results from an increase in muscle mass, rather than fat, mediated by hormones (e.g. Faucheux *et al.* 1978). This finding may also explain why the exponent in static adult allometric analyses (long bone length against body mass) is significantly lowered, relative to the ontogenetic exponent, in *Gorilla* but not in *Pan* or *Homo* (Hartwig-Scherer 1992), since body mass continues to increase after cessation of long bone growth in the former. When humerus length is used as a proxy for body mass, however, curvilinearity is not observed. For example, when radius length is regressed against humerus length, a linear model fits the scatter best, obscuring the true curvilinear nature of the relationship between radius length and body mass in *Gorilla*. Hence, analyses of proportional differences between bones throughout ontogeny are unlikely to reveal shape changes during skeletal development and maturation, relative to overall size (see also Fig. 9 in Dainton and Macho 1999). Thus, even long bones, which fuse last, provide a poor yardstick against which to assess differences in ontogenetic development between species, since their cessation of growth is not necessarily synchronised with the attainment of full adult body mass. This situation seems further confounded when the ontogenetic trajectories of small, irregular bones, such as carpals (Fig. 2.2), are assessed against either body mass or an intervening variable (Fig. 2.3, Table 2.2, 2.3).

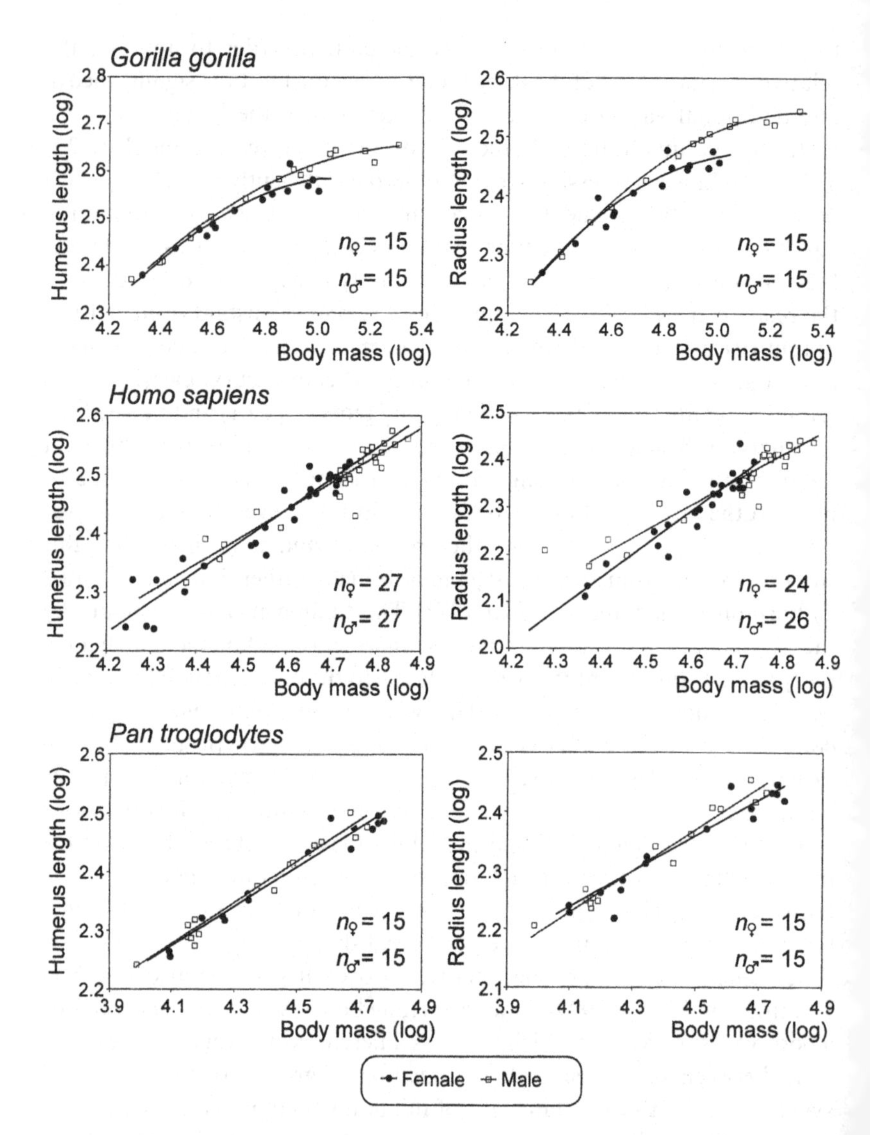

Fig. 2.1. Ontogenetic changes in humerus length (mm) and radius length (mm) relative to body mass (g) in each species (*Gorilla*, *Homo*, *Pan*) are shown, by sex. Only those regressions that fit the data plot best, i.e. linear or quadratic, are indicated.

Table 2.1. *Coefficient of determination and F-statistics are given for linear, quadratic and cubic least square regression analyses of either humerus or radius against body mass, or radius length against humerus length, for each species separately. The significance of additional variance explained by the respective regressions is also given (Sokal and Rohlf 1995)*

(a) Males

		Humerus/ mass		Radius/ mass		Radius/ humerus	
	Curve	r^2	F	r^2	F	r^2	F
Gorilla	Linear	0.938	859.81**	0.953	2022.78**	0.990	1064.37**
	Quadratic	0.988	46.02**	0.995	86.93**	0.990	0.08
	Cubic	0.988	0.21	0.995	0.46	0.990	0.00
Homo	Linear	0.887	202.37**	0.871	156.65**	0.966	666.23**
	Quadratic	0.890	0.63	0.872	0.26	0.967	0.53
	Cubic	0.890	0.04	0.872	0.01	0.967	0.05
Pan	Linear	0.965	331.02**	0.964	324.00**	0.992	1508.77**
	Quadratic	0.965	0.00	0.964	0.20	0.992	0.69
	Cubic	0.965	0.00	0.964	0.00	0.992	0.08

**$p < 0.01$.

(b) Females

		Humerus/ mass		Radius/ mass		Radius/ humerus	
	Curve	r^2	F	r^2	F	r^2	F
Gorilla	Linear	0.886	133.33**	0.873	102.82**	0.912	134.60**
	Quadratic	0.926	6.03*	0.906	3.87	0.925	1.99
	Cubic	0.927	0.11	0.907	0.04	0.925	0.06
Homo	Linear	0.908	250.38**	0.828	105.98**	0.905	183.26**
	Quadratic	0.909	0.45	0.844	2.04	0.906	0.13
	Cubic	0.909	0.01	0.844	0.02	0.906	0.00
Pan	Linear	0.959	328.54**	0.941	188.72**	0.962	278.62**
	Quadratic	0.968	3.14	0.945	0.75	0.962	0.19
	Cubic	0.968	0.06	0.945	0.04	0.962	0.06

**$p < 0.01$.
*$p < 0.05$.

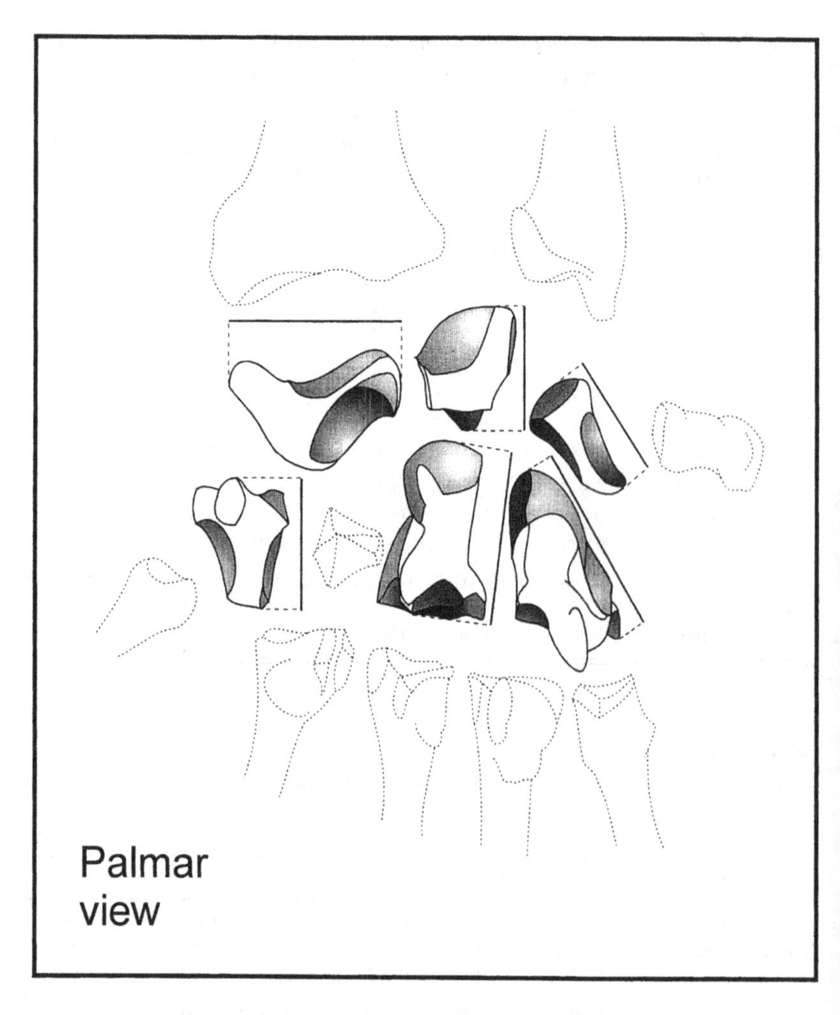

Fig. 2.2. Schematic view of the wrist in palmar view showing the measurements discussed in the present study. Each dimension conforms essentially to the bone's longest axis and was measured using digital callipers to the nearest 0.1 mm. Neither pisiform nor trapezoid dimensions are analysed because their relatively late ossification would have resulted in sample sizes being too small.

Body mass and carpals

Of the 36 comparisons presented in Table 2.2 for humerus length and body mass, respectively, linear regressions adequately describe the relationships between carpal and humerus length in 23 instances. This is the case for only 14 analyses against body mass where quadratic equations explain

Table 2.2. *Coefficient of determination and F-statistics are given for linear (L), quadratic (Q) and cubic (C) least square regression analyses of carpal dimensions against either body mass or humerus length, for each species separately. The significance of additional variance explained by the respective regressions is also given (Sokal and Rohlf 1995)*

(a) Male

		Body mass		Humerus length		Body mass		Humerus length	
		r^2	F	r^2	F	r^2	F	r^2	F
		Scaphoid				Lunate			
Gorilla	L	0.893	173.21**	0.931	12.08**	0.800	68.65**	0.882	67.26**
	Q	0.952	11.43**	0.931	0.00	0.893	7.96*	0.882	0.00
	C	0.954	0.26	0.931	0.00	0.895	0.19	0.882	0.00
Homo	L	0.418	16.09**	0.541	30.60**	0.551	34.29**	0.643	88.00**
	Q	0.430	0.46	0.611	3.96	0.614	3.93	0.825	24.86**
	C	0.430	0.00	0.611	0.00	0.614	0.04	0.825	0.02
Pan	L	0.795	73.32**	0.744	40.52**	0.778	47.61**	0.696	25.53**
	Q	0.869	6.88*	0.78	1.96	0.820	2.57	0.697	0.12
	C	0.870	0.09	0.78	0.00	0.820	0.00	0.700	0.01
		Triquetral				Hamate			
Gorilla	L	0.893	249.04**	0.953	231.33**	0.686	40.64**	0.768	37.05**
	Q	0.959	18.35**	0.955	0.29	0.809	7.30*	0.772	0.18
	C	0.961	0.34	0.955	0.03	0.814	0.28	0.772	0.03
Homo	L	0.518	24.90**	0.622	54.33**	0.771	8.77**	0.829	144.58**
	Q	0.541	1.10	0.748	10.95**	0.779	0.09	0.856	4.75*
	C	0.542	0.02	0.748	0.00	0.779	0.00	0.857	0.09
Pan	L	0.854	125.77**	0.792	56.31**	0.854	105.96**	0.799	56.40**
	Q	0.918	9.44**	0.831	2.77	0.892	4.76*	0.830	2.15
	C	0.919	0.09	0.831	0.03	0.892	0.07	0.830	0.03
		Capitate				Trapezium			
Gorilla	L	0.817	69.37**	0.859	67.31**	0.892	80.16**	0.869	84.70**
	Q	0.862	4.71*	0.859	0.00	0.899	0.59	0.908	3.81
	C	0.863	0.11	0.859	0.00	0.900	0.07	0.908	0.08
Homo	L	0.741	87.11**	0.780	139.97**	0.712	67.69**	0.780	248.73**
	Q	0.787	5.49*	0.861	14.46**	0.758	4.35*	0.928	47.17**
	C	0.787	0.00	0.861	0.03	0.758	0.04	0.928	0.00
Pan	L	0.841	69.78**	0.802	46.13**	0.880	81.95**	0.856	65.58**
	Q	0.875	6.24*	0.809	0.40	0.882	0.19	0.856	0.05
	C	0.875	0.00	0.809	0.01	0.882	0.02	0.856	0.01

**$p < 0.01$.
*$p < 0.05$.

Table 2.2. (*continued*)
(b) Female

		Body mass		Humerus length		Body mass		Humerus length	
		r^2	F	r^2	F	r^2	F	r^2	F
		Scaphoid				Lunate			
Gorilla	L	0.717	34.46**	0.667	24.85**	0.682	35.14**	0.691	29.60**
	Q	0.791	3.55**	0.730	2.36	0.805	6.30*	0.766	3.22
	C	0.792	0.05	0.731	0.05	0.806	0.06	0.767	0.04
Homo	L	0.763	66.92**	0.744	68.75**	0.869	173.98**	0.802	92.33**
	Q	0.772	0.79	0.794	4.65*	0.889	4.14	0.817	1.68
	C	0.772	0.01	0.794	0.00	0.89	0.05	0.818	0.07
Pan	L	0.745	131.76**	0.788	132.49**	0.812	114.21**	0.815	126.90**
	Q	0.937	34.08**	0.934	24.49**	0.922	15.47**	0.929	17.69**
	C	0.938	0.00	0.934	0.10	0.922	0.00	0.929	0.02
		Triquetral				Hamate			
Gorilla	L	0.777	91.14**	0.800	66.25**	0.722	46.02**	0.749	44.94**
	Q	0.914	16.03**	0.879	6.52*	0.827	6.66*	0.817	4.09
	C	0.915	0.12	0.879	0.06	0.827	0.00	0.817	0.00
Homo	L	0.715	50.76**	0.719	49.69**	0.843	149.37**	0.824	128.13**
	Q	0.718	0.27	0.719	0.00	0.853	1.72	0.832	1.35
	C	0.719	0.01	0.719	0.00	0.853	0.05	0.833	0.05
Pan	L	0.767	86.41**	0.799	76.80**	0.821	116.46**	0.853	228.83**
	Q	0.902	15.20**	0.885	8.26*	0.923	14.36**	0.959	28.47**
	C	0.902	0.00	0.885	0.01	0.923	0.01	0.959	0.10
		Capitate				Trapezium			
Gorilla	L	0.621	28.75**	0.644	26.91**	0.530	18.97**	0.541	14.84**
	Q	0.718	4.76*	0.737	3.91	0.721	6.82*	0.541	0.00
	C	0.718	0.00	0.737	0.00	0.721	0.00	0.635	2.59
Homo	L	0.875	178.24**	0.878	181.21**	0.586	27.42**	0.699	43.38**
	Q	0.877	0.49	0.879	0.16	0.594	0.40	0.699	0.00
	C	0.877	0.00	0.879	0.01	0.594	0.01	0.710	0.69
Pan	L	0.731	67.27**	0.768	100.65**	0.885	136.44**	0.911	179.86**
	Q	0.881	13.78**	0.915	19.37**	0.935	7.66*	0.949	7.45*
	C	0.881	0.00	0.916	0.07	0.935	0.07	0.949	0.06

**$p < 0.01$.
*$p < 0.05$.

Table 2.3. *Contingency table showing the observed and expected frequencies of cases where linear or quadratic regression analysis describe the scatter plot sufficiently, by species, and when carpel dimensions are regressed either against humerus length or body mass. The resulting chi-squares and significance levels are also shown*

	Carpal/humerus		Carpal/mass	
Species	Linear	Quadratic	Linear	Quadratic
Gorilla				
Observed	11	1	2	10
Expected	7.7	4.3	4.7	7.3
	+	–	–	+
Homo				
Observed	6	6	10	2
Expected	7.7	4.3	4.7	7.3
	–	+	+	–
Pan				
Observed	6	6	2	10
Expected	7.7	4.3	4.7	7.3
	–	+	–	+
		$\chi^2 = 6.02$*		$\chi^2 = 14.96$**

**$p < 0.01$.
*$p < 0.05$.
Plus signs indicate more 'observed' than 'expected' cases and minus signs the reverse.

a significantly higher proportion of the variances; these differences are significant at the 5% probability level ($\chi^2 = 4.5$; $p < 0.5$). Hence, cessation of carpal growth appears to predominantly occur prior to attainment of full adult body mass in hominoids and, in a significant number of instances, even prior to skeletal maturity as judged by growth of long bone length. Since both carpals and long bones seem to stop growing relatively early in *Gorilla*, the relatively higher percentage of adequate linear models with regard to carpal lengths against humerus length is unsurprising. Detailed inspection of the results indicates, however, that there are differences between species and, in a few instances, sexes (Fig. 2.3, Table 2.2). In both *Gorilla* and *Pan*, cessation of carpal growth occurs prior to that of body mass, but this is significantly less frequent in *Homo* (Table 2.3). In other words, the African great apes differ from modern humans with regard to the timing of carpal growth cessation relative to body mass. The high frequency of quadratic best fits in *Homo* males with regard to carpal lengths against humerus length suggests yet a different pattern: humerus length may continue to grow even after increase in body mass has declined.

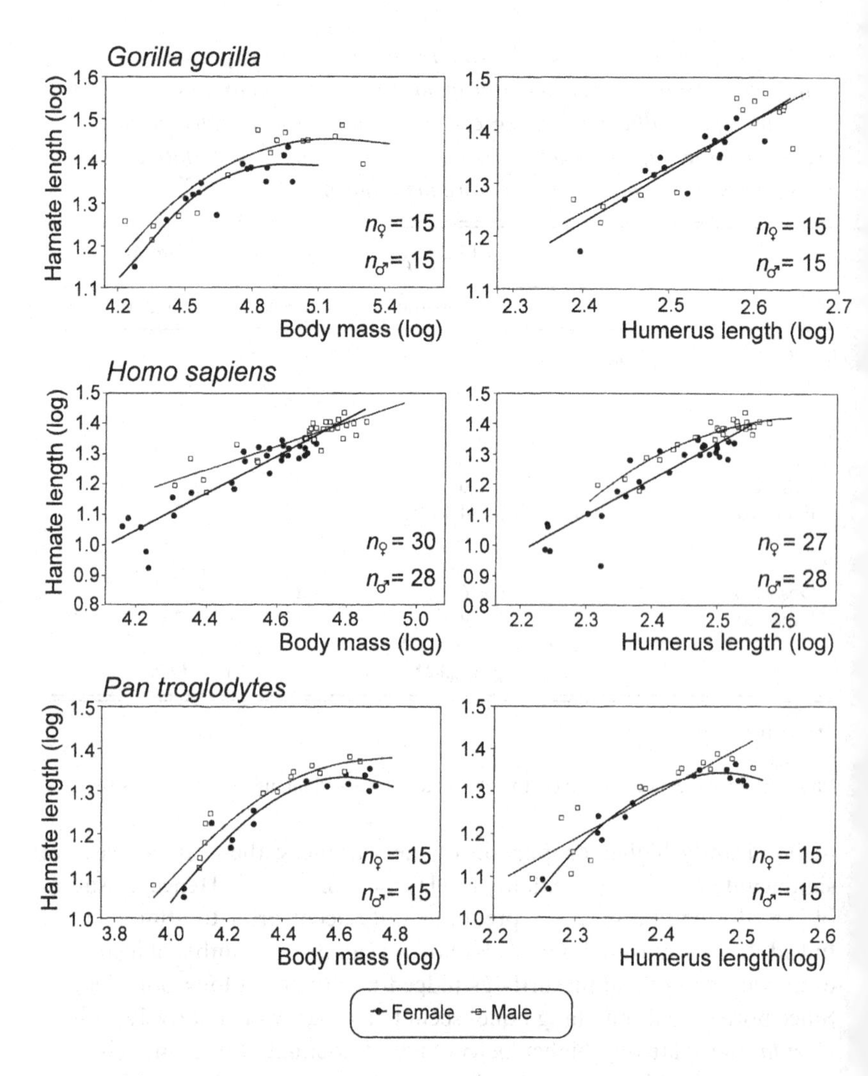

Fig. 2.3. Ontogenetic changes in hamate length (mm) in *Gorilla*, *Homo* and *Pan* relative to body mass (g) or humerus length (mm) are shown, by sex. Only those regressions that fit the data plot best, i.e. linear or quadratic, are indicated.

Considering the growth patterns of both long bones and carpals relative to body mass, overall somatic (body mass) and skeletal (long bones, carpals) maturity are clearly dissociated within species, and there appear to be species-specific differences even between these closely related taxa. If corroborated further, this will have important implications for investigating evolutionary changes in ontogenies, especially with regard to heterochronic processes. These curvilinear relationships also confound the interpretation of ontogenetic allometry.

Ontogenetic allometry

Although it has been observed that, even when intervening variables are used, curvilinear relationships are not uncommon (Laird 1965; Cock 1966; Gould 1966; Shea 1981, 1992a), many researchers still advocate ontogenetic scaling as a 'criterion of subtraction', whereby subadult and adult individuals are analysed simultaneously (e.g. Ravosa 1991; Shea 1981, 1986, 1992a, 1995; Inouye 1992, 1994; Taylor 1995). This approach attempts to distinguish between features that differ between species as a result of overall differences in body size (ontogenetic scaling) and those resulting from a more fundamental shift in biological organisation (Shea 1995). This has been taken even further in assessing ancestor–descendant relationships, such that some proxy for size is taken to adequately reflect age, allowing observations to be made of the heterochronic processes that may have played a part in the descendant's evolution (e.g. McKinney 1986). Godfrey and Sutherland (1995) have recently criticised this approach and have demonstrated that the simple allometric power function does not always capture the underlying biological relationship between size and shape such that 'when growth trajectories are not simple power functions, the same heterochronic perturbation can yield very different ancestral and descendant growth allometries. Conversely, very similar growth allometries can be produced by very different heterochronic perturbations . . .' (*ibid.*, p. 59). We concur with these conclusions in light of the fact that both carpals and long bones sometimes cease to grow prior to attainment of full body mass (adulthood), and demonstrate below the kinds of errors that may occur even in a simple comparison of ontogenetic scaling. Nonetheless, the identification either of shifts in growth trajectories or of heterochronic changes (i.e. in onset timing, rate of growth and offset timing) (Hall 1984) from study of patterns alone is not precluded.

In a careful study, Hartwig-Scherer (1992) noted that, in many cases, hominoid long bone dimensions scale differently to body mass in subadults and adults, with the cut-off point at about the time of eruption of the last permanent molar. When all individuals with a fully erupted permanent dentition (*Gorilla*, $n = 11$; *Pan*, $n = 10$; *Homo*, $n = 24$) are removed from regression analyses, the best fit polynomial is consistently linear for the dimensions examined here (Dainton and Macho 1999). This begs the question of what effect the inclusion of adults has on interspecific

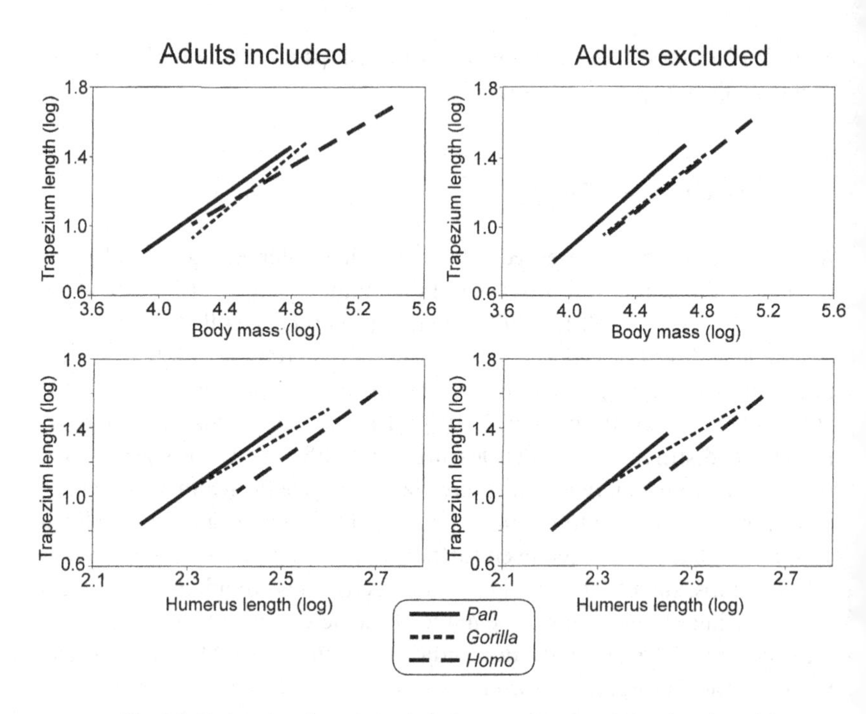

Fig. 2.4. Reduced major axis analysis for trapezium length (mm) against either body mass (g) or humerus length (mm) were performed for each species separately. In one set of analyses, adults are included whereas in the other they are excluded. Note the differences in ontogenetic trajectories.

ontogenetic analyses when ontogenetic trajectories are assessed either against body mass or an intervening variable. In order to shed light on this question, separate reduced major axis (RMA) analyses are carried out with adults either included or excluded, and using both body mass and humerus length, as an indicator of size/maturity. Allocation of sex to subadult specimens is difficult and records may not always be available which could explain why studies of ontogenetic allometries rarely distinguish between the sexes. Moreover, a previous analysis of carpal ontogenetic trajectories revealed that there are no statistically significant differences between sexes (Dainton and Macho 1999). Hence, for the present purpose sexes are pooled.

Figure 2.4 graphically displays the different RMA results for trapezium length (Table 2.4). If these results are interpreted within the framework of ontogenetic enquiry, one would inevitably draw different conclusions, depending on research design. Not all interpretations can be valid! Moreover, the measurement shown in Fig. 2.4 is no exception, and Table 2.4 highlights differences in results with regard to slope (Clarke 1980) and line position (Tsutakawa and Hewett 1978) between the four different approaches, for a series of long bone and carpal dimensions. When bone dimensions are assessed against body mass, inclusion of adults yields different results in either slope, line position or both for 74% of cases (Table 2.4). Even in long bones, which are more highly correlated with body mass than carpals, four out of six RMA comparisons change with the inclusion of adults. Conversely, only 38% of the RMA regressions of carpal and radius measurements against humerus length differ between 'adults included' and 'adults excluded' (Table 2.4). The high frequency of discrepancy (71%) observed between subadult analyses using either humerus length or body mass as an indictor of size further highlights the inaccuracy inherent in using this intervening variable. Hence, for comparative ontogenetic studies, not only is (estimated) body mass a much more reliable variable against which to assess skeletal development, but caution must be exercised in combining adult and subadult specimens when only linear regression models are employed. It logically follows that inclusion of adults in ontogenetic analyses often obscures the true growth rates (i.e. artificially lowering the allometric exponent) and may lead to misinterpretations regarding differences between taxa. Furthermore, the cessation of growth of some skeletal element relative to body mass, which is important for elucidating heterochronic processes, cannot be determined if their interrelationship does, or is assumed to, remain unchanged into adulthood. Some issues regarding the methods employed, however, require further consideration.

As regards *Homo*, too many data points have probably been removed by choosing third molar eruption as the cut-off point between adults and subadults. Perhaps a more suitable method for delineation between individuals of a single species, who have or have not attained full growth of a particular structure would be to remove data points, beginning with the oldest individual, until a curvilinear equation no longer explains a significantly greater portion of the variance than a linear fit. This approach would eliminate problems associated with different offset points, and would ensure that ontogenetic growth trajectories (rates) are comparable, while allowing for the determination of the timing of offset of growth relative to body mass. Termination of growth is an important variable if the true nature of heterochronic processes are to be elucidated, particularly where progenesis or hypermorphosis (Alberch *et al.* 1979) of the element in question is concerned. Alternatively, appropriate line-fitting methods could be employed (see e.g. Jolicoeur 1989), although at present, we know of no statistical technique that tests for statistically significant differences between such curves, the results of which are interpretable in terms of

Table 2.4. *Results are given for ontogenetic analyses of carpal and long bone dimensions against the body size estimators (X) body mass (bm) and humerus length (hl), when adults are either included or excluded. Sexes are pooled. In species comparisons statistically significant differences in slope (Clarke 1980) or line position (Tsutakawa and Hewett 1978) are indicated. Whenever differences in results are observed in potentially the 'same' analyses this discrepancy is highlighted (light grey, adults included or not included; dark grey, size variable is body mass or humerus length)*

Bone	Species comparison	X	Slope differences		Intercept differences	
			Included	Excluded	Included	Excluded
Humerus	*Homo/Gorilla*	bm	**	**	**	**
	Homo/Pan	bm	**	**	—	**
	Pan/Gorilla	bm	*	—	**	**
Radius	*Homo/Gorilla*	bm	**	**	*	**
		hl	◇◇	◇	◇	◇
	Homo/Pan	bm	**	**	**	**
		hl	◇◇	◇◇	◇◇	◇◇
	Pan/Gorilla	bm	—	*	*	—
		hl	◇	◇◇	◇◇	◇◇
Scaphoid	*Homo/Gorilla*	bm	**	*	*	*
		hl	—	—	—	—
	Homo/Pan	bm	**	*	**	**
		hl	◇	—	◇◇	◇◇
	Pan/Gorilla	bm	—	—	—	*
		hl	—	—	◇◇	◇◇
Lunate	*Homo/Gorilla*	bm	**	—	—	—
		hl	—	—	—	—
	Homo/Pan	bm	**	—	**	**
		hl	—	—	◇◇	◇◇
	Pan/Gorilla	bm	—	—	—	**
		hl	—	—	**	**
Triquetral	*Homo/Gorilla*	bm	—	—	—	*
		hl	◇◇	—	—	—
	Homo/Pan	bm	—	—	**	**
		hl	◇	—	◇◇	◇◇
	Pan/Gorilla	bm	—	—	*	**
		hl	—	—	◇◇	◇◇
Hamate	*Homo/Gorilla*	bm	**	**	—	—
		hl	◇	◇◇	—	—
	Homo/Pan	bm	**	**	**	**
		hl	—	—	◇◇	◇◇
	Pan/Gorilla	bm	—	—	—	**
		hl	—	—	◇◇	◇◇

Table 2.4. (*continued*)

Bone	Species comparison	X	Slope differences: Included	Slope differences: Excluded	Intercept differences: Included	Intercept differences: Excluded
Capitate	*Homo/Gorilla*	bm	**	—	—	*
		hl	—	—	◇◇	◇
	Homo/Pan	bm	**	—	**	**
		hl	—	—	◇◇	◇◇
	Pan/Gorilla	bm	—	—	*	**
		hl	—	—	◇◇	◇◇
Trapezium	*Homo/Gorilla*	bm	**	—	**	—
		hl	—	—	◇◇	◇◇
	Homo/Pan	bm	—	—	*	**
		hl	◇	◇	◇◇	◇
	Pan/Gorilla	bm	—	—	**	*
		hl	—	—	◇	◇◇

** $P < 0.01$, body mass (bm). ◇◇ $P < 0.01$, humerus length (hl),
* $P < 0.5$, body mass (bm). ◇ $P < 0.05$, humerus length (hl).

heterochronic processes. In an interesting paper, Rice (1997) strongly advocated full examination of the shape, as well as endpoints, of ontogenetic trajectories, which, although requiring the fitting of curves rather than simple linear lines, allows a full appreciation of the transformations between two growth trajectories, even if these cannot always be described within the traditional heterochronic framework (Alberch *et al.* 1979; McNamara 1986).

A fundametal objection that may be raised with regard to the analyses presented above concerns the use of two different equations for the estimation of body mass in adults and subadults (Hartwig-Scherer and Martin 1992). The use of two equations could create a shift in ontogenetic trajectories at about third molar eruption, resulting in the appearance of a curvilinear relationship that does not reflect the real situation. Differences in the timing of growth cessation relative to body mass in different species (Tables 2.1 and 2.2) would caution against such an interpretation. The following section will illustrate further the relationship between tooth eruption and age with body mass cessation and skeletal maturity in order to address, among other things, propositions that the results may be artefacts.

Age and maturation

Sufficient published data are available on the relationship between body mass and age (Leigh and Shea 1996) and dental eruption (Kuykendall *et al.* 1992; Willoughby 1978, cited in Smith *et al.* 1994) in the African great apes, to examine the relationship between M_3 eruption and somatic maturity. For modern humans, a growth curve was calculated, based on the double Gompertz equation (German *et al.* 1994) fit to cross-sectional data consisting of age and body mass measurements for 197 White European subadults (courtesy of Dr Li Yu) and the known age skeletal sample of modern humans (present study) for whom body mass was estimated (Hartwig-Scherer and Martin 1992). Only in *Homo* does M_3 eruption and somatic growth cessation coincide (Fig. 2.5; age–body mass axis), whereas in both *Gorilla* and *Pan* M_3 erupts prior to full somatic maturity. In order to determine how carpal and long bone growth may relate to body mass and age in general, their dimensions at given body masses are added to the plots as a third dimension (Fig. 2.5). Before interpreting these curves, however, it must be appreciated that our African ape body mass estimates are probably lower, at a given age, than those published for hominoids (Leigh and Shea 1996).

In Fig. 2.6, body mass estimates and alveolar tooth emergence status for the present sample are superimposed onto the JPPS growth curves (after the authors of the paper; Jolicoeur *et al.* 1988) provided by Leigh and Shea (1996). The range of permanent molar eruption for *Gorilla* (Willoughby 1978, cited in Smith *et al.* 1994) and *Pan* (Kuykendall *et al.* 1992) are also indicated. Individuals at any given dental eruption stage tend to cluster together along the age–body mass curves, although there is a systematic trend for these individuals to apparently erupt their teeth earlier than expected. It needs to borne in mind, however, that both Leigh and Shea's (1996) study on body mass increase and Kuykendall and colleagues' (1992) documentation of tooth eruption are based on captive individuals, whereas those for *Gorilla* are somewhat dubious (Smith *et al.* 1995). Captive individuals are on average larger than their wild counterparts (Smith and Jungers 1997), and erupt their teeth relatively earlier (Phillips-Conroy and Jolly 1988). Nonetheless, dental development is influenced less by environmental factors than is skeletal development (Demirjian 1978; Watts 1990; Marzke *et al.* 1996). Furthermore, in the present study, alveolar rather than gingival eruption is recorded and the data sets on eruption are thus not strictly comparable. Taking these caveats into account, it is not surprising that molar eruption in the present sample appears to occur too early: the wild animals measured are simply smaller relative to similarly aged captive

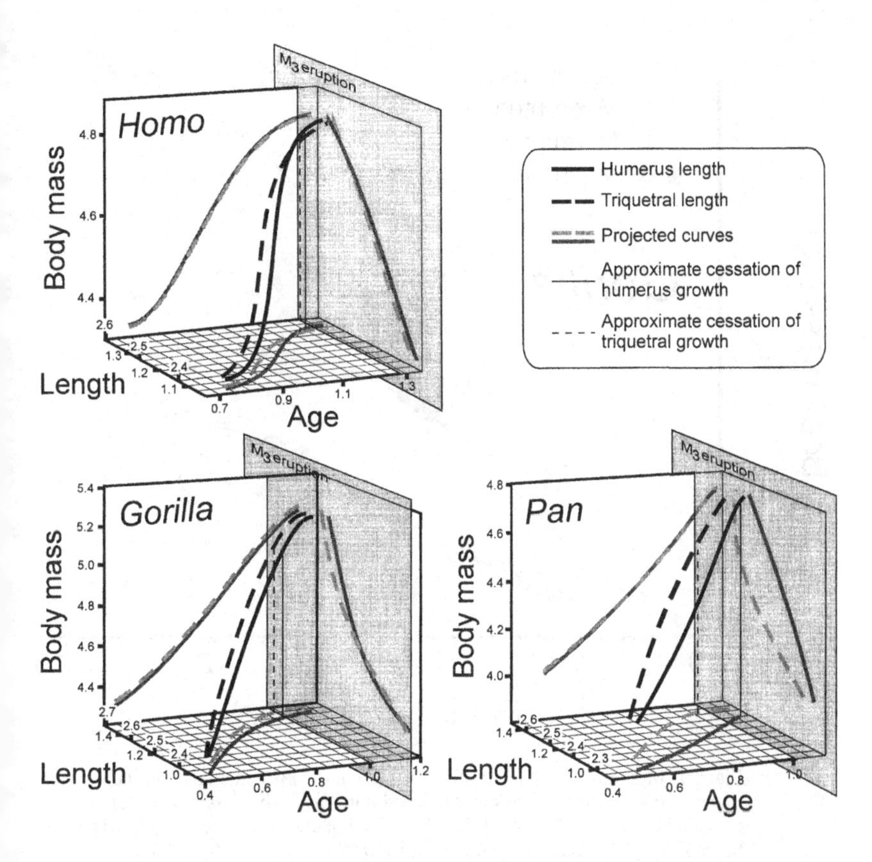

Fig. 2.5. In three-dimensional plots, the growth curves for male humerus length (log mm) and triquetral length (log mm) are superimposed onto the known (Leigh and Shea 1996, for *Gorilla* and *Pan*) and calculated (this study for *Homo*) body mass/age curves. Average M_3 eruption times (taken from Smith *et al.* 1994) are also shown.

animals and are therefore shifted towards the *y*-axis (Fig. 2.6). This has consequences for interpreting Fig. 2.5 in that both the carpal and long bone growth trajectories of the African apes should be shifted to slightly older ages. Inspection of the eruption status of permanent molars and the carpal and humerus growth trajectories relative to body mass confirms this suggestion (Fig. 2.7).

Trends exhibited in Fig. 2.5 and 2.7 confirm that in both *Gorilla* and *Homo* the cessation of carpal and humerus growth occurs at, or just after, the eruption of third permanent molars. However, in *Gorilla* cessation of overall skeletal growth occurs substantially earlier than body mass termination, whereas in *Homo* skeletal maturation apparently coincides with full body mass attainment. Notably, in *Pan* the endpoints of growth

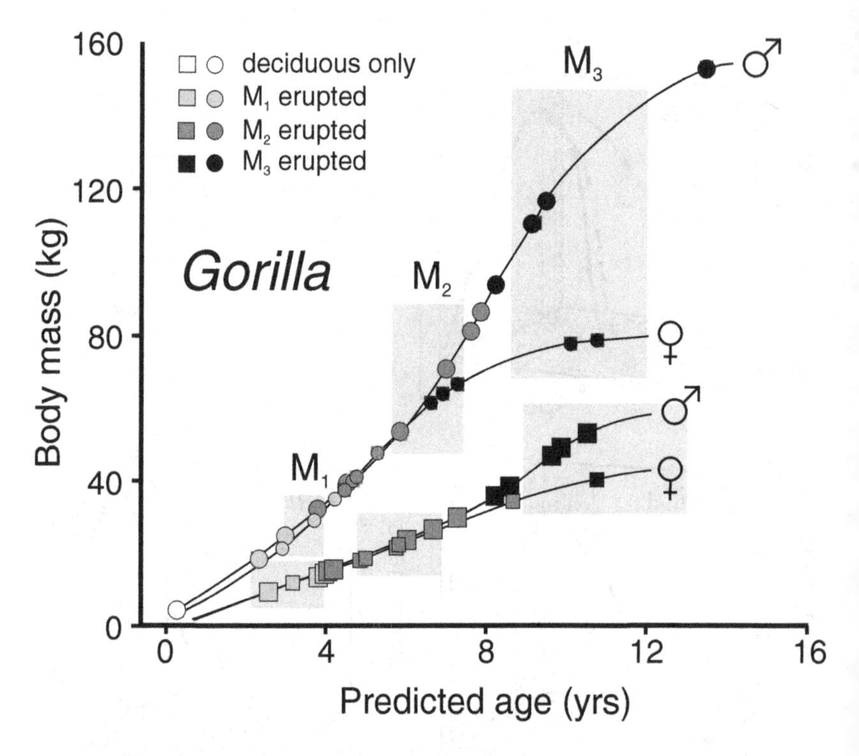

Fig. 2.6. Individuals of *Pan* and *Gorilla*, used in the present study, at different stages of tooth eruption are superimposed onto the body mass/age curves provided by Leigh and Shea (1996). The range of eruption times for M_1, M_2 and M_3 is shown (Kuykendall *et al.* 1992, for *Pan*; Smith *et al.* 1994, for *Gorilla*).

trajectories for humerus and carpals are dissociated, with carpals ceasing growth at a much earlier age, again just before the eruption of M_3. This finding sheds further light on the proposition that use of two different equations for the determination of adult and subadult body masses, with full third molar eruption as the cut-off point, has significantly influenced the results presented above. If this were the case, (1) different bones would show a change in growth trajectory relative to body mass at similar ages within a species and (2) growth curves would tail off at about the eruption of M_3. Clearly, the data for *Pan* refute such a proposal, since different bones show a change in growth trajectory at different ages. It is, however, surprising that cessation of carpal growth appears to occur consistently close to the eruption of the last permanent molar in all species studied here, although it seems to be dissociated from other aspects of growth. This trend warrants further exploration.

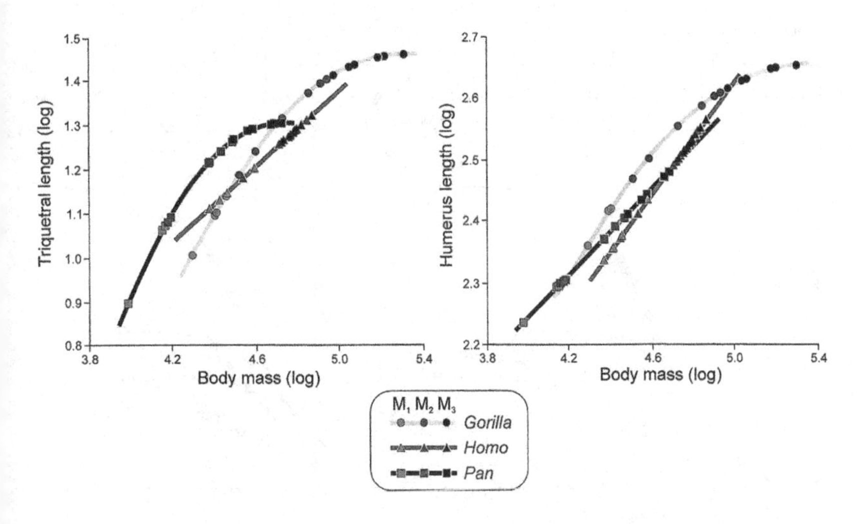

Fig. 2.7. Growth curves for triquetral length (mm) and humerus length (mm) of male *Gorilla*, *Homo* and *Pan* against estimated body mass (g) are shown. The eruption status of each individual, projected onto these growth curves (at the respective body mass), is highlighted.

Phylogenetic considerations

Hominid evolution is characterised by a prolongation of ontogeny (Schultz 1936; Beynon and Dean 1988), an increase in brain (McHenry 1994; Aiello and Wheeler 1995) and body size (McHenry 1992, 1994), and a delay of life history events, such as sexual and somatic maturity (Schultz 1960; Gould 1977; Smith 1989). Carpals similarly follow this overall trend toward delayed maturation, but growth cessation of modern human carpals is shifted later relative to sexual and somatic maturity, as compared to the African great apes (Fig. 2.8a). Despite this difference in carpal growth cessation relative to life history events, carpal growth cessation consistently occurs close to M_3 eruption in all three taxa. In other words, both dental development and carpal maturation apparently follow a similar growth trend, but are dissociated from other life history events in modern humans when compared with African great apes. This observation is not foreshadowed by early carpal development in hominoids, since the appearance of primary ossification centres of carpals is not strongly related to M_1 calcification (Winkler 1996), probably because dental eruption is less likely to be affected by sex and/or environmental factors than is wrist maturation (Marzke *et al.* 1996). While it is improbable that carpal and dental development are directly linked, they may be subject to the same developmental

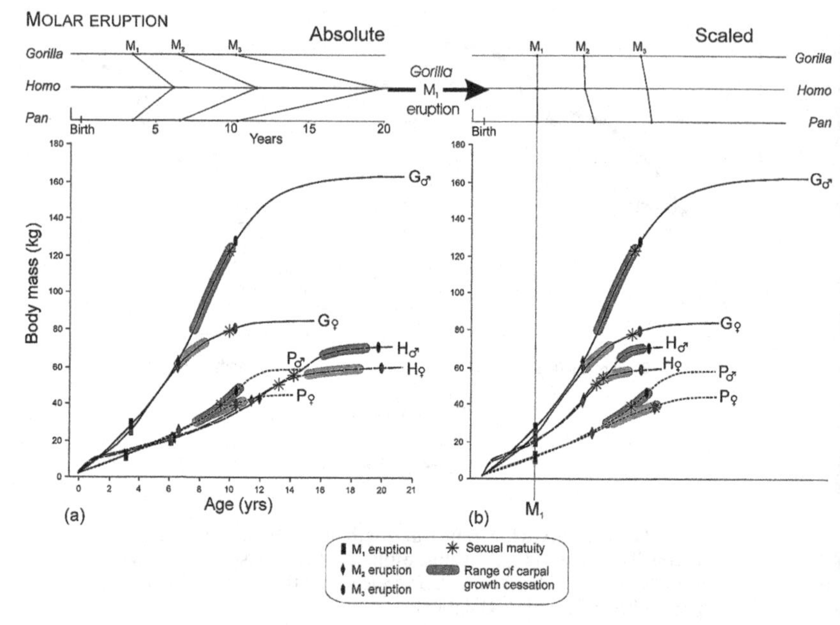

Fig. 2.8. (a) The average cessation of carpal growth, relative to estimated body mass is superimposed onto the growth curves for *Gorilla* and *Pan* (Leigh and Shea 1996), and *Homo* (Tanner and Whitehouse 1976). Average molar eruption times (Smith *et al.* 1994, for *Gorilla* and *Homo*; Kuykendall *et al.* 1992, for *Pan*) and reported timing of sexual maturity (Stewart and Harcourt 1987 (*Gorilla*); Nishida and Hiraiwa-Hasegawa 1987 (*Pan*); Eveleth and Tanner 1990 (*Homo*)) are also indicated. (b) The growth curves shown in (a) are scaled to the same length of infancy (i.e. 3.5 years).

biases, although the observed patterns could also be coincidental and merely due to chance. The following sections explore these alternatives.

Mechanisms of growth

In an innovative study, Vogl and colleagues (1993) appraised the effects of growth hormone on skeletal elements in mice that were transgenic for growth hormone, with special reference to long bones and the mandible. They found that excess growth hormone is particularly influential in promoting mandibular growth via muscle action, while growth of long bone length is stimulated largely through the effects of growth hormone at the growth plates. Hence, two main routes can be postulated by which growth hormone mediates bone growth. Firstly, it promotes long bone elongation by stimulating the proliferation of cartilage in the epiphyseal growth plates (Vogl *et al.* 1993) (Fig. 2.9a), which is then subject to en-

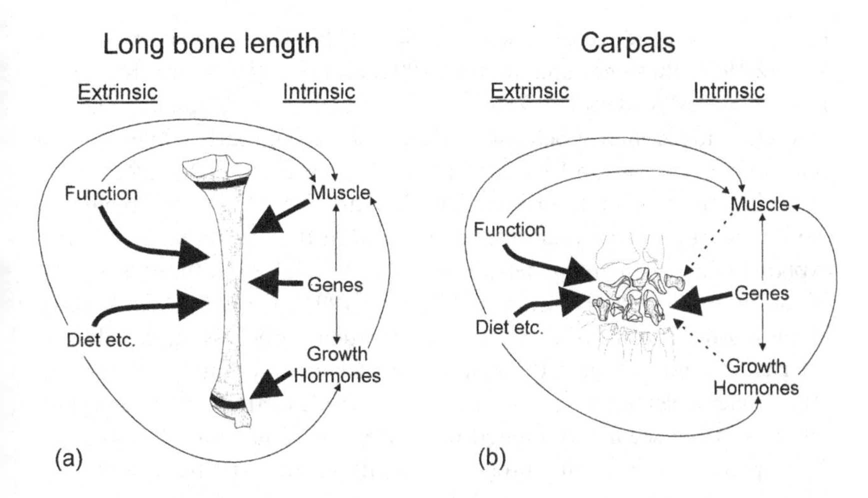

Fig. 2.9. Path diagram illustrating the factors influencing growth of (a) long bones in length and (b) carpals. Thick arrows indicate major influences on the bone, thin arrows indicate indirect effects on bones, while dotted arrows are those that have apparently little direct affect on bone growth.

dochondral ossification (Moss-Salentijn 1992). Specifically, growth hormone promotes hypertrophy of cartilage cell columns, enhancing growth activity and prolonging bone growth by delaying epiphyseal fusion (Oberbauer *et al.* 1992). The allometric relationship of bone length to body mass during ontogeny, however, remains unaltered (Oberbauer *et al.* 1992). Secondly, growth hormone acts to enlarge muscle tissue, which subsequently influences bone growth epigenetically (Herring 1993).

Carpals lack epiphyseal growth plates and, with the exception of the pisiform, trapezium and hook of the hamate, lack muscle attachments too (Kaplan 1965; Kapandji 1982; Lewis 1989). Prior to ossification, carpals exist as cartilagenous models, resembling their future bone shape and, similar to long bones, the process of ossification proceeds from the centre of these models (Roche 1978). While the articular cartilage of carpals does have a germinal layer, allowing some growth to occur once the initial cartilage model is ossified (Siffert 1966), articular cartilage chondrocytes are much less responsive to growth factors than those of epiphyseal plates (Masoud *et al.* 1986; Moss-Salentijn 1992) (Fig. 2.9b). The absence both of growth plates and muscle attachments renders it unlikely, therefore, that carpals are strongly influenced by changes in the concentration of growth hormone in the body postnatally and, hence, the size and shape of carpals is seemingly more strongly determined by intrinsic (genetic) mechanisms than long bones (Bryant and Simpson 1984; Hall 1984) (Fig. 2.9b).

However, extrinsic factors such as function (loading etc.) (Bertram and Swartz 1991; Biewener and Bertran 1993) and diet (Pratt and McCance 1964; Stuart-Macadam 1989) also play a role in bone development. Despite differences in locomotor behaviour during infancy and early adolescence in *Gorilla* and *Pan* (Doran 1996, 1997), carpals appear to be little affected by direct functional stimuli (Dainton and Macho 1999). Only with regard to the lunate can behavioural differences between the species be invoked to explain difference in ontogenetic trajectories, but this is not the case for any of the other carpals (Dainton and Macho 1999). On the basis of these observations, it follows that carpals are apparently less susceptible to external stimuli (e.g. growth hormone, muscle action, function) than are other skeletal elements, and thus behave more like teeth in this regard (Fig. 2.9). Given the seemingly limited influence of external stimuli on carpal development throughout ontogeny, carpals promise to be usefully exploited for taxonomic purposes. Furthermore, absence of the influence of external stimuli on carpal growth may explain why carpal and M_3 maturation remain correlated, despite other shifts in life history variables between the species examined; the development of teeth is, after all, also little affected by extrinsic factors (Smith 1989). It is proposed that a comparison of carpals and teeth, which exhibit predominantly intrinsically determined growth patterns, with those structures that have a greater extrinsic determinant of growth may contribute to an understanding of evolutionary processes. The extended ontogeny in *Homo* and the early cessation of long bone growth in *Gorilla* could thus be explained.

What do carpals tell us about modern human brains?

In exploring difference in ontogenies between modern humans and the African great apes (Fig. 2.7a), not only does the delayed maturation of *Homo* deserve mention, but also the dissociation of certain life history events, i.e. that of dental and carpal maturation from somatic and sexual maturity. Vrba (1994) suggested that heterochronic changes in human evolution affect mainly the growth rates of earlier, postnatal phases of development, which facilitated the extension of prenatal growth rates of neural tissue to be carried through into the first year of life. Following completion of brain growth, modern humans and chimpanzees share similar growth rates (Vrba 1994). However, this does not take into account the magnitude of the significantly extended period of brain growth (or overall infancy) in modern humans. In a stimulating study, Riska and Atchley (1985) investigated the relationship between brain and body size in mammals. They advocate that a distinction be made in interpreting evolutionary changes between hyperplastic and hypertrophic growth. The former is brought about by an increase in cell numbers and characterises early ontogeny, when selection for larger body size is also likely to result in increased brain size and vice versa through the influence of strong pleiotropic gene effects. Hypertrophic growth occurs later in ontogeny when body size increase can be selected for only through an increase in the size of cells, muscle fibres etc.; it will not affect structures that have already ceased growing, i.e. the brain. Given the fundamental shift of all modern human life history events to later ages, it seems probable that selection has acted on either brain size during pre- and early postnatal growth, when pleiotropic gene effects are greatest, delaying key life history events and allowing early, increased, growth rates to be active for longer. Whether the target for selection was body size, brain size or, perhaps, even extended infancy, is uncertain. Hafner and Hafner (1984) have, for example, argued, that encephalisation in certain geomyoid rodent genera may have been an incidental by-product of developmental changes in an ancestral lineage. Whatever the impetus for the shift in brain size, body size and life history variables observed in hominid evolution, pleiotropic effects would mean that selection early in ontogeny for any of these variables would be likely to result in a fundamental delay of life history events.

Smith and colleagues (1995) used the relationship between brain size and M_1 eruption across all higher primates to predict the timing of M_1 eruption in hominoids on the basis of cranial capacity. They suggested that the extended maturation of modern humans has developed progressively during hominid evolution. In all living primates, about 95% of adult brain size is attained at the age when M_1 erupts (Beynon and Dean 1988; Bogin 1995). If the propositions regarding extension of human ontogeny are correct, little difference in the timing of later life history variables would be expected between species when the interval between birth and brain growth cessation (estimated from M_1 eruption) is scaled to an equal time interval. In fact, this is only the case for dental eruption and carpal growth, while sexual maturation and body mass cessation actually occurs relatively earlier in modern humans (Fig. 2.8b). Since both dental eruption and carpal growth are less influenced by growth hormone, secondary selection for relatively smaller body size and/or a shorter maturation period may, therefore, be postulated for human evolution. In this case selection would have acted on the endocrine system in later ontogeny, for a hormonally mediated reduction of body size (the hypertrophic growth suggested by Riska and Atchley (1985)) leaving brain, carpal and tooth developmental timing relatively unaltered. Indirect support for this proposition is provided by the fossil record. African *Homo erectus* was of a similar or even greater body size than modern humans (Ruff and Walker 1993), although its cranial capacity suggests a brain size of about 30–35% smaller than that of modern humans (Rightmire 1990; Wood 1991; McHenry 1994). On the basis of brain size, the timing of M_1 eruption is, therefore, intermediate between that of modern humans and African apes (Smith *et al.* 1995). Bearing in mind the similarities in body size, infancy in African *H. erectus* would have been somewhat shorter, while the juvenile and adolescent periods would be similar to those of modern humans. If our hypothesis is correct, it may be concluded that African *H. erectus* and its immediate ancestors were subject to selection for extended growth in early ontogeny, but not yet to secondary selection for decreased body size in later ontogeny. Whatever initially sparked an increase in brain/body size, the advantages a larger brain conferred were probably, by this time, exerting direct selection pressure to maintain and even enlarge it further. Excessive body size is not necessarily advantageous, however, owing to the metabolic requirements needed to build and maintain it. Moreover, postponing sexual maturation lengthens the vulnerable juvenile period while delaying the ability to reproduce. Since selection for brain size inevitably results in body size increase due to pleiotropic effects early in ontogeny (Riska and Atchley 1985), body size is likely to have been secondarily selected *against* after infancy. Selection on hormonal control is the simplest way to accomplish the latter, since a simple shift in hormonal levels can produce a co-ordinated response in numerous target organs (Shea 1992b). We thus preliminarily conclude that the delay of carpal (and dental) maturation, relative to other long bones and overall somatic maturation in modern humans, indicates that two processes have occurred during this species recent evolutionary history: (1) selection for larger brain size prenatally and early postnatally and (2) secondary selection against the large body size this entails during later stages of ontogeny. These differential selection pressures may explain why there is a high correlation between M_1 eruption and brain size, yet a lower correlation between either M_1 eruption or brain size and later life history events in primates (Smith 1989; Smith *et al.* 1995). An integrated analysis of ontogenetic changes can similarly be used to interpret the ontogenetic patterns found in *Gorilla.*

Are gorillas too big for their bones?

Gorilla long bones tend to cease growing relatively early, in relation to body mass, than those of either *Pan* or *Homo*. The final length of a long bone is apparently achieved when its growth plate(s) exhaust the supply of mitotically active cells. This variable is, however, determined at the onset of growth and is controlled by genes that dictate the size of the initial cellular condensation early in development (Hall 1984; Oster *et al.* 1988; Atchley and Hall 1991). While an increase in circulating growth hormone mediates increased hypertrophy of cells in long bone growth plates and results in prolonged growth (Oberbauer *et al.* 1992; Vogl *et al.* 1993), it cannot overcome the initial limiting factor of cell number and mitotic potential set early in ontogeny (Hall 1984). In other words, this limit on growth of long bone length cannot be surmounted by selection for larger body size later in ontogeny. The increased body size of *Gorilla* is almost certainly the result of selection for hypertrophic growth (Riska and Atchley 1985) mediated by selection on endocrine control (Shea 1992b; Leigh and Shea 1996). Whatever the impetus for selection for larger body size (e.g. environmental degradation, sexual selection) *Gorilla* has apparently increased its body size beyond the limits of long bone growth potential, such that long bone epiphyses run out of mitotically active cells before body mass has ceased to grow. This proposition is supported by the fact that gorillas have relatively smaller brains for their body size, than chimpanzees, which would not be the case if selection for larger body size acted on the initial condensation events early in ontogeny, where pleiotropic effects would also cause an expansion of brain size (Riska and Atchley 1985). In light of this explanation, it is unsurprising that *Gorilla* and *Pan* long bones exhibit similar growth trajectories (Jungers and Susman 1984; Hartwig-Scherer 1992), given that similar allometric trajectories characterise long bone growth in mice with both normal and heightened growth hormone levels (Shea *et al.* 1990; Oberbauer *et al.* 1992). What is missing from these observations, however, is an appreciation of when the offset of growth occurs and therefore a sound explanation of why adult long bone allometry differs between these species. It is not necessary to invoke biomechanical explanations for these adult differences, in fact they are in some ways forced upon *Gorilla* as a consequence of enlarged body size. From this, it may even be inferred that enlarged body size in *Gorilla* may be a recent acquisition in evolutionary time.

This chapter has tried to illustrate that system-specific developmental patterns in *Gorilla*, *Homo* and *Pan* are not only comprehensible with regard to their underlying biological mechanisms, which determine their growth, but are also likely to reveal evolutionary processes. It would seem that carpal and dental growth *are* subject to the same developmental biases; they do not covary due to chance. This has important implications for the study of life histories in extant and extinct species.

Conclusions

This study aimed to assess the ontogenetic changes of three different structures, i.e. body mass, carpal and tooth development, in skeletal samples of *Gorilla*, *Pan* and *Homo*. Contrary to common contentions, it demonstrates that heterochronic processes can, to a certain extent, be determined from analyses of patterns alone. However, care must be taken in interpreting the trajectories when (estimated) body mass is substituted by an intervening variable. Moreover, we caution against inclusion of adults in studies of ontogenetic allometry, since different structures stop growing at different times during ontogeny, thus resulting in curvilinear relationships between variables. Inclusion of adults can, therefore, markedly affect the slope of regression analyses and will lead to misinterpretations.

Careful analyses of ontogenetic changes can reveal information otherwise unobtainable from methods commonly employed in studies of ontogenetic allometry. In particular, an integrated investigation of different body systems has much to offer in elucidating biological processes underlying the evolution of morphology and life histories. Modern humans and gorillas are a case in point. The integrated approach toward the study of ontogenetic trajectories of carpals, long bones and teeth has indicated that at least two heterochronic changes are necessary to account for the evolution of modern human body size, whereas only one change needs to be invoked to explain changes in body mass and life histories in gorillas. Although the hypotheses thus formulated are, at present, speculative, the approach toward studying ontogenetic changes advocated here is likely to contribute to a better understanding of the processes that operated during evolution in the past.

Acknowledgements

For access to the skeletal material we thank Mr Derek Howlett, The Powell-Cotton Museum, Birchington, and Dr Bruce Latimer and Lyman Jellema, Hamman–Todd Collection, Cleveland Museum of Natural History, Cleveland. Dr Li Yu kindly provided growth data for a British population. We are also grateful to Drs Charles FitzGerald and Robert Hoppa for inviting us to contribute to this volume. The NERC (GR9/01635) and RDF (University of Liverpool) provided financial support (G.M.). M.D. is funded by a NERC studentship (GT4/97/159).

References

Aiello LC and Wheeler P (1995). The expensive tissue hypothesis: the brain and the digestive system in human and primate evolution. *Current Anthropology* **36**, 199–221.

Alberch P, Gould SJ, Oster GF and Wake DB (1979). Size and shape in ontogeny and phylogeny. *Paleobiology* **5**, 296–317.

Atchley WR and Hall BK (1991). A model for development and evolution of complex morphological structures. *Biological Review* **66**, 101–157.

Bertram JEA and Swartz SM (1991). The law of bone transformation, a case of crying Wolff? *Biological Review* **66**, 245–273.

Beynon AD and Dean MC (1988) Distinct dental development patterns in early fossil hominids. *Nature* **335**, 509–514.

Biewener AA and Bertram JEA (1993) Mechanical loading and bone growth in vivo. In *Bone* vol. 7 *Bone Growth – B*, ed. B.K. Hall, pp. 1–36. Boca Raton, FL: CRC Press.

Bogin B (1995) Growth and development: recent evolutionary and biocultural research. In *Biological Anthropology. The State of the Science*, ed. N.T. Boaz and L.D. Wolfe, pp. 49–70. Oregon: International Institute for Human Evolutionary Research, Oregon State University Press.

Bryant PJ and Simpson P (1984) Intrinsic and extrinsic control of growth in developing organs. *Quarterly Review of Biology* **59**, 387–415.

Clarke MRB (1980) The reduced major axis of a bivariate sample. *Biometrika* **67**, 441–446.

Cock AG (1966) Genetical aspects of metrical growth and form in animals. *Quarterly Review of Biology* **41**, 131–190.

Dainton M and Macho GA (1999) Did knuckle walking evolve twice? *Journal of Human Evolution* **36**, 171–184.

Demirjian A (1978) Dentition. In *Human Growth*, 2: *Postnatal Growth*, ed. F. Falkner and J.M. Tanner, pp. 413–444. New York: Plenum Press.

Doran DM (1996) Comparative positional behaviour of the African apes. In *Great Ape Societies*, ed. W.C. McGrew, L.F. Marchant and T. Nishida, pp. 213–224. Cambridge: Cambridge University Press.

Doran DM (1997) Ontogeny of locomotion in mountain gorillas and chimpanzees. *Journal of Human Evolution* **32**, 323–344.

Eveleth PB and Tanner JM (1990) *Worldwide Variation in Human Growth*, 2nd edn. Cambridge: Cambridge University Press.

Faucheux B, Bertrand M and Bourlière F (1978) Some effects of living conditions upon the pattern of growth in the stumptail macaque (*Macaca arctoides*). *Folia Primatologica* **30**, 220–236.

German RZ, Hertweck JE, Sirianni JE and Swindler DR (1994) Heterochrony and sexual dimorphism in the pigtailed macaque (*Macaca nemestrina*). *American Journal of Physical Anthropology* **93**, 373–380.

Godfrey LR and Sutherland MR (1995) Flawed inference: why size-based tests of heterochronic processes do not work. *Journal of Theoretical Biology* **172**, 43–61.

Gould SJ (1966) Allometry and size in ontogeny and phylogeny. *Biological Review* **41**, 587–640.

Gould SJ (1977) *Ontogeny and Phylogeny*. Cambridge, MA: Belknap Press of Harvard University Press.

Greulich WW and Pyle SI (1959). *Radiographic Atlas of Skeletal Development of the Hand and Wrist*. 2nd edn. Stanford: Stanford University Press.

Hafner MS and Hafner JC (1984) Brain size, adaptation and heterochrony in geomyoid rodents. *Evolution* **38**, 1088–1098.

Hall BK (1984) Developmental processes underlying heterochrony as an evolutionary mechanism. *Canadian Journal of Zoology* **62**, 1–7.

Hall BK (1992) *Evolutionary Developmental Biology*. London: Chapman & Hall.

Hartwig-Scherer S (1992) Allometry in hominoids: a comparative study of skeletal growth trends. PhD thesis, University of Zurich.

Hartwig-Scherer S and Martin RD (1992) Allometry and prediction in hominoids: a solution to the problem of intervening variables. *American Journal of Physical Anthropology* **88**, 37–57.

Herring SW (1993) Formation of the vertebrate face: epigenetic and functional influences. *American Zoologist* **33**, 472–483.

Humphrey LT (1998) Growth patterns in the modern human skeleton. *American Journal of Physical Anthropology* **105**, 57–72.

Huxley JS (1932) *Problems of Relative Growth*. London, Methuen/MacVeagh.

Inouye SE (1992) Ontogeny and allometry of African ape manual rays. *American Journal of Physical Anthropology* **23**, 107–138.

Inouye SE (1994) Ontogeny of knuckle-walking hand postures in African apes. *Journal of Human Evolution* **26**, 459–485.

Jolicouer P (1989) A simplified model for bivariate complex allometry. *Journal of Theoretical Biology* **140**, 41–49.

Jolicouer P, Pontier J, Pernin MO and Sempe M (1988) A lifetime asymptotic growth curve for human height. *Biometrics* **44**, 995–1003.

Jungers WL (1985) Body size and scaling of limb proportions in primates. In *Size and Scaling in Primate Biology*, ed. W.L. Jungers, pp. 345–381. New York: Plenum Press.

Jungers WL and Cole MS (1992) Relative growth and shape of the locomotor skeleton in lesser apes. *Journal of Human Evolution* **23**, 93–105.

Jungers WL and German RZ (1981) Ontogenetic and interspecific skeletal allometry in non-human primates: bivariate versus multivariate analysis. *American Journal of Physical Anthropology* **55**, 195–202.

Jungers WL and Hartman SE (1988) Relative growth of the locomotor skeleton in orang-utans and other large-bodied hominoids. In *Orang-utan Biology*, ed. J.H. Schwartz, pp. 347–359. New York: Oxford University Press.

Jungers WL and Susman RL (1984) Body size and skeletal allometry in African apes. In *The Pygmy Chimpanzee. Evolutionary Biology and Behaviour*, ed. R.L. Susman, pp. 131–177. New York: Plenum Press.

Kapandji IA (1982) *The Physiology of the Joints*, vol. 1 *The Upper Limbs*, 2nd edn. New York: Churchill Livingstone.

Kaplan EB (1965) *Functional and Surgical Anatomy of the Hand*, 2nd edn. London: Pitman Medical Publishing Company Ltd.
Kuykendall KL, Mahoney CJ and Conroy CC (1992) Probit and survival analysis of tooth emergence ages in a mixed longitudinal sample of chimpanzees (*Pan troglodytes*). *American Journal of Physical Anthropology* **89**, 379–399.
Laird AK (1965) Dynamics of relative growth. *Growth* **29**, 249–263.
Leigh SR and Shea BT (1996) Ontogeny of body size variation in African apes. *American Journal of Physical Anthropology* **99**, 43–65.
Lewis OJ (1989) *Functional Morphology of the Hand and Foot*. Oxford: Clarendon Press.
Lumer H (1937) The consequences of sigmoid growth for relative growth functions. *Growth* **1**, 140–154.
Macho GA and Wood B (1995) The role of time and timing in hominid dental evolution. *Evolutionary Anthropology* **4**, 17–31.
Marzke MW, Young DL, Hawkey DE, Su SM, Fritz J and Alford PL (1996) Comparative analysis of weight gain, hand/wrist maturation, and dental emergence rates in chimpanzees aged 0–24 months from varying captive environments. *American Journal of Physical Anthropology* **99**, 175–190.
Masoud I, Shapiro F and Moses A (1986) Tibial epiphyseal development: a cross sectional histologic and histomorphic study in the New Zealand white rabbit. *Journal of Orthopaedic Research* **4**, 212–228.
McHenry HM (1992) Body size and proportions in early hominids. *American Journal of Physical Anthropology* **87**, 407–431.
McHenry HM (1994) Behavioural ecological implications of early hominid body size. *Journal of Human Evolution* **27**, 77–87.
McKinney ML (1986) Ecological causation of heterochrony: a test and implications for evolutionary theory. *Paleobiology* **12**, 282–289.
McNamara KJ (1986) A guide to the nomenclature of heterochrony. *Journal of Palaeontology* **60**, 4–13.
Moss-Salentijn L (1992) Long bone growth. In *Bone*, vol. 6 *Bone Growth – A*, ed. B.K. Hall, pp. 185–208. Boca Raton, FL: CRC Press.
Mosimann J and James F (1979) New statistical approaches for allometry with application to Florida red winged black birds. *Evolution* **33**, 444–459.
Nishida T and Hiraiwa-Hasegawa M (1987) Chimpanzees and bonobos: cooperative relationships among males. In *Primate Societies*, ed. B.B. Smuts, D.L. Cheney, R.M. Seyfarth, R.W. Wrangham and T.T. Struhsaker, pp. 165–177. Chicago: University of Chicago Press.
Oberbauer AM, Currier TA, Nancarrow CD, Ward KA and Murray JD (1992) Linear bone growth of oMT1a-oGH transgenic male mice. *American Journal of Physiology* **262**, E936–E942.
Oster GF, Shubin N, Murray JD and Alberch P (1988) Evolution and morphogenetic rules: the shape of the vertebrate limb in ontogeny and phylogeny. *Evolution* **42**, 862–884.
Phillips-Conroy JE and Jolly CJ (1988) Dental eruption schedules of wild and captive baboons. *American Journal of Primatology* **15**, 17–29.

Pratt CWM and McCance RA (1964). Severe undernutrition in growing and adult animals. 14. The shafts of the long bones in pigs. *British Journal of Nutrition* **18**, 613–624.

Ravosa MJ (1991) Ontogenetic perspective on the mechanical and nonmechanical models of primate circumorbital morphology. *American Journal of Physical Anthropology* **85**, 95–112.

Rice SH, (1997) The analysis of ontogenetic trajectories. When a change in size or shape is not heterochrony. *Proceedings of the National Academy of Sciences, USA* **94**, 907–912.

Rightmire GP (1990) *The evolution of* Homo erectus. Cambridge: Cambridge University Press.

Riska B and Atchley WR (1985) Genetics of growth predict patterns of brain size evolution. *Science* **229**, 668–671.

Roche AF (1978) Bone growth and maturation. In *Human Growth. 2. Postnatal Growth*, ed. F. Falkner and J.M. Tanner, pp. 25–60. New York: Plenum Press.

Ruff CB and Walker A (1993) Body size and body shape. In *The Nariokotome* Homo erectus *skeleton*, eds. A. Walker and R. Leakey, pp. 234–265. Cambridge, MA: Harvard University Press.

Schultz AH (1936) Characters common to higher primates and characters specific to man. *Quarterly Review of Biology* **11**, 259–283 and 425–455.

Schultz AH (1960) Age changes in primates and their modification in man. In *Human Growth*, ed. J. M. Tanner, pp. 1–20. Oxford: Pergamon.

Shea BT (1981) Relative growth of the limbs and trunk in the African apes. *American Journal of Physical Anthropology* **56**, 179–201.

Shea BT (1986) Scapula form and locomotion in chimpanzee evolution. *American Journal of Physical Anthropology* **70**, 475–488.

Shea BT (1992a) Ontogenetic scaling of skeletal proportions in the talapoin monkey. *Journal of Human Evolution* **23**, 283–307.

Shea BT (1992b) Developmental perspective on size change and allometry in evolution. *Evolutionary Anthropology* **1**, 125–134.

Shea BT (1995) Ontogenetic scaling and size correction in the comparative study of primate adaptations. *Anthropologie* **33**, 1–16.

Shea BT, Hammer RE, Brinster RL and Ravosa MR (1990) Relative growth of the skull and postcranium in giant transgenic mice. *Genetical Research* **56**, 21–34.

Siffert RS (1966) The growth plate and its affections. *Journal of Bone and Joint Surgery* **48**, 546–563.

Simpson SW, Russell KF and Lovejoy CO (1996) Comparison of diaphyseal growth between the Libben population and the Hamman-Todd chimpanzee sample. *American Journal of Physical Anthropology* **99**, 67–78.

Sirianni JE and Swindler DR (1985) *Growth and Development of the Pigtailed Macaque*. Boca Raton, FL: CRC Press.

Smith BH (1989) Dental development as a measure of life history in primates. *Evolution* **43**, 683–688.

Smith BH (1991) Dental development and the evolution of life history in the Hominidae. *American Journal of Physical Anthropology* **86**, 157–174.

Smith BH (1993) The physiological age of KNM-WT 15 000. In *The Nariokotome*

Homo erectus *Skeleton*, eds. A. Walker and R. Leakey, pp. 195–220. Cambridge, MA: Harvard University Press.

Smith BH, Crummett TL and Brandt KL (1994) Ages of eruption of primate teeth: a compendium for ageing individuals and comparing life histories. *Yearbook of Physical Anthropology* **37**, 177–231.

Smith RJ (1996) Biology and body size in human evolution. *Current Anthropology* **37**, 451–481.

Smith RJ and Jungers WL (1997) Body mass in comparative primatology. *Journal of Human Evolution* **32**, 523–559.

Smith RJ, Gannon PJ and Smith BH (1995) Ontogeny of australopithecines and early *Homo*: evidence from cranial capacity and dental eruption. *Journal of Human Evolution* **29**, 155–168.

Sokal RR and Rohlf FJ (1995) *Biometry*, 3rd edn. New York: W. H. Freeman and Company.

Sprent P (1972) The mathematics of size and shape. *Biometrics* **28**, 23–37.

Stewart KJ and Harcourt AH (1987) Gorillas: variation in female relationships. In *Primate Societies*, ed. B.B. Smuts, D.L. Cheney, R.M. Seyfarth, R.W. Wrangham and T.T. Struhsaker, pp. 155–164. Chicago: University of Chicago Press.

Stuart-Macadam PL (1989) Nutritional deficiency diseases: a survey of scurvy, rickets and iron-deficiency anaemia. In *Reconstruction of Life from the Skeleton*, ed. M.T. Iscan and K.A.R. Kennedy, pp. 201–222. New York: Alan R. Liss.

Tanner JM and Whitehouse RH (1976) Clinical longitudinal standards for height, weight, velocity, weight velocity and the stages of puberty. *Archives of Disease in Childhood* **51**, 170–179.

Tanner JM, Whitehouse RH, Cameron N, Marshall WA, Healy MJR & Goldstein H (1983). *Assessment of Skeletal Maturity and Prediction of Adult Height* (*TW2 Method*), 2nd edn. London: Academic Press.

Taylor AB (1995) Effects of ontogeny and sexual dimorphism on scapula morphology in the mountain gorilla (*Gorilla gorilla beringei*). *American Journal of Physical Anthropology* **98**, 431–445.

Tsutakawa RK and Hewett JE (1978) Comparison of two regression lines over a finite interval. *Biometrics* **34**, 391–398.

Vogl C, Atchley WR, Cowley DE, Crenshaw P. Murray JD and Pomp D (1993) The epigenetic influence of growth hormone on skeletal development. *Growth, Development and Ageing* **57**, 163–182.

Vrba ES (1994) An hypothesis of heterochrony in response to climatic cooling and its relevance to early hominid evolution. In *Integrative Paths to the Past – Paleoanthropological Advances in Honor of F. Clark Howell*, ed. R.S. Corruccini and R.L. Ciochon, pp. 345–376. Englewood Cliffs, NJ: Prentice Hall.

Watts ES (1990) Evolutionary trends in primate growth and development. In *Primate Life History and Evolution*, ed. C.J. DeRousseau, pp. 89–104. New York: Wiley-Liss.

Winkler LA (1996) Appearance of ossification centres of the lower arm, wrist, lower leg, and ankle in immature orangutans and chimpanzees with an assessment of the relationship of ossification to dental development. *American Journal of*

Physical Anthropology **99**, 191–203.

Wintheiser JG, Clauser DA and Tappen NC (1977) Sequence of eruption of permanent teeth and epiphyseal union in three species of African monkeys. *Folia Primatologica* **27**, 178–197.

Wood B (1991) *Koobi Fora Research Project*, vol. 4, *Hominid Cranial Remains.* Oxford: Oxford University Press.

3 *Relative mandibular growth in humans, gorillas and chimpanzees*

LOUISE T. HUMPHREY

Introduction

Recent years have seen a large number of comparative developmental studies among apes, employing a variety of methodologies. These include studies of the developing dentition (Dean and Wood 1981; Smith 1994; Smith *et al.* 1994), cranial and postcranial skeleton (Shea 1981, 1983; Hartwig-Scherer and Martin 1992; Simpson *et al.* 1996; Taylor 1997), skeletal maturation (Watts 1986, 1990) and body weight (Leigh 1992; Leigh and Shea 1995, 1996).

The allometric method has been widely used in comparative analysis of primate skeletal growth. This method enables the comparison of growth trajectories to be carried out without reference to dental or chronological age. Instead, the growth of a given set of dimensions is examined relative to a reference variable, such as body weight (Hartwig-Scherer and Martin 1992), skeletal weight (Taylor 1997) or basicranial length (Ravosa 1991; Shea 1983). Subadult crania may be grouped into broad developmental classes on the basis of tooth eruption to allow comparisons to be made at equivalent developmental stages. This may be useful for comparisons within a species (e.g. between males and females) or between species with similar patterns of dental development, but it is not appropriate in a broader comparative context.

An alternative method for ontogenetic comparisons is to describe the relationship between size and chronological age (Dean and Wood 1981; Simpson *et al.* 1996; Humphrey 1998). Direct comparisons of skeletal size against chronological age are rare, owing to the scarcity of individuals of known life history in human and primate skeletal collections. As an increasing amount of information about the relationship between chronological age and dental development has become available (e.g. Smith *et al.* 1994) it is now possible to carry out this type of growth study using skeletal collections of unknown individuals.

This chapter presents a comparative study of the mandibular growth of modern humans, common chimpanzees and lowland gorillas and examines the relationships between mandibular growth, tooth emergence and chronological age in the three species. The growth patterns of a series of mandibular dimensions are summarised by a set of equations, each describing a curve fitted to size plotted against estimated chronological age. The equations can be used to determine the actual size or percentage of adult size attained at a specified chronological age or stage of dental development.

The first part of this chapter compares the amount of skeletal growth accomplished at different stages of tooth emergence and different estimated chronological ages in humans, gorillas and chimpanzees. The timing of tooth emergence has been shown to be correlated with brain size and body size across primate species and may be a useful predictor of the rate of growth in a species (Smith 1989; Smith *et al.* 1994). The growth of the mandible is expected to be more closely integrated with the process of tooth emergence than other parts of the skeleton since the mandible houses the lower dentition and provides attachment for the masticatory muscles. This chapter examines whether growth of the mandible is more tightly constrained relative to stages of tooth emergence than relative to chronological age. The second purpose is to assess the variation in the relative timing of growth of different mandibular dimensions within and among humans, gorillas and chimpanzees.

Sample and methods

The African ape mandible sample comprises wild-caught specimens of *Gorilla gorilla* (29 adults and 38 subadults) and *Pan troglodytes* (28 adults and 32 subadults) from the collections held by the Anthropological Institute and Museum of the University of Zurich. The modern human sample comprises 53 adults and 25 subadults from the Spitalfields collection, held by the Natural History Museum, London, and 41 adults and 17 subadults from the St Bride's collection, held in the crypt of St Bride's Church, London. Both human groups represent a relatively affluent section of an urban population and are of individuals who were buried between 1750 and 1850. Fourteen mandibular dimensions were measured. Only individuals in which fusion of the mandibular symphysis had occurred were measured. Individuals with extensive antemortem tooth loss were excluded from the sample.

The human sample used here comprises subadult and adult skeletons of individuals of known age at death (Bowman *et al.* 1992; Molleson and Cox 1993). As a result it is possible to avoid the error and possible bias introduced by the use of dental age estimates. This is particularly useful for adolescents and adults for whom accurate age estimation is problematic. From an analytical perspective the further advantage of working with individuals of known age at death is that it is unnecessary to make *a priori* assumptions about the age at which growth is complete or mean final size, which may differ from the mean size of individuals exhibiting complete dental maturity. These advantages have allowed for detailed analyses of human skeletal growth to be carried out (Humphrey 1998). The present study involves a comparison of the growth of three species, and in order to provide comparable results it is necessary to standardise the methods used. This analysis aims to replicate the techniques used previously (Humphrey 1998) as far as is possible with skeletal samples of unknown individuals. In practise this means assuming that the samples of all three species are randomly selected from populations comprising individuals of unknown chronological age, since the chronological ages of the African apes used here are not known. The methodology had to be adapted in several ways.

Ageing the subadult sample

For this study, tooth emergence was used as an indicator of chronological age. There are several disadvantages in using tooth emergence for estimating age. Firstly, gingival eruption may be difficult to define in skeletal material, since few teeth have a clear stain line indicating how much of the cusp had pierced the gum. Secondly, there is a relatively large time interval between the age at which the emergence of the deciduous dentition is complete and the age at which the first permanent teeth emerge during which individuals cannot be aged accurately on the basis of tooth emergence. Additionally in modern humans the emergence of the third permanent molar does not give an accurate indication of age because it is highly variable between individuals. An advantage of tooth emergence is that data can be collected using non-invasive methods. An impressive data base on tooth emergence in extant primates has recently been compiled and summarised by Smith and co-workers (1994). Furthermore, while studies of known individuals have led to the development of numerous techniques for estimating the chronological age of subadult humans and chimpanzees using tooth formation (e.g. Moorrees, Fanning and Hunt 1963; Demirjian *et al.* 1973; Demirjian and Goldstein 1976; Anemone *et al.* 1991; Kuykendall 1996), the gorilla has been less widely studied and comparable ageing techniques are not readily available.

For the present analysis each individual was assigned an age at death on the basis of the state of emergence of the mandibular and maxillary deciduous and permanent teeth, using the dental emergence charts presented by Smith and colleagues (1994). For each species the sequence of tooth emergence was determined from the mean age of emergence of each tooth. Where separate values are available for males and females, mean age of emergence was calculated as the midpoint of the male and female values. In most cases age at death was estimated from an upper and lower age boundary indicated by the state of emergence of the dentition as a whole. The lower age boundary was calculated as the mean age of emergence of the latest tooth to emerge, and the upper boundary was calculated as the mean age of emergence of the earliest tooth that had not emerged. Occasionally the upper and lower boundaries were reversed as a result of minor variations in the sequence of dental emergence. Age at death was calculated as half way between the lower and upper boundaries. If a tooth of an individual had clearly only just emerged, age was calculated as a quarter of the way between the lower and the upper boundary. In a few cases where a particular tooth was in the process of emerging on one or both sides, individuals were assigned the mean age of emergence of that tooth. African apes in which all of the permanent teeth are fully erupted but basi-occipital synchondrosis is incomplete were aged at 12 years.

Curve fitting

The growth of each mandibular variable was described by fitting a modified Gompertz equation to the data points. Previous work has shown that the Gompertz equation provides an accurate description of human skeletal growth and that the modified Gompertz equation, in which the lower asymptote is forced to be zero, is less severely perturbed than the standard Gompertz equation by the removal of young individuals from the sample (Humphrey 1998). The curves were fitting using SPSS. The technique requires initial values for each of the parameters to be suggested. For each variable in each species the suggested initial values were the mean adult size for the upper asymptote (a), 0.2 for the slope (b) and 0 for the point of inflexion (m). The percentage of adult size attained at successive dental ages was calculated by dividing size, as determined from the fitted curve, into mean adult size.

Calculation of mean adult size and the distribution of adult data points

All individuals in which the permanent third molars were in occlusion and basioccipital synchondrosis was complete were used to estimate adult size. Mean adult size was calculated as the mean size of individuals who had achieved this state of maturity. The estimate of mean adult size is used to estimate the upper asymptote during the curve fitting procedure. In this study it is also used to simulate the distribution of adult data points, since the age of adult individuals is unknown. The distribution of adult data points was simulated by incorporating a string of mean adult values into the data set for each variable. Mean adult values were added at one year age intervals between the ages of 15 and 30 years in chimpanzees and gorillas and one and a half year intervals between the ages of 22.5 and 45 years in *Homo*.

Normalisation

Under some circumstances it is appropriate to use normalised data for interspecific developmental comparisons. Various developmental system can be used to normalise the duration of development between species, including the skeleton, dentition and sexual maturation (Simpson *et al.* 1996). For example, Smith (1986) extended the total duration of pongid dental development to equal that of humans, with 10 pongid years equalling 18 human years. Simpson and co-workers (1996) have argued that it is more appropriate to determine a normalising function from a cumulative developmental process rather than single (usually terminal) events such as age of menarche, age at completion of epiphyseal fusion, or the age of occurrence of a particular aspect of dental development. Accordingly, their technique, which yielded a ratio of 1.62 between *Pan* and *Homo*, is based on the timing of the development of all of the permanent postcanine teeth.

The normalising ratio used in this study is calculated from the mean age of emergence of each of the permanent teeth using data from Smith and co-workers (1994). It is therefore based on the same criteria as those used for estimating the age of the mandible samples. The mean age of emergence of eight maxillary and eight mandibular permanent teeth is 10.57 years for modern humans, 7.13 years for chimpanzees and 6.98 years for gorillas. These figures yield an adjustment ratio of 1.48 between humans and chimpanzees and 1.51 between humans and gorillas. For the current analysis a normalising ratio of 1.5 was used in both cases.

Outstanding differences between the human and African ape studies

This study aimed to standardise the conditions under which the human and African ape material were studied and analysed, but several factors could not be fully controlled. The most obvious difference between the samples is that the African apes were wild-caught and can be assumed to represent a random sample of the population from which they were drawn. The human sample is a mortality sample comprising individuals whose normal developmental trajectories were permanently interrupted by death, which followed an unknown period of developmental disruption (Johnston 1968).

Many of the human children used in this study suffered severe dental pathology, whereas this was rare amongst the wild-caught African apes. Caries can result in premature loss of the deciduous teeth, which can affect the timing of the emergence of the underlying permanent teeth (Posen 1965; Bailit 1976). Early loss of the deciduous molars can result in delayed emergence of the permanent premolars as a result of alveolar remodelling. Slightly later, but still premature, loss of the deciduous molars can result in advanced emergence of the permanent premolars because they have a clear passageway through which to erupt. These processes were clearly present among the Spitalfields children, who exhibited an irregular pattern of premolar emergence including a high level of asymmetry.

A third difference between the African ape and human samples is the state of preservation. The African ape mandibles were mainly complete and undamaged and most of the mandibular and maxillary teeth were retained n the tooth rows. The human bones were frequently broken, damaged or distorted. Breakage of the mandible at the ramus or mandibular condyle was particularly common, resulting in incomplete sets of measurements. Many of the individuals had loose or missing teeth, resulting in partially completed dental emergence charts, and less precise information for age estimation. This problem helped to counterbalance the differences in the composition of the African ape and human samples. The samples of African ape mandibles used here are heavily biased against individuals who had not fully erupted their deciduous dentition. Contrastingly this age range is well represented in many archaeological samples, including the Spitalfields sample, but individuals in this age range were most likely to be rejected because of breakage or because the state of tooth emergence could not be defined accurately.

Mandibular growth in humans, gorillas and chimpanzees

Modern humans are said to be characterised by a prolonged period of growth and maturation compared to chimpanzees and gorillas. This aspect of human life history has been widely discussed in relation to encephalisation, social structure and the development of culture (Smith 1986). The extent to which different developmental systems might vary in terms of relative advancement or retardation, and the implications of this variation have not been fully explored (Watts 1990). The present contribution is part of ongoing work examining the relationships between skeletal growth, skeletal maturation (e.g. epiphyseal fusion of the long bones), tooth formation and emergence and chronological age in anthropoid primates.

In this chapter, the relationship between mandibular growth, tooth emergence and chronological age in humans, common chimpanzees and lowland gorillas is examined. The results are discussed in two sections. The first section evaluates growth attainment, at successive dental stages and relative to estimated chronological age and normalised dental age in humans, chimpanzees and gorillas. The second section examines differences in the growth patterns of the individual mandibular variables and considers the sequence in which different parts of the mandible attain their adult size in the three species. Mandibular growth is assessed in terms of the percentage of adult size attained at different stages of growth.

Relative growth

Figure 3.1 illustrates the percentage of adult size attained in humans, chimpanzees and gorillas for each mandibular variable at the species-specific ages of emergence of the first, second and third permanent mandibular molars (i.e. at equivalent dental stages). The variables are arranged in ascending order of the mean percentage of adult size attained across the three species at the age of first molar emergence. At the age of first molar emergence (Fig. 3.1a), humans have attained an average of 78.4% of adult size compared to 62.4% in chimpanzees and 60.1% in gorillas. The differences between the three species diminish with the emergence of successive molars. By the age of second molar emergence (Fig. 3.1b), the figures are 89.4% for humans, 78.2% for chimpanzees and 77.0% for gorillas and by the age of third molar emergence (Fig. 3.1c), they are 96.9%, 90.2% and 89.7%, respectively. Humans have attained a higher percentage of adult size in all of the variables at the first two stages of tooth emergence. At the age of third molar emergence, humans have attained a higher percentage of

(a)

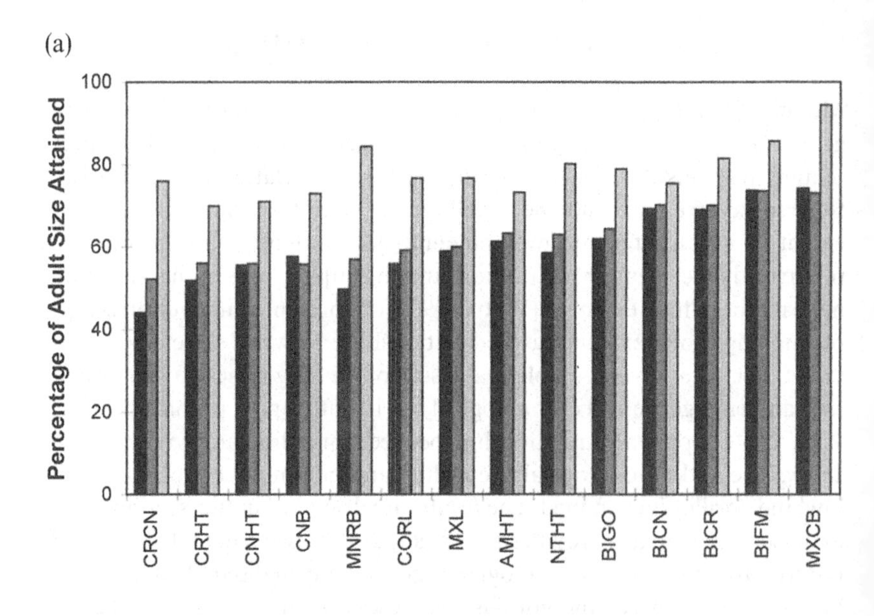
Percentage of Adult Size Attained
100
80
60
40
20
0
CRCN
CRHT
CNHT
CNB
MNRB
CORL
MXL
AMHT
NTHT
BIGO
BICN
BICR
BIFM
MXCB

(b)

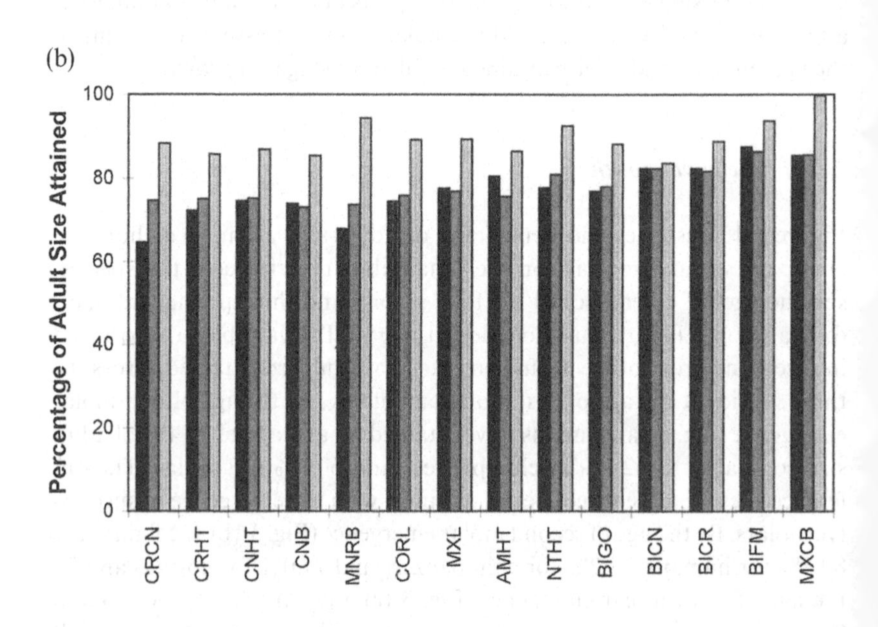
Percentage of Adult Size Attained
100
80
60
40
20
0
CRCN
CRHT
CNHT
CNB
MNRB
CORL
MXL
AMHT
NTHT
BIGO
BICN
BICR
BIFM
MXCB

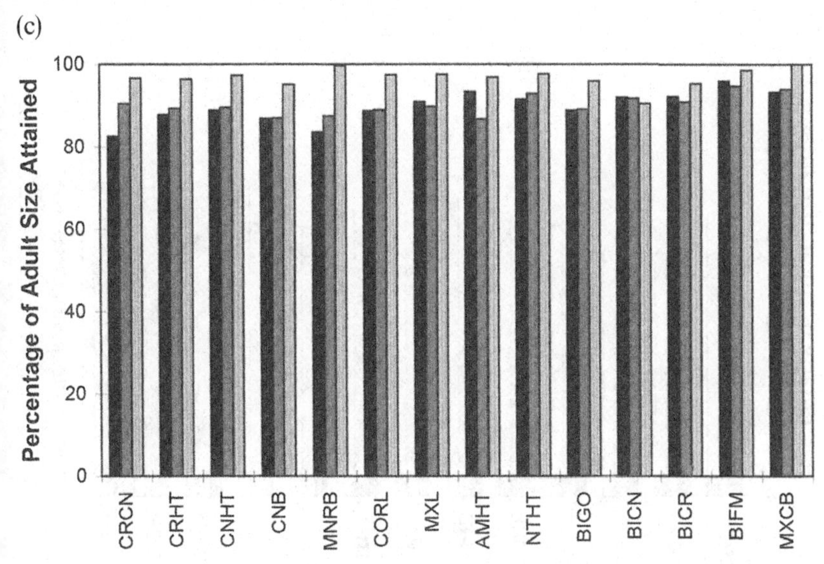

Fig. 3.1. The percentage of adult size attained in humans (pale grey bars), chimpanzees (mid-grey bars) and gorillas (black bars) for each mandibular variable at the species-specific ages of emergence of the first (a), second (b) and third (c) permanent mandibular molars. The ages for dental emergence are from Smith *et al.* (1994). CRCN, coronoid–condyle distance; CRHT, coronoid height; CNHT, condyle height; CNB, condyle breadth; MNRB, minimum ramus breadth; CORL, corpus length; MXL, maximum mandibular length; AMHT, anterior mandible height; NTHT, mandibular notch depth; BIGO, bigonial breadth; BICN, bicondylar breadth; BICR, bicoronoid breadth; BIFM, distance between mentale foramina; MXCB, maximum corpus breadth.

adult size in all of the mandibular dimensions apart from bicondylar breadth. At the age of first molar emergence, chimpanzees have attained a higher percentage of adult size than gorillas in 11/14 variables, declining to 9/14 variables at the age of second and third molar emergence, but the average difference between chimpanzees and gorillas is small at every stage.

Figure 3.2 compares the percentage of adult size attained in humans, chimpanzees and gorillas for each mandibular variable at estimated chronological ages of 3 years, 6 years and 9 years. The variables are arranged in ascending order of the mean percentage of adult size attained across the three species at age 3 years. At 3 years (Fig. 3.2a), humans have attained an average of 68% of adult size in the 14 mandibular variables, compared to 60.8% in chimpanzees and 56.8% in gorillas. At 6 years (Fig. 3.2b), the differences between humans and the other species are considerably smaller, with humans having attained an average of 77.7% of adult size compared to 76.3% in chimpanzees and 74.3% in gorillas. By 9 years (Fig. 3.2c), chimpanzees and gorillas have overtaken humans in terms of the average

(a)

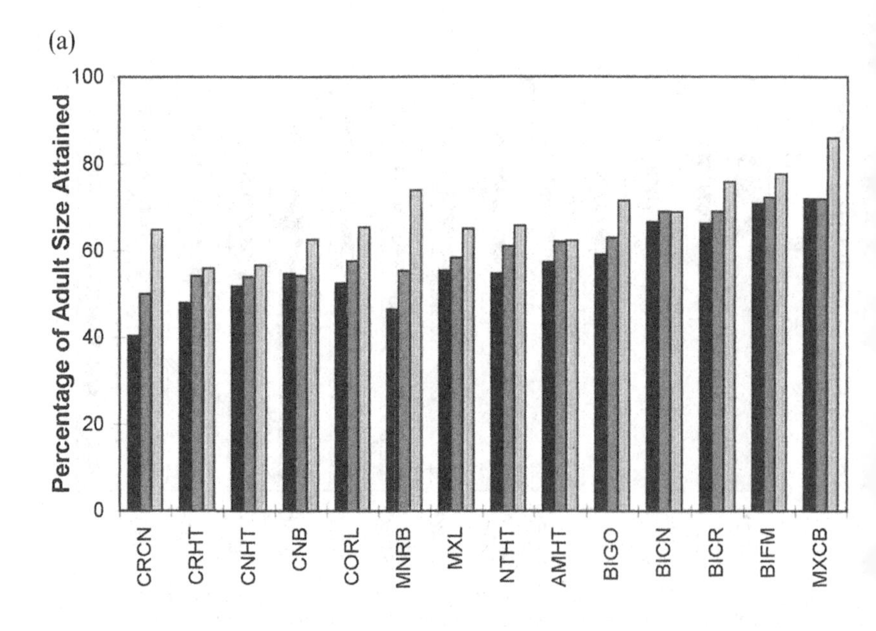
Percentage of Adult Size Attained
100
80
60
40
20
0
CRCN
CRHT
CNHT
CNB
CORL
MNRB
MXL
NTHT
AMHT
BIGO
BICN
BICR
BIFM
MXCB

(b)

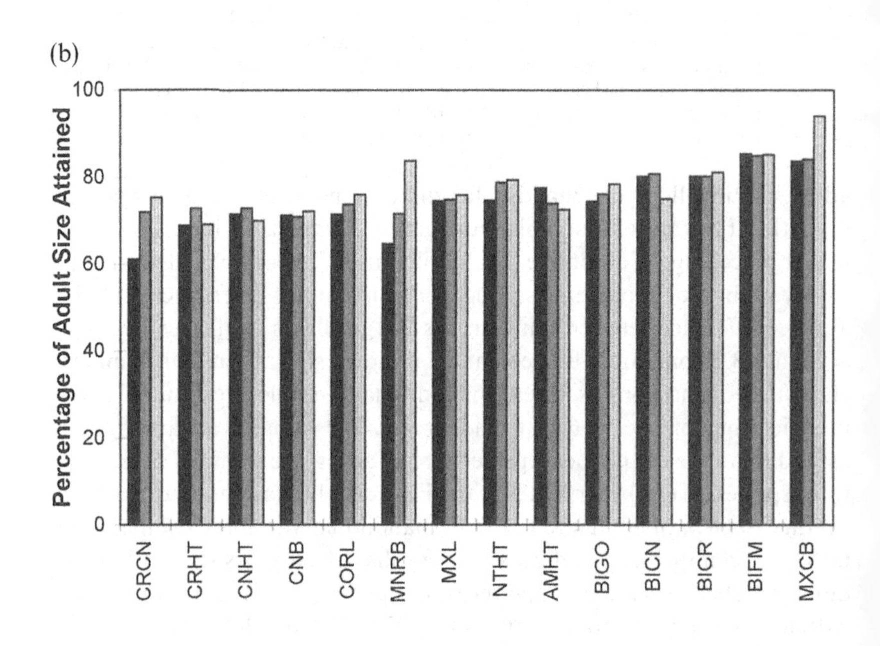
Percentage of Adult Size Attained
100
80
60
40
20
0
CRCN
CRHT
CNHT
CNB
CORL
MNRB
MXL
NTHT
AMHT
BIGO
BICN
BICR
BIFM
MXCB

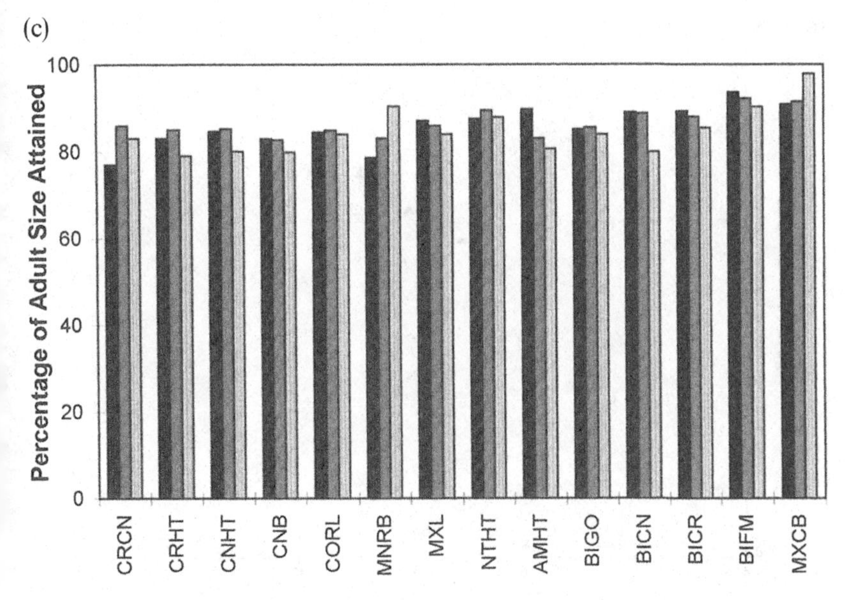

Fig. 3.2. The percentage of adult size attained in humans (pale grey bars), chimpanzees (mid-grey bars) and gorillas (black bars) for each mandibular variable at estimated chronological ages of 3 years (a), 6 years (b) and 9 years (c). The labels for the variables are the same as those in Fig. 3.1.

percentage of adult size attained across the 14 mandibular dimensions. At this age, humans have attained an average of 84.8% of adult size compared to 86.6% in chimpanzees and 86.0% in gorillas. The average figures mask considerable variation in the extent to which the 14 mandibular dimensions differ in terms of their progress toward adult size between the three species, but they reflect a general trend that is echoed in the diminishing number of variables in which humans are advanced at successive chronological ages. At 3 years humans have attained a higher percentage of adult size in all 14 variables. By 6 years this has been reduced to nine variables and at 9 years humans are advanced in the growth of only two variables.

The results can be summarised by superimposing plots showing the age at which 70% of adult size is attained against the age at which 90% of adult size is attained for each of the mandibular variables in the three species. Figure 3.3a shows a plot of the age at which 70% of adult size is attained against the age at which 90% of adult size is attained, using the unadjusted estimated chronological ages. Humans attain 70% of adult size earlier than chimpanzees or gorillas in most of the dimensions. The exceptions are mandibular condyle height, coronoid height and anterior mandible height. Humans attain 90% of adult size later than the chimpanzees or gorillas in

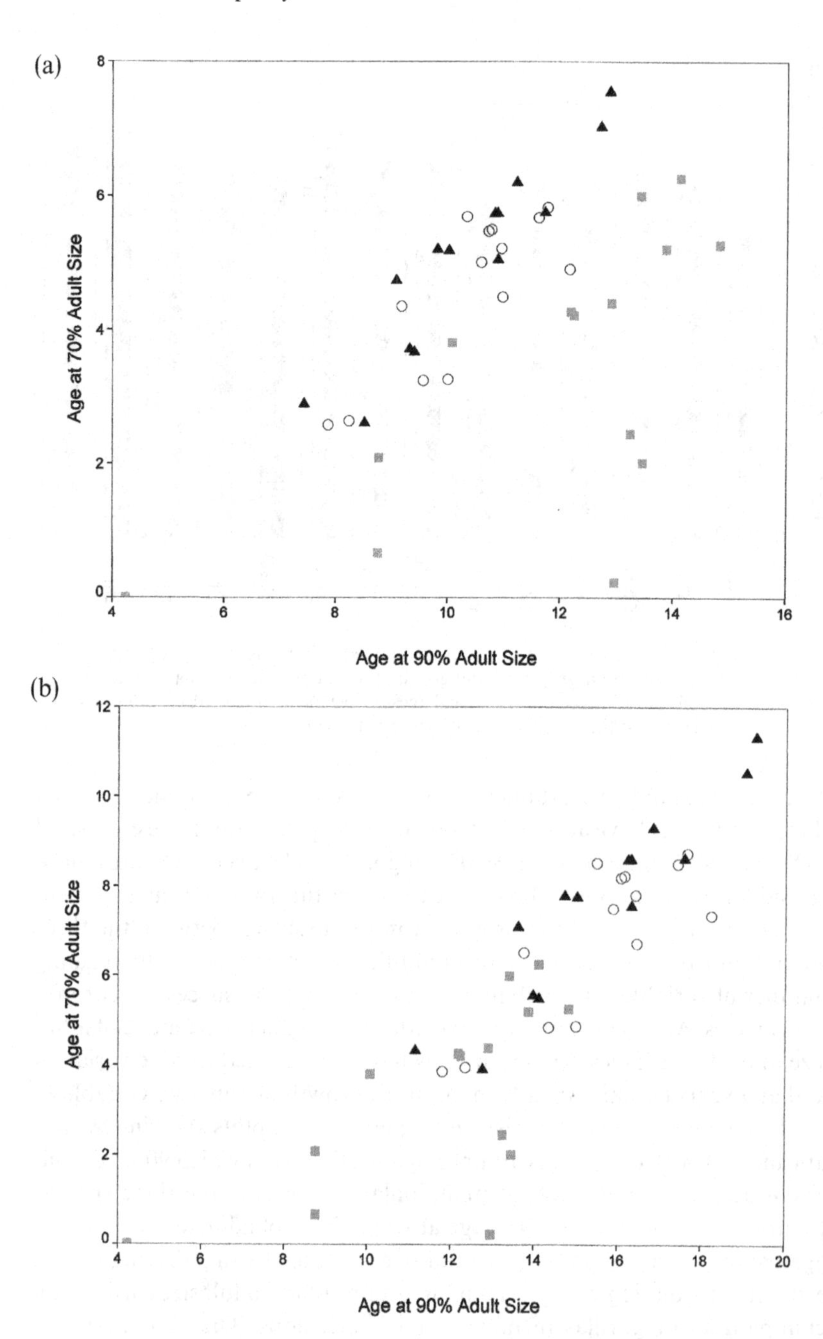

Fig. 3.3. Age at which 70% of adult size is attained against the age at which 90% of adult size is attained in 14 mandibular variables in humans (pale grey squares), chimpanzees (open circles) and gorillas (black triangles), based on estimated chronological age (a) and normalised dental age (b).

all of the dimensions except minimum ramus breadth and maximum condyle breadth. This reflects a clear difference in the growth pattern of the mandibular dimensions, which will be described in more detail below.

Figure 3.3b shows an equivalent plot using the normalised dental ages. Humans are advanced in the normalised dental age at which both 70% and 90% of adult size are attained in all of the mandibular dimensions except anterior mandible height, for which the gorilla is the earliest species to attain 90% of adult size. The normalised plot indicates that mandibular growth in gorillas and chimpanzees is delayed relative to dental development in comparison with modern humans throughout the growth period. The spread of data points on both the unadjusted and normalised plots (Figs. 3.3a and 3.3b) illustrates that all three species show considerable variation in the progress of different mandibular dimensions toward adult size.

The preceding results demonstrate that humans, gorillas and chimpanzees differ in their rate of attainment of adult mandibular size relative to both estimated chronological age and the emergence of individual teeth. In both cases the largest differences are between humans and the African apes. Chronological age is estimated from the average species values for the age of emergence of individual teeth using data from Smith and colleagues (1994). During the first few years of growth humans are advanced over chimpanzees and gorillas in the percentage of adult size attained in all of the mandibular dimensions. The average difference decreases until approximately 9 years, when humans are overtaken by the African apes in their progress toward the attainment of adult size, although the actual age at which this transition occurs varies quite considerably between different mandibular dimensions. Compared to chimpanzees and gorillas, humans attain 70% of adult size early and 90% of adult size late in most dimensions. This reflects a marked slowing down in the rate at which adult size is attained in the human mandibles compared to chimpanzees and gorillas.

The differences between humans and the African apes in the percentage of adult size attained are higher when comparisons are made at equivalent stages of tooth emergence, with the greatest differences occurring at the age of first molar emergence. The average percentage of adult size attained at the time of emergence of the first, second and third permanent molars is very similar in chimpanzees and gorillas, whereas humans are considerably advanced over chimpanzees and gorillas in the average percentage of adult size attained at each of these stages. Interestingly, humans have attained roughly the same percentage of adult size at the age of first molar emergence as the African apes have attained at the age of second molar emergence, and roughly the same percentage of adult size at the age of second molar emergence as the African apes have attained at the age of third molar emergence.

The differences between chimpanzees and gorillas are generally low, whether measured at equivalent chronological ages or tooth emergence stages, but they are lowest for comparisons made at tooth emergence stages. In all examples, chimpanzees are, on average, slightly advanced over gorillas in terms of their attainment of adult size. The overall similarity in the patterns of attainment of adult size described here indicate that differences in adult size can be attributed largely to differences in growth rate between gorillas and chimpanzees rather than differences in the duration of growth. This supports earlier reports (Shea 1983; Leigh and Shea 1996) that differences in adult cranial size and body weight between gorillas and chimpanzees are mainly the result of variation in growth rates rather than in the duration of growth.

The third type of growth comparison described above examines the percentage of adult size attained relative to normalised dental age. The normalised dental age is an adjustment based on the mean age of emergence of the permanent teeth in each species. Mandibular growth occurs later relative to normalised dental age in chimpanzees and gorillas than in humans. The normalising ratio of 1.5 used here is low compared to some previous suggestions (Smith 1986; Simpson *et al.* 1996). A higher ratio would yield an even greater discrepancy between the relative timing of tooth emergence and mandibular growth in humans and African apes. For comparative purposes a normalisation ratio can be calculated for mandibular growth. The average ages at which 95% of adult size is attained in the 14 mandibular variables are 16.35 years in humans, 13.34 years in gorillas and 13.29 years in chimpanzees. This yields a normalising ratio of 1.23 between humans and both species of African ape. This ratio is based on terminal events (Simpson *et al.* 1996) and does not reflect the whole process of growth, nor the growth of the individual mandibular dimensions. However, any single ratio would fail to reflect the fact that humans are generally advanced over chimpanzees and gorillas in the early stages of growth and delayed in the later stages.

Growth patterns and sequences

Examination of the growth patterns of the individual mandibular dimensions demonstrates that chimpanzees grow up more like gorillas than like humans, even in those dimensions in which humans and chimpanzees are more alike in adult size. A typical example is bicondylar breadth (Fig. 3.4).

(a)

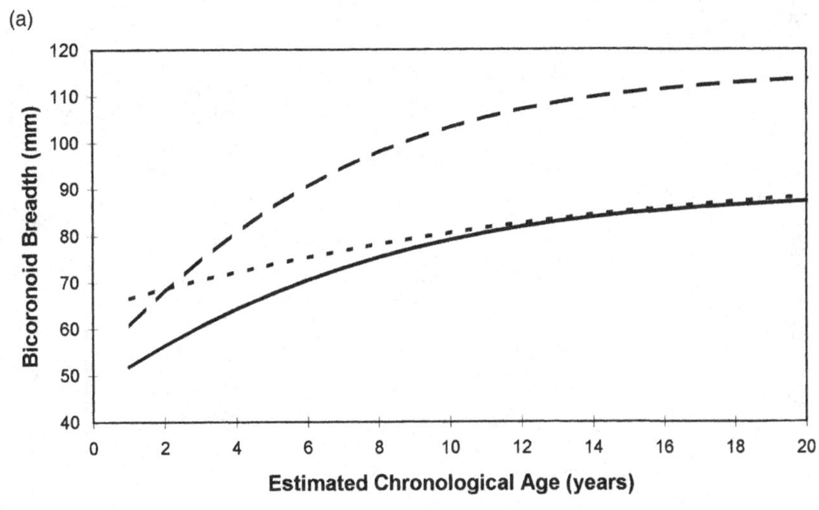

(b)

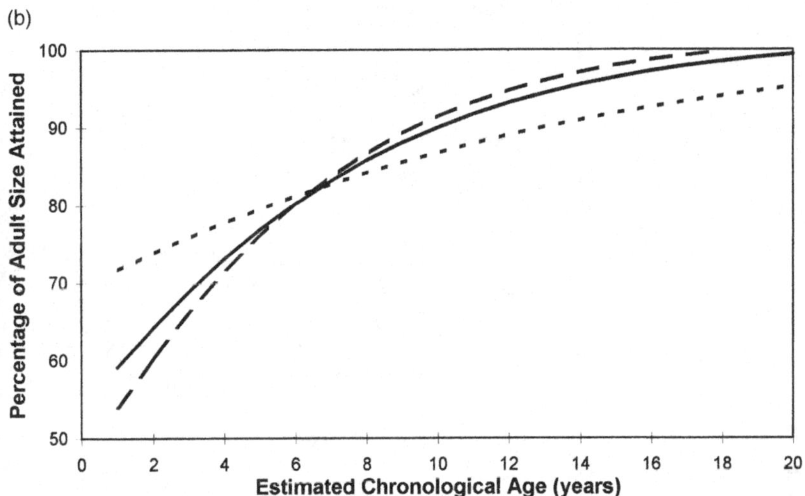

Fig. 3.4. Gompertz curves describing the growth pattern for bicoronoid breadth (a) and the percentage of adult size attained (b) relative to estimated chronological age in humans (dotted line), chimpanzees (dashed line) and gorillas (continuous line).

Adult humans have a slightly larger bicoronoid breadth than chimpanzees and both species are much smaller than gorillas. During infancy, humans are larger than chimpanzees of the same estimated chronological age, but between approximately 9 years and the completion of growth in the chimpanzee, the two species are virtually indistinguishable in size (Fig. 3.4a). In terms of percentage of adult size attained (Fig. 3.4b), chimpanzees

(a)

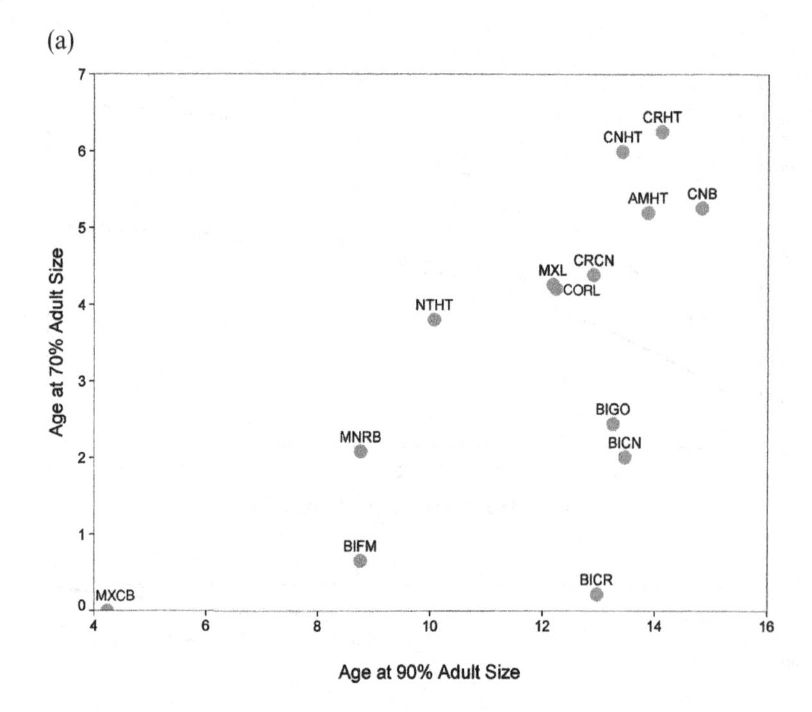
CRHT
CNHT
AMHT
CNB
CRCN
MXL
CORL
NTHT
BIGO
MNRB
BICN
BIFM
BICR
MXCB
Age at 70% Adult Size
Age at 90% Adult Size
0
1
2
3
4
5
6
7
4
6
8
10
12
14
16

(b)

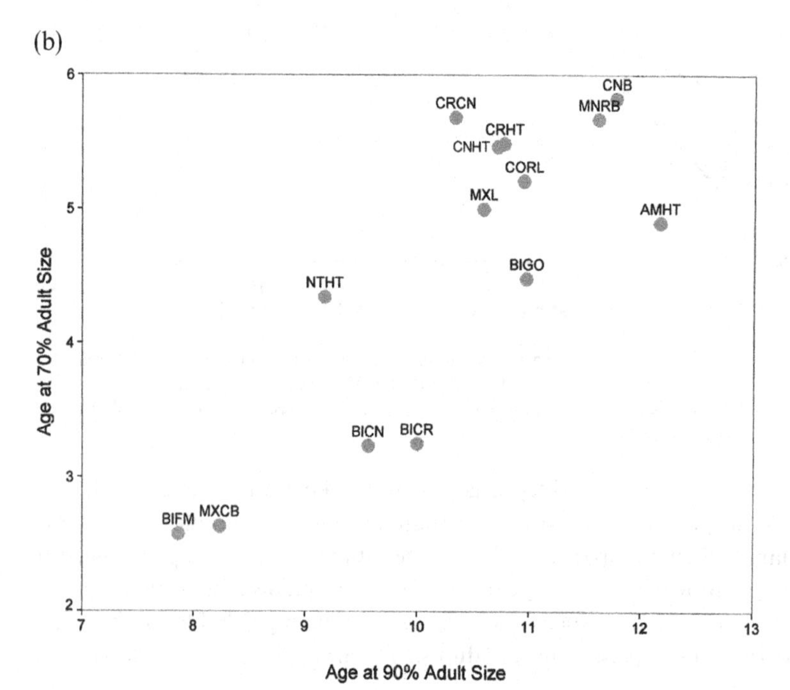
CNB
CRCN
MNRB
CRHT
CNHT
CORL
MXL
AMHT
BIGO
NTHT
BICN
BICR
BIFM
MXCB
Age at 70% Adult Size
Age at 90% Adult Size
2
3
4
5
6
7
8
9
10
11
12
13

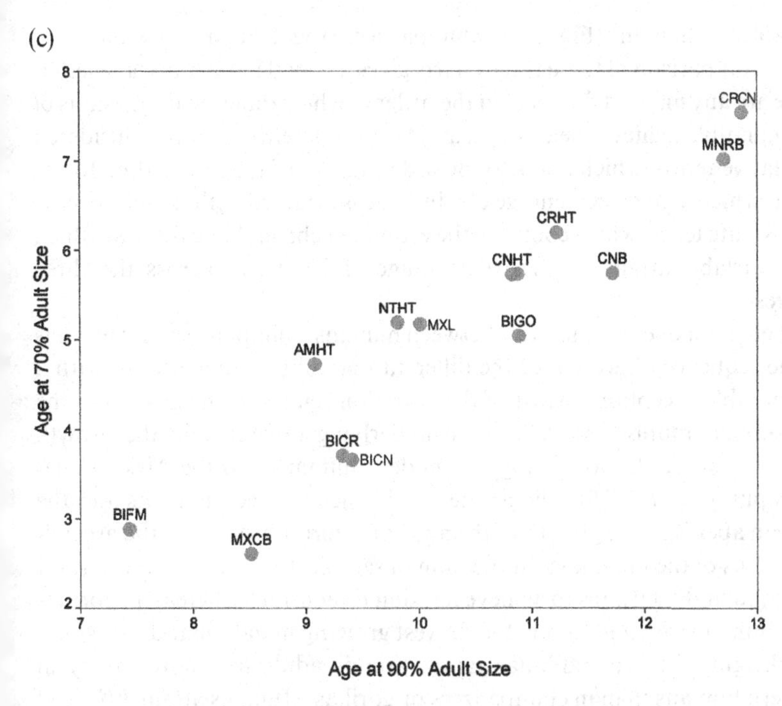

Fig. 3.5. Age at which 70% of adult size is attained against the age at which 90% of adult size is attained in 14 mandibular variables in humans (a), chimpanzees (b) and gorillas (c). The labels for the variables are the same as those in Fig. 3.1.

are similar to gorillas throughout the growth period. Humans are advanced over both species until approximately 5 years, when the situation is reversed. While most variables follow the same general pattern, there are a few exceptions. Humans are advanced in the percentage of adult size attained in the maximum breadth of the mandibular corpus throughout growth, while chimpanzees and gorillas follow a virtually indistinguishable growth pattern. Humans are advanced over chimpanzees in the percentage of adult size attained in the minimum breadth of the mandibular ramus, and chimpanzees in turn are advanced over gorillas. Chimpanzees and humans follow a similar pattern of attainment of adult size of anterior mandible height and they are advanced over gorillas until approximately 5 years. Chimpanzees and gorillas show a similar shape of growth pattern for the distance between the mandibular condyle and coronoid process, but chimpanzees are advanced over gorillas in terms of the percentage of adult size attained throughout growth.

Figure 3.5 shows the age at which 70% of adult size is attained against the age at which 90% of adult size is attained for each of the mandibular variables in humans (Fig. 3.5a), chimpanzees (Fig. 3.5b) and gorillas (Fig. 3.5c). Comparison of these three charts gives a broad indication of whether there are any major differences in the order in which different dimensions of the mandible achieve their adult size. These differences can be considered in relative terms, which refers to the ordering of variables according to the age at which a given percentage of adult size is attained within each species, or absolute terms which compare the estimated chronological age at which each variable attains a given percentage of adult size across the three species.

There is an overall similarity between humans, chimpanzees and gorillas in the sequence of growth of the different mandibular dimensions, with a few notable exceptions. Most of the variation relates to the growth of the mandibular ramus, and reflects an underlying difference in the growth patterns and adult morphology of modern humans and the African apes (Humphrey *et al.* 1999). The greatest difference between humans and the African apes is in the growth of the anteroposterior breadth of the ascending ramus of the mandible. In the human sample, this is one of the earliest mandibular dimensions to achieve its adult dimensions whereas, in gorillas and chimpanzees, it is among the slowest growing mandibular dimensions. The height of the mandibular ramus attains adult size more slowly in modern humans than in chimpanzees or gorillas. Humans attain 70% and 90% of adult size in the heights of the coronoid process and mandibular condyle relatively and absolutely later than chimpanzees and gorillas. The difference is particularly evident in the early stage of growth. The heights of the coronoid process and mandibular condyle are the last two dimensions to attain 70% of adult size in modern humans and are two of only three variables in which humans attain 70% of adult size at a later estimated chronological age than chimpanzees or gorillas. In chimpanzees and gorillas the breadth of the mandibular ramus continues to increase in size after the completion of growth of the height of the mandibular ramus, whereas in humans this situation is reversed. Additionally, there is a marked difference between modern humans, chimpanzees and gorillas in the relative timing of the growth of the breadth of the mandible measured at the mandibular condyles and coronoid processes. All three species display rapid early growth in bicondylar and bicoronoid breadth, but humans are marked by a slowing down in the rate of growth and attain 90% of adult size relatively late.

In humans, the maximum length of the mandible continues to grow after the mandibular ramus has achieved its maximum anteroposterior dimensions. This is achieved in part by a shifting of the position of the mandibular ramus through deposition of bone on the posterior border and resorption on the anterior border (Enlow 1990). Later growth of the mandibular ramus occurs predominantly though bone deposition on the mandibular condyle and coronoid process, and results in a gradual increase in the height of the mandibular ramus and breadth of the mandible measured across the coronoid processes and mandibular condyles, according to the 'enlarging V principle' (Enlow 1990). The early widening of the mandible, particularly the increase in bicondylar breadth, in all three species is necessary to keep pace with the expansion of the basicranium. The delayed completion of growth of bicondylar and bicoronoid breadth in humans compared to chimpanzees and gorillas occurs because the mandible continues to widen gradually in association with the superior growth of the mandibular condyle and coronoid process. In gorillas and chimpanzees, the anteroposterior growth of the mandibular ramus continues after the height of the mandibular ramus and bicoronoid and bicondular breadth have reached their maximum dimensions, indicating that the rate of bone deposition on the posterior border of the mandibular ramus exceeds the rate of resorption on the anterior border for the entire duration of mandibular growth.

The maximum breadth of the mandibular corpus is among the earliest growing variables in all three species, but humans are advanced by several years over chimpanzees and gorillas in the age at which 90% of adult size is attained. The mandibular corpus houses the lower dentition, and is expected to grow in order to accommodate the widest dimensions of the developing teeth. The maximum postcanine breadth of the mandibular corpus needs to accommodate the buccolingual breadth of the largest molar. The early completion of the growth of the maximum breadth of the mandibular corpus in humans compared to gorillas and chimpanzees may reflect differences in relative tooth size between the three species. In humans, 90% of adult size is attained a few months after the formation of the first and largest permanent molar crown is complete. In chimpanzees and gorillas the second and third permanent molars are larger than the first permanent molar, and 90% of adult size is attained a few months after the age at which formation of the third permanent molar crown is complete.

The overall sequences in which different parts of the mandible attain adult size are remarkably similar in chimpanzees and gorillas, with only three dimensions showing markedly different patterns of growth. The anterior height of the mandible attains adult size relatively and absolutely early in gorillas. This difference is already evident at the age at which 70% of adult size is attained and becomes even more pronounced in the later stages of growth. In contrast, chimpanzees are advanced in the growth of the minimum breadth of the mandibular condyle and in the distance between the mandibular condyle and the coronoid process. The distance between the coronoid process and mandibular condyle attains adult size relatively late in gorillas in comparison with both chimpanzees and humans.

Conclusions

Humans, gorillas and chimpanzees differ in the percentage of adult mandible size attained at equivalent stages of dental development. Humans are more advanced than chimpanzees or gorillas in the overall progress of mandibular growth relative to tooth emergence. This is true whether an assessment is made at the species-specific ages of emergence of the first, second and third permanent molars or by using a normalising ratio to adjust for the average age of emergence of the permanent dentition. Humans, gorillas and chimpanzees also differ in the percentage of adult mandible size attained at the same estimated chronological age but the differences between humans and the African apes are lower for comparisons made at the same chronological age than for those made at equivalent stages of tooth emergence. During the early period of growth, humans have, on average, attained a higher percentage of adult size than have chimpanzees and gorillas, but they are overtaken by chimpanzees and gorillas in the later stages of growth. The overall differences between gorillas and chimpanzees are small at equivalent stages of tooth emergence and estimated chronological ages.

The findings presented in this chapter reflect variation in the relative timing of mandibular growth and tooth emergence in humans, gorillas and chimpanzees. Furthermore, all three species show variation in the pattern of growth of the different parts of the mandible and, despite an overall similarity in the sequence in which different parts of the mandible attain their adult size and function, there are notable differences between humans and the African apes in particular. Given that there is good reason to assume that the growth of the mandible should be closely integrated with the process of dental development, other aspects of skeletal growth and maturation may prove to be even more variable relative to dental development. These results indicate that it is not possible to predict skeletal growth attainment on the basis of chronological age or dental emergence.

There is convincing evidence that the pattern and rate of dental development of early hominids, including *Homo habilis*, is more like that of the African apes than that of modern humans (Bromage and Dean 1985; Smith 1986, 1994; Dean 1987). Together with analyses describing the relationship between tooth emergence and life history variables (Smith 1989; Smith *et al.* 1994), these studies suggest that the prolonged life history pattern seen in modern humans evolved subsequent to the appearance of the genus *Homo*. The present analysis demonstrates that mandibular growth is less delayed in modern humans relative to the African apes than dental emergence. This suggests that there is a certain amount of independence in the extent to which different developmental systems have been affected by an overall lengthening of the developmental period during human evolution. We must therefore be cautious about drawing inferences about the rate of development of fossil taxa based solely on dental evidence.

Acknowledgements

I thank Theya Molleson, Louise Scheuer, Jacqui Bowman, Bob Martin, Nina Bahr and Thomas Geissmann for permission to use, and help with access to, the skeletal collections, and Leslie Aiello, Chris Dean, Rob Foley and Bob Martin and Christophe Soligo for their helpful insights into various aspects of this work. I would also like to thank Charles FitzGerald and Rob Hoppa for inviting me to contribute to this volume and for their comments on the manuscript.

References

Anemone RL, Watts ES and Swindler DR (1991) Dental development of known-age chimpanzees, *Pan troglodytes* (Primates, Pongidae). *American Journal of Physical Anthropology* **86**, 229–241.

Bailit HL (1976) Variations in tooth eruption: a field guide. In *The Measures of Man*, ed. E. Giles and J.S. Friedlander, pp. 321–337. Cambridge, MA: Peabody Museum Press.

Bowman JE, MacLaughlin SM and Scheuer JL (1992) The relationship between biological and chronological age in the juvenile remains from St Bride's Church, Fleet Street, *Annals of Human Biology* **19**, 216.

Bromage TG and Dean MC (1985) Re-evaluation of the age at death of immature fossil hominids. *Nature* **317**, 525–527.

Dean MC (1987) The dental development status of six East African fossil hominids. *Journal of Human Evolution* **16**, 197–213.

Dean MC and Wood BW (1981) Developing pongid dentition and its use for ageing individual crania in comparative cross-sectional growth studies. *Folia Primatologica* **36**, 111–127.

Demirjian A and Goldstein H (1976) New systems for dental maturity based on 7 and 4 teeth. *Annals of Human Biology* **3**, 411–421.

Demirjian A, Goldstein H and Tanner JM (1973) A new system of dental age assessment. *Human Biology* **45**, 211–227.

Enlow DH (1990) *Facial Growth*, 3rd edn. Philadelphia: W.B. Saunders Company.

Hartwig-Scherer S and Martin RD (1992) Allometry and prediction in hominoids: a solution to the problem of intervening variables. *American Journal of Physical Anthropology* **88**, 37–57.

Humphrey LT (1998) Patterns of growth in the modern human skeleton. *American Journal of Physical Anthropology* **105**, 57–72.

Humphrey LT, Dean MC and Stringer CB (1999) Morphological variation in great ape and modern human mandibles. *Journal of Anatomy*, in press.

Johnston FE (1968) The growth of the skeleton in earlier peoples. In *The Skeletal Biology of Earlier Human Populations*, ed. D.R. Brothwell, pp. 57–66. Oxford: Pergamon Press.

Kuykendall KL (1996) Dental development in chimpanzees (*Pan troglodytes*): the timing of tooth calcification stages. *American Journal of Physical Anthropology* **99**, 135–158.

Leigh SR (1992) Patterns of variation in the ontogeny of primate body size dimorphism. *Journal of Human Evolution* **23**, 27–50.

Leigh SR and Shea BT (1995) Ontogeny and the evolution of adult body size dimorphism in apes. *American Journal of Primatology* **36**, 37–60.

Leigh SR and Shea BT (1996) Ontogeny of body size variation in African apes. *American Journal of Physical Anthropology* **99**, 43–66.

Molleson T and Cox M (1993) *The Spitalfields Project*, vol. 2 *The Middling Sort*. Council for British Archaeology Research Report no. 86.

Moorrees CFA, Fanning EA and Hunt EE (1963) Age variation of formation stages for 10 permanent teeth. *Journal of Dental Research* **42**, 1490–1502.

Posen AL (1965) The effect of premature tooth loss of deciduous molars on premolar eruption. *Angle Orthodontist* **35**, 249–252.

Ravosa MJ (1991) The ontogeny of cranial sexual dimorphism in two old world monkeys: *Macaca fascicularis* (Cercopithecinae) and *Nasalis larvatus* (Colobinae). *International Journal of Primatology* **12**, 403–426.

Shea BT (1981) Relative growth of the limbs and trunk in the African apes. *American Journal of Physical Anthropology* **56**, 179–201.

Shea BT (1983) Allometry and heterochrony in the African apes. *American Journal of Physical Anthropology* **62**, 275–289.

Simpson SW, Russel KF and Lovejoy CO (1996) Comparison of diaphyseal growth between the Libben population and the Hamann-Todd chimpanzee sample. *American Journal of Physical Anthropology* **99**, 67–78.

Smith BH (1986) Dental development in *Australopithecus* and early *Homo*. *Nature* **323**, 327–330.

Smith BH (1989) Dental development as a measure of life history in primates. *Evolution* **43**, 683–688.

Smith BH (1994) Patterns of dental development in *Homo*, australopithecines, *Pan* and *Gorilla*. *American Journal of Physical Anthropology* **94**, 307–325.

Smith BH, Crummett TL and Bradt KL (1994) Ages of eruption of primate teeth: a compendium for ageing individuals and comparing life histories. *Yearbook of*

Physical Anthropology **37**, 177–232.

Taylor AB (1997) Relative growth, ontogeny, and sexual dimorphism in *Gorilla* (*Gorilla gorilla gorilla* and *G. g. beringei*:) evolutionary and ecological considerations. *American Journal of Primatology* **43**, 1–31.

Watts ES (1986) Evolution of the human growth curve. In *Human Growth: A Comprehensive Treatise*, vol. 1, ed. F. Falkner and J.M. Tanner, pp. 153–165. New York: Plenum Press.

Watts ES (1990) Evolutionary trends in primate growth and development. In *Primate Life History and Evolution*, ed. C.J. De Rousseau, pp. 89–104. New York: Wiley-Liss.

4 *Growth and development in Neandertals and other fossil hominids:*

implications for the evolution of hominid ontogeny

ANDREW J. NELSON AND JENNIFER L. THOMPSON

Introduction

Most studies of fossil hominid morphology focus on the analysis of adult skeletal remains. However, an understanding of growth and development is crucial to the complete interpretation of the adult form (Thompson 1942). Many recent publications have focused on overall body size (see below) and sexual dimorphism (e.g. McHenry 1991a), which are clearly the end product of growth trajectories. Others have focused on subtle details of cranial (e.g. Tobias 1991) and postcranial (e.g. Trinkaus 1976) morphology. The interrelationships and correlations of individual traits can be fully discerned only when the initial appearance and ontogenetic development of these morphological features are established.

The outcome of growth is, of course, the attainment of full body size. This subject has come under some scrutiny in the recent literature because of implications for understanding evolutionary processes, particularly with regard to the question of how modern humans attained their highly encephalised (large brain relative to body mass) condition. The classic work in this regard is that of Pilbeam and Gould (1974), who described a pattern of gradual increase in both brain size and body size from *Australopithecus africanus* through *Homo habilis* and *H. erectus* to modern humans. Brain size was thought to increase faster than body mass, yielding a progressive increase in encephalisation, presumably due to a strong positive selection on brain size. However, recent research (e.g. Kappelman 1996; Ruff *et al.* 1997; Nelson *et al.* 1992) has challenged the body mass estimates used by Pilbeam and Gould (1974). There is considerable disagreement among these studies with regard to the absolute body mass estimates, but the pattern that is emerging contrasts sharply with that presented by Pilbeam and Gould (1974). It is clear that both brain size and body mass increased until the Upper Pleistocene, after which body mass (and to a lesser extent brain size) decreased, moving toward modern values. This decrease has left modern *H. sapiens* in our highly encephalised position as a result of a body mass reduction event, rather than a substantial increase in brain size (Kappelman 1996; Mai *et al.* 1992). The mechanism for this shift presented in Mai and colleagues (1992) relies on the relationship between brain growth and overall somatic growth. The end of the childhood development stage is marked by the cessation of brain growth (Bogin 1997). However, the majority of somatic growth takes place in the later adolescent growth stage (Bogin 1997). Thus, a shift from a large-bodied, large-brained Upper Pleistocene hominid to a smaller bodied, but still large-brained modern hominid could be achieved by an alteration in the pattern of adolescent somatic growth (Mai *et al.* 1992). The focus of this study is on when the modern adolescent growth stage appeared in the course of hominid evolution, with a particular focus on the Neandertals.

The study of growth and development in fossil hominids, like similar studies in archaeological populations, is plagued by the problem of a lack of absolute chronological referents. Studies of growth and development in archaeological (modern *H. sapiens*) populations can estimate chronological age of skeletonised individuals with some degree of precision from techniques of ageing developed on the basis of modern known age samples. Examples of these are the changing morphology of the public symphysis (McKern and Stewart 1957), epiphyseal fusion (Krogman and Iscan 1986), dental development (Moorrees *et al.* 1963; Anderson *et al.* 1976), and long bone growth (Maresh 1970) (for a detailed review, see Saunders 1992). Unfortunately, techniques of ageing developed on the basis of modern reference samples are almost certainly not directly applicable to hominid populations, particularly with regard to earlier taxa. By the same token, techniques of ageing modelled on modern hominoids (e.g. Dean and Wood 1981) are probably directly applicable only to the earliest hominid taxa.

Despite these difficulties, there have been many important contributions made on the basis of the study of these various techniques when applied to fossil hominids. Examples of such studies include growth of the cranial base in modern humans and apes (Dean and Wood 1984), australopithecine dental formation and development (e.g. Mann 1975; Bromage and Dean 1985; Smith 1986), relative dental and skeletal development in *H. erectus* (Smith 1993), and craniofacial growth (Minugh-Purvis, 1988), determination of the rate of growth (e.g. Skinner 1978; Tompkins 1996a,b), and the timing of the appearance of adult morphological features in archaic *H. sapiens* (e.g. Vlček 1964, 1970; Tillier 1986; Trinkaus 1986; Tompkins and Trinkaus 1987; Stringer *et al.* 1990). The study by Smith (1993), which considered several techniques, obtained conflicting results for a *H. erectus* juvenile (see below), highlighting the fact that any single system of assessing the ontogenetic status of a juvenile individual is unlikely to apply directly to any given juvenile fossil hominid. Thus, a more complete reconstruction of the growth and development programme must rely on the analysis of several different indicators of maturity, seeking concordance or patterns of disagreement with the modern human ontogenetic programme.

Human growth

A consideration of patterns of growth and development in fossil hominids must begin with a detailed understanding of the pattern of growth and development in modern *H. sapiens.* The following section owes a great deal to the work of Barry Bogin (e.g. Bogin 1988, 1997; Bogin and Smith 1996).

Modern humans grow very slowly, with most of their growth taking place over a period of 18 to 20 years, instead of approximately 12 years, as in apes. In general, the development of the modern human pattern of growth and development has involved the addition and extension of developmental stages and the delay of somatic growth, relative to the ape pattern (Schultz 1960; Watts 1986). The programme of postnatal human growth and development demonstrates five identifiable stages. These are infancy, childhood, juvenile, adolescent and adult (Smith and Tompkins 1995; Bogin and Smith 1996). Old World monkeys and apes have only three basic stages in their growth and development: infancy, juvenile and adulthood (Smith and Tompkins 1995; Bogin and Smith 1996). Infancy in general is defined as the period from immediately after birth to the time of weaning (Bogin and Smith 1996). Human infancy is unique in terms of its continuation of foetal brain growth rates for a year after birth (Dienske 1986), which is accompanied by secondary altriciality and an extreme delay in motor development. Childhood is defined by Bogin (1997) as the time when the individual has been weaned, but remains dependent on adults for food and protection. Among modern primates this period is unique to modern humans, allowing for a delay in sexual development and an opportunity for further brain and somatic growth, but at a rate much reduced from infancy. The juvenile stage begins with the eruption of the first molar and the cessation of brain growth and ends with puberty (Bogin and Smith 1996). This period further delays sexual maturity in modern humans, but represents the single intervening phase between infancy and puberty in modern primates (Bogin 1988). Adolescence is marked by the onset of puberty and ceases with the termination of growth. It is characterised by the attainment of full sexual maturity, the development of secondary sexual characteristics, a growth spurt, and the adoption of adult behaviours (Tanner 1962, 1990; Bogin 1993, 1994a,b, 1995; Bogin and Smith 1996). Puberty in African apes seems to begin at the age of 7–10 years, with the onset of menarche and the appearance of other sexual characteristics in females and the acceleration of testes growth in males (Leigh and Shea 1996). Puberty occurs in human girls and boys about the age of 10 and 12 years, respectively. While both human and non-human primates demonstrate some sort of acceleration in growth velocity during this time (Laird 1967; Leigh 1996), the magnitude and late timing of the human growth spurt (and thus the nature of adolescence) appear to be unique to modern humans (Smith 1993). However, this is an area of some controversy (for references, see Leigh 1996). Adulthood in all primates begins with the cessation of growth and the attainment of full sexual maturity, and ends with the death of the individual (Bogin 1997).

Morphology, growth and development in fossil hominids

Early hominids

The oldest well-documented species of early hominid is *A. afarensis*, which has been proposed and generally accepted as an ancestor of modern humans (Johanson and White 1979). Specimens of this species indicate that it closely resembles the chimpanzee in both body size and brain size (McHenry 1992). From the beginning, there has been controversy over whether they more closely resembled modern humans or apes their pattern of dental growth. The early consensus was that they were human-like in terms of the order of eruption of their teeth (Clark 1950; Broom and Robinson 1952; Clements and Zuckerman 1953; Koski and Garn 1957), with regard to the timing of dental developmental events (Clark 1967; Robinson 1970), and in terms of their life history characteristics, including extended and delayed growth and development, and longer parental care (Mann 1968, 1975).

This view of australopithecines and early *Homo* as very human-like in their growth and development was challenged in the 1980s by research demonstrating that incremental growth markings on teeth could be used to estimate the timing of dental developmental events (Bromage and Dean

1985; Dean 1985a,b, 1987a,b, 1989; Beynon and Dean 1988, 1991). This work revealed that the dental development and growth of early hominids was very rapid, being much more ape-like than human-like (Smith 1986; Beynon and Dean 1988; but see Mann *et al.* 1987, 1990; Mann 1988). If one assumes that early hominids grew more quickly than modern humans, and for a shorter time (Bromage and Dean 1985; Smith 1986; Conroy and Vannier 1987; Beynon and Dean 1988, 1991; Dean 1989; Smith and Tompkins 1995; but see also Mann *et al.* 1987, 1990; Mann 1988), then it is likely that the early hominids, like the apes, went through a juvenile phase immediately after infancy, without the intervening childhood phase.

Homo habilis is differentiated from the rest of the early hominids primarily on the basis of an increase in cranial capacity (Tobias 1991). Martin (1983) argued that any hominid with adult cranial capacity of 850 cm^3 or greater would require a human-like pattern of continuing rapid foetal growth for some time after birth. Bogin (1997), on the basis of this analysis, argued that a childhood phase of growth would also be needed in order to complete brain growth. He suggested that a brief childhood phase may have been present even in *H. habilis*. However, it is unlikely that they and other early hominids underwent a childhood and adolescence similar to that of modern humans, as the early hominids probably lacked the long delay of linear growth and sexual maturation that characterises modern humans.

Homo erectus

The origin of *H. erectus* heralded the appearance of an adult morphology that is much more reminiscent of that of modern humans. *Homo erectus* was very close to, and possibly larger than, modern humans in terms of body size (Kappelman 1996; Ruff *et al.* 1997; Nelson *et al.* 1999) and stature (McHenry 1991b). The range of cranial capacity recorded for this taxon (750 cm^3 to 1231 cm^3, data from Aiello and Dean 1990: 192) reaches into the lower end of the range of variation for modern humans (Aiello and Dean 1990: 193). Overall limb proportions appear to lie within the modern human range of variation (McHenry 1978; Walker and Leakey 1993).

The study of growth and development in *H. erectus* is greatly aided by the presence of an almost complete immature skeleton, KNM-WT 15 000. This skeleton dates to about 1.6 million years ago and has been carefully documented by Walker and Leakey (1993). Its ontogenetic status has been studied by Smith (1993). The results of Smith's (1993) work suggest that *H. erectus* was more human-like in its growth and development than earlier hominids, but had not yet fully achieved the modern human rate and pattern of growth. In an analysis of dental and postcranial elements of the KNM-WT 15 000 specimen, Smith (1993) determined that it had a dental age of 11 years old or less, but a stature comparable to that of a modern human youth of about 15 years of age. In other words, the size and maturation of the skeleton was too advanced relative to the dentition for this individual to have followed a modern human growth pattern. This implies that it is unlikely that *H. erectus* underwent the long, slow growth of childhood (*contra* Bogin 1997, see above), followed by the rapid adolescent growth spurt that characterised modern humans. It is more likely that *H. erectus* underwent a more constant rate of growth after infancy to adulthood.

Archaic **Homo sapiens/*Neandertals***

Archaic *H. sapiens* are characterised by a combination of primitive traits such as cranial superstructures and modern traits such as a large cranial capacity (Trinkaus 1982; Klein 1989). The European late archaic group, the Neandertals, had very large cranial capacities, with a group mean that exceeds that of modern *H. sapiens* (data in Aiello and Dean 1990). They had a massive, but compact body, weighing on average 75 kg or more (Kappelman 1996; Ruff *et al.* 1997; Nelson *et al.* 1992). They possessed a suite of characteristics that includes distal limb shortening and large noses, traits that have been interpreted as adaptations to life during the last glacial period (Coon 1962; Trinkaus 1981).

The state of secondary altriciality that characterises infancy in modern humans was apparently already established in Neandertals, on the basis of estimates of the ratio of adult/newborn brain weight (Rosenberg 1992; Smith and Tompkins 1995). Mann and co-workers (1996) have argued that the childhood developmental stage was also present in Neandertals, on the basis of a statis period in dental eruption. This stasis occurs between the eruption of the lateral incisors and the second molars in modern humans (Hellman 1943) and Neandertals, but not in apes (Mann *et al.* 1996). However, the length of this stage, relative to that seen in modern humans and relative to the Neandertal stages of infancy and adolescence, is not known and may be difficult to determine, since the duration of this stage varies between upper and lower jaws, between males and females, and between populations (Jaswal 1983). Whether the Neandertals demonstrated the modern pattern of slowed childhood growth accompanied by rapid adolescent growth is not currently known (Thompson 1998).

The appearance of modern human pattern of growth and development

The various stages of the human growth cycle, as outlined above, are defined on the basis of criteria such as duration of dependency, timing of cessation of brain growth, and so on. Thus, these stages are not all amenable to delineation on the basis of dental studies or postcranial studies alone. However, a comparison of patterns of maturation of the two systems can allow the determination of whether any given hominid is following the modern human dental/linear trajectory, or whether they are retaining primitive elements in their growth pattern.

This comparison can be affected by determining the dental age and height age of a fossil hominid on the basis of modern human criteria. If the two estimates are in accord with each other, then the individual was probably following a modern human pattern of dental development and linear growth, although we cannot know for certain whether the estimated chronological age is correct. If the two estimates do not agree with each other, then the individual was departing from the modern human pattern. In particular, if estimated height age exceeds estimated dental age as seen above for *H. erectus*, then the individual was probably following a primitive growth pattern of early linear growth, as seen in African apes. It is the aim of this chapter to test whether a primitive or derived pattern of growth and development existed in a later hominid group, the Neandertals.

Dental age and height age in Neandertals

There are two Neandertal juvenile males, Teshik-Tash 1 and Le Moustier 1, of the appropriate age to assist in addressing the question of whether Neandertals experienced an extended period of slow childhood growth and a rapid adolescent growth spurt. However, any attempt to assess their position along a growth trajectory is subject to the difficulties inherent in skeletal analysis discussed earlier, particularly as applied to fossil hominids, namely the lack of any absolute chronological referent, and the problem of choosing the appropriate sample on which to base estimates of age. For the purposes of this study, we elected to employ ageing and stature estimation models based on modern Euro-American samples. We recognise fully that a Neandertal is unlikely to track the absolute trajectory of a modern Euro-American child, but this strategy allows us to make comparisons to the largest available reference data bases, which are of Euro-American origin. The conversion of femoral length into stature, while introducing a possible source for error, allows the individuals to be readily compared with growth curves not only for modern Euro-American but also for many other extant populations. Thus stature is ultimately more useful for broad-based comparisons than femoral length alone. In this study we assess stature as a proportion of the mean adult outcome, rather than in absolute terms. The use of 100% (which is the sex-specific mean adult stature) as the termination of growth allows for the examination of growth patterns relative to population-specific outcomes. This is preferable to the use of an absolute outcome target from a reference population, which may differ (in this case exceed) the mean for the population under examination.

In this study, dental age was estimated on the basis of crown calcification and root formation using the standards outlined by Anderson and colleagues (1976), who provided data on individuals of Euro-American origin. This system of ageing is applied tooth by tooth, with a summary age calculated on the basis of all available teeth. On the basis of these criteria, these two individuals are estimated to be 9.9 and 15.5 years of age, respectively. The dental age of Le Moustier 1 was based on the study of the original specimen (Thompson 1995) while that of Teshik-Tash 1 was based on the description of developing dentition outlined by Minugh-Purvis (1988). Estimates from multiple teeth were in close concordance.

The assessment of height age for Le Moustier 1 and Teshik-Tash 1 must begin by estimating their stature. Traditionally, researchers have used regression models to calculate stature (e.g. Trotter and Gleser 1952), but these have been criticised as population specific (Feldesman 1992). Feldesman (Feldesman *et al.* 1990; Feldesman and Fountain 1996) has developed a simple ratio to estimate stature on the basis of femoral length using data collected from 13 000 adults from 51 skeletal populations from various parts of the world. Feldesman maintains that this ratio is very conservative across modern human groups and, by extension, across fossil groups (Feldesman *et al.* 1990; but see Meadows and Jantz 1995; Holliday and Ruff 1997).

The estimation of stature in subadults is particularly problematic because very few prediction models exist for this age group. This is due to the fact that the proportion of total stature that can be attributed to the length of the long bones changes during the adolescent growth spurt. Thus any prediction model based on adults will not accurately predict the stature of a subadult. Feldesman (1992) addressed this problem by examining femur length and stature in samples from four growth studies (Tupman 1962; Anderson *et al.* 1963, 1964; Maresh 1970). He calculated separate femur/stature ratios for each year from 8 to 18 years of age, and concluded that the ratio for children under 12 years of age differed significantly from those aged between 12 and 18, which he subsumed into a single group.

We elected to use the age-specific femur/stature ratios (Feldesman 1992) to estimate stature in the two Neandertal juveniles (see Table 4.1), Le Moustier 1 and Teshik-Tash 1. These individuals have both been sexed as males (Klaatsch and Hauser 1909; Thompson 1995, for Le Moustier 1; Oakley *et al.* 1971, for Teshik-Tash 1). Using the ratio method, the femoral length of 380 mm for Le Moustier 1 (Klaatsch and Hauser 1909) yields a stature estimate of 138.5 cm. This figure contrasts with earlier published estimates of 145–150 cm (Klaatsch and Hauser 1909; Hrdlička 1930). However, even this estimate must be considered to be generous, given the shortening of the tibia already present in Le Moustier 1 (Thompson and Nelson 1997), and his low vault height (Thompson and Bilsborough 1997), relative to the modern humans in Feldesman's (1992) sample. The stature calculated for Teshik-Tash 1 (see Table 4.1) is 112.9 cm.

In order to compare this stature estimate to an average stature for an adult Neandertal, we applied Feldesman's adult stature ratio (Feldesman *et al.* 1990; Feldesman and Fountain 1996) to an adult male Neandertal sample (see Table 4.1). In order to be consistent with the juvenile sample, the Neandertal adult sample was restricted to continental Europe and to males. This calculation yielded an average stature of 162.8 cm. A comparison of the stature calculated for the Le Moustier boy with the mean adult Neandertal sample suggests that Le Moustier 1 had achieved 85.1% of the mean adult stature. Since most children stay approximately at the same centile during preadolescence and return to this centile when growth is completed (Tanner *et al.* 1966), and since adult outcome is not greatly influenced by the earliness or lateness of maturation (Tanner and Whitehouse 1976), it is possible to examine the proportion of adult stature achieved by an average modern boy of comparable age. An average modern human boy of 15 years of age will have achieved about 96.9% of adult stature (see Table 4.1), and will be ending his growth spurt (Largo *et al.* 1978; Gasser *et al.* 1985; Frisancho 1990). The figure calculated for Le Moustier 1 more closely matches the proportion of adult stature achieved by a modern 11–12 year old male (see Table 4.1). The proportion of adult stature achieved by Teshik-Tash 1 closely approximates to that of a modern human male of 6–7 years of age. Note that the height age discussed here is *proportional height age*.

Several archaeological studies have documented an apparent lag in postcranial development in ancient populations relative to the healthy, tall, Euro-American reference samples (e.g. Ubelaker 1984; Saunders and Hoppa 1993; Molleson 1995). For that reason we have included data from

Table 4.1. *Absolute and proportional stature for the modern reference, Inuit, and Neandertal samples*

Specimen/sample/Ref.	Age (yr)	Stature (cm)	Adult sature (cm)	Proportion of adult stature
Euro-American sample		50th ‰	50th ‰	
Frisancho 1990	6	119.4	176.6	67.6
	7	125.4	176.6	71.0
	8	130.1	176.6	73.7
	9	135.8	176.6	76.9
	10	140.9	176.6	79.8
	11	146.4	176.6	82.9
	12	151.4	176.6	85.7
	13	159.3	176.6	90.2
	14	166.9	176.6	94.5
	15	171.2	176.6	96.9
	16	174.1	176.6	98.6
Inuit sample				
XIV-C:002	12[a]	132.3[b]	163.8[b]	80.8
XIV-C:198	14.8[a]	149.8[b]	163.8[b]	91.5
Neandertal sample				
Teshik-Tash 1	9.9[a]	112.9[b,c]	162.8[b,d]	69.3
Le Moustier 1	15.5[e]	138.5[b]	162.8[b,d]	85.1

[a]Ages for the archaeological sample were determined using Anderson *et al.*'s (1976) standards for crown calcification and root formation. Age for Teshik-Tash 1 was determined on the basis of dental description in Minugh-Purvis 1988, using Anderson *et al.*'s (1976) standards.
[b]Statures calculated using age appropriate femur/stature ratios from Feldesman (1992) for the juveniles and the adult femur/stature ratio from Feldesman *et al.* (1990).
[c]Femur length calculated from diaphyseal length using the regression formula epiphyseal length (cm) = diaphyseal length (cm) × 1.096 + 5.68, $N = 24$, $r^2 = 0.992$. Formula calculated as a least-squares regression on the basis of the Sadlermiut sample.
[d]Mean male Neandertal stature calculated on the basis of femoral lengths of La Chapelle 1, La Ferrassie 1, Fond de Foret 1, Neandertal 1, and Spy 2.
[e]Thompson 1995.

two Inuit juveniles of roughly similar age (see Table 4.1). The Arctic-dwelling Inuit are subject to many of the same environmental pressures as were the Neandertals, and are considered to have been under considerable resource stress (Merbs and Wilson 1962; Szathmary 1984; Merbs 1996). Thus they may be a more appropriate modern analogue for the Neandertals. The examination of the two Inuit males demonstrates that while they are indeed absolutely short for their ages when compared to the Euro-American reference sample, they are still much more advanced in proportion to adult growth achieved than were the Neandertals of a similar dental age.

The salient conclusion offered by the examination of dental age and proportional height age as assessed for these two Neandertal juveniles is that they demonstrate a dental age that exceeds their proportional height age. As such, they contrast with *H. erectus*, and therefore do not demonstrate a primitive pattern of growth and development. However, because the Neandertal age estimates based on the two developmental standards are not in accord, these hominids also contrast with modern humans.

A consideration of the discordance between dental and proportional height age leads us to consider two different reconstructions of the pattern of Neandertal adolescent growth. The first reconstruction involves the assumption that the dental age estimates for Le Moustier 1 and Teshik-Tash 1 are correct, following those of modern human populations, but that Neandertals experienced a substantial delay in skeletal maturity relative to their dental age. The converse is that Neandertals experienced a programme of skeletal growth and development like that of modern humans, accompanied by an accelerated dental eruption programme.

Delayed skeletal growth

Teshik-Tash 1 represents an individual who still needed to complete about 30% of statural growth, while a typical Euro-American boy of similar dental age would have only about 20% of statural growth to complete (see Table 4.1). Since preadolescence is a period of very slow growth in modern humans, this delay of growth in Teshik-Tash 1 may mean that the Neandertal childhood and/or juvenile stage of growth was extended over and above that seen in modern humans. This would mean that Neandertals were delaying a substantial amount of their growth until later in the growth period.

Judging by his dental age, Le Moustier 1 represents a 15.5 year old who still needed to complete approximately 15% of his statural growth. An average modern human boy of the same age would need to grow only the final 3% of his ultimate stature (Table 4.1). Again, this implies that a substantial amount of Neandertal growth was delayed until the end of the growth period. If Neandertals grew for the same length of time as modern humans do (from 0 to *ca.* 18–20 years of age) including a growth spurt, and if the teeth give us the appropriate chronological age estimate, then this would require a very late, accelerated and compressed growth period between 15.5 and 19.5 (or so) years of age in order for Neandertals to reach the mean adult stature at the time when their M3s are completing their formation. While this would be unusual in humans, it is not impossible.

For example, Singh (1980) reported that Indian children with chronic poor nutrition undergo peak spurts in weight and height at 15–16 years of age.

It is possible, however, that Neandertals did not grow for the same length of time as modern humans, but instead continued to grow substantially in stature into their 20s. Humans experience a small increase in stature after their epiphyses have fused, often until the age of 30, but the amount of growth is minimal and takes place on the bodies of the vertebrae (Tanner 1990). If Neandertal long bones did not fuse until well after an equivalently aged modern human adolescent, and their vertebrae continued to grow beyond 20 years of age, it would suggest that in Neandertals there was no pronounced growth spurt, but rather an extended, gradual growth until full stature had been achieved.

Again this situation is unusual, but not unknown in modern humans. Little and co-workers (1983) reported that Turkana adolescents undergo slow and prolonged growth, continuing to grow into their 20s. This is probably due to the interaction of seasonal food shortages combined with a high protein/low energy diet (Little and Johnson 1987). Other nutritionally poor populations have been reported to show a pattern of prolonged growth (Stini 1972).

The modern examples presented above represent malnourished samples that depart from the norms of the species. Thus, these children cannot be thought of as typical of modern *H. sapiens* at large. Furthermore, there is no evidence that Le Moustier 1 was malnourished at the time of his death. This individual demonstrates no pathological conditions that would have been active at the time of his death, nor does he demonstrate lines of arrested growth in the diaphyses of his long bones (as observed by examination by the authors, of radiographs of the preserved long bones).

Advanced dental growth

If the postcrania of Neandertals followed a modern human schedule of development, the implication would be that their dental growth was substantially accelerated relative to that of modern humans. On the basis of the proportion of adult stature achieved by the two juvenile Neandertals, which aligns Teshik-Tash 1 and Le Moustier 1 respectively with a 6–7 year old and an 11–12 year old Euro-American, this would require their dental growth be accelerated by about 3.5–4 years and place them at the extreme end of the modern human range.

A number of authors have argued that Neandertals did have accelerated dental growth relative to modern humans (e.g. Bolk 1926; Wolpoff 1978,

1979, 1996). Tompkins' (1996b) work demonstrated contrasts between Neandertals and modern human aspects of dental growth. When compared with recent humans, Neandertals and other Mid–Upper Pleistocene hominid fossils showed a relative delay of I1 and P3 formation but advances in their M2 formation compared with a modern sample of European descent and their M3 formation compared with a modern sample of Southern Africans. Tompkins' (1996b) results revealed that while the rate of dental growth in Neandertals and other Mid–Upper Pleistocene hominids may be somewhat accelerated over that of populations of European descent, it may be similar to that seen in recent human populations from Southern Africa. The exception is the M3, which may form more rapidly in Neandertals.

However, studies of non-European populations have also shown an advance of dental development over that seen in children of European descent, including the M3s (Friedlaender and Bailit 1969; Mayhall *et al.* 1978; Singh 1980; Jaswal 1983; Owsley and Jantz 1983). It must be noted that these studies which advocate advanced eruption of the dentition in Neandertals are based entirely on dental analyses, and do not consider external development referents. In the absence of a comparison with different chronological indicators, the conclusion can only be that some aspects of the dentition are advanced relative to other aspects of the dentition. An absolute chronological advance, or an advance relative to any other developmental system, cannot be supported on the basis of this evidence alone.

The possible acceleration of M3 development has important implications for the expression of the growth spurt in Neandertals. If Le Moustier 1 was actually 11–12 years of age (as indicated by proportional stature), and he was to finish his postcranial growth at an age comparable to that in modern humans (*ca.* 18 years), then he would have at least 6 years to complete the development of his third molars (Le Moustier's M3 roots are only 25% complete), a task which normally takes a modern human adolescent fewer than 3 years. This would obviate the need for the accelerated postcranial growth that characterises the modern human growth spurt by allowing the same amount of growth to take place over a longer period of time. However, if skeletal growth ceased with the completion of the third molar, and if Le Moustier 1 took the same amount of time to complete his M3 development as a modern human (*ca.* 3 years), then he would have completed his growth at approximately 14–15 years of age (3 years from 11–12 years of age). This early termination of growth would have incorporated an early growth spurt. Given that in both apes and humans dental and skeletal growth normally ceases at approximately the same time (Tanner 1990; Anemone *et al.* 1996; Kuykendall and Conroy 1996; Leigh and Shea

1996; Simpson *et al.* 1996), the latter scenario is the more likely of the two, if the dentition is indeed advanced.

Discussion

The study of Neandertal ontogeny has focused on the estimation of dental age at death of individual specimens (e.g. Legoux 1966; Wolpoff 1979; Dean *et al.* 1986; Thompson 1995), the determination of the rate of growth (e.g. Skinner 1978; Tompkins 1996a,b), the pattern of craniofacial growth (Minugh-Purvis 1988), and the timing of the appearance of adult morphological features (e.g. Vlček 1964, 1970; Tillier 1986; Trinkaus 1986; Tompkins and Trinkaus 1987; Stringer *et al.* 1990; for a fuller discussion of this research, see Trinkaus and Tompkins 1990). For instance, Tillier (1986 and *in lit.*) has examined immature Neandertal and archaic hominids with the aim of understanding their pattern and rate of development using primarily morphological descriptions of these specimens. Minugh-Purvis (1988, 1993, 1995) has taken a morphometric and qualitative approach, examining the craniofacial measurements of hominids in order to compare patterns of ontogenetic development between Neandertals and other immature Mid–Upper Pleistocene populations. Other researchers (e.g. Dean *et al.* 1986; Tompkins 1996b) have used the dentition to determine the rate of growth and timing of eruption of the teeth. With respect to the rate of growth, a number of authors have argued that Neandertals had accelerated dental growth relative to modern humans (e.g. Bolk 1926; Wolpoff 1978, 1979, 1996) or that the timing of the formation and eruption of certain teeth differed from that of modern humans (e.g. Thoma 1963; Dean *et al.* 1986; Tompkins 1996b), or that they were close to modern humans in their rate of dental growth (e.g. Smith 1991).

Relatively little work has been done on postcranial growth in Neandertals. Heim (1982), in his study of the La Ferrasie children, found some indicators of advanced skeletal development. However, as Tompkins (1996b) suggested, Heim's findings need to be placed in a broader comparative framework to assess their significance. In his paper on relative dental growth, Tompkins predicted that Neandertals not only had accelerated postcanine dental growth, but that they also had 'precocious' development of the postcrania due to the pleiotrophic effect of common regulatory genes (Tompkins 1996b: 112–113). This would mean that both dental and skeletal growth in Neandertals are accelerated relative to modern human standards. In terms of estimates of age, it also implies that the evaluation of the dental and skeletal maturity of any Neandertal juvenile using modern human standards should produce age estimates that are in accord (both accelerated). However, on the basis of the results presented in this chapter, this hypothesis can be rejected.

On the basis of both relative dental and postcranial development of the Le Moustier 1 and Teshik-Tash 1 juveniles, it would appear that Neandertals experienced either delayed skeletal growth or advanced dental development. Unfortunately, these Neandertal juveniles do not allow us to choose between these two possibilities. A detailed consideration of maturation as measured by dental and postcranial development has demonstrated that Neandertals do not follow a programme of growth and development typical of modern humans. We have presented a variety of possible reconstructions of the Neandertal growth and development programme which can be subsumed under two major headings, skeletal retardation or dental advancement relative to modern humans. The former has two possible expressions: a late compressed growth spurt, or an elongated period of growth (beyond 18 years). The latter also has two possible expressions: early dental maturation accompanied by a long, slow adolescent growth period without a growth spurt; or early dental maturation accompanied by the cessation of linear growth early, incorporating an early growth spurt. Unfortunately, the nature of age estimation on ancient skeletal material, means that we cannot know which maturation system best reflects true chronological age. Thus we cannot choose objectively between these alternatives. However, given that the pattern expressed in the Neandertals is to some extent 'hyper-human' rather than primitive (as seen in *H. erectus*), it is highly likely that the Neandertals did indeed express an elongated childhood/juvenile period of slow linear growth (cf. Mann *et al.* 1996), although the duration of this period is still unclear. That there was a period of stasis between the eruption of the lateral incisors and the eruption of the second molars is borne out by the wear patterns on the teeth of Le Moustier 1 (as observed on the original by one of the authors, J.T.). The childhood and juvenile periods were probably followed by an adolescent growth spurt, although when that occurred cannot yet be determined.

Conclusions

The programme of growth and development expressed in modern *H. sapiens* is clearly derived relative to African apes and other primates. The relatively simple ape sequence from infancy, through the juvenile stage to adulthood has been elaborated over the 5 million years of hominid evolution. The early hominids clearly retained much of the primitive scheme and remained quite ape-like in this regard until the brain size increase of *H. habilis* required the addition of the childhood stage (Bogin 1997). The duration of that childhood stage is unclear, as *H. erectus* also retained the primitive pattern of rapid early linear growth.

The detailed examination of both dental and height maturation in Neandertals reveals several important points about their growth and development. In contrast to *H. erectus*, Neandertals do not appear to retain the primitive pattern of early linear growth. In fact, relative to dental maturation, linear growth in Neandertals appears to be greatly delayed. Thus, it is clear that the Neandertals did indeed demonstrate the slow linear growth that characterises the childhood stage. Unfortunately, we cannot yet determine the precise duration of the childhood and juvenile stages. However, this conclusion has important implications with regard to both the large brain and body size noted for Neandertals. Furthermore, we cannot definitively confirm the existence of the adolescent growth spurt. If the growth spurt did occur either it would have occurred quite early, with adult size being achieved at an early age, or it would have occurred very late.

It is clear that the key to understanding modern human growth and development, particularly with regard to our unique, highly encephalised condition, will require further investigation of the adolescent growth period in other archaic and early modern *H. sapiens* groups. Juvenile fossils that retain both cranial and postcranial remains will be crucial to the resolution of these questions.

Acknowledgements

Both authors express their thanks to Dr J. Cybulski and Dr C. Merbs for permission to examine the Inuit material at the Museum of Civilization, Hull, with special appreciation of the efforts of Janet Young, who helped make our visit a successful one. We would like to thank Professor Berndt Hermann for giving us copies of his radiographs of the preserved elements of the Le Moustier 1 postcrania. Thanks also to Chris Nelson and Barry Bogin for invaluable assistance in the preparation of this manuscript.

J.L.T. thanks Professor W. Menghin, Director of the Museum für Vor- und Frühgeschichte, Berlin, for permission to examine the original of Le Moustier 1 and Mrs Hoffmann for her assistance.

A.J.N. acknowledges the support of the Department of Anthropology, University of Western Ontario, and the generosity of Anne Foley.

References

Aiello L and Dean MC (1990) *An Introduction to Human Evolutionary Anatomy.* London: Academic Press.

Anderson DL, Thompson GW and Popovich F (1976) Age of attainment of mineralization stages of the permanent dentition. *Journal of Forensic Sciences* **21**, 191–200.

Anderson M, Green WT and Messner MB (1963) Growth and prediction of growth in the lower extremities. *Journal of Bone and Joint Surgery* **45A**, 1–4.

Anderson M, Messner MB and Green WT (1964) Distribution of lengths of the normal femur and tibia in children from 1–18 years of age. *Journal of Bone and Joint Surgery* **46A**, 1197–1202.

Anemone RL, Mooney MP and Siegel MI (1996) Longitudinal study of dental development in chimpanzees of known chronological age: implications for understanding the age at death of Plio-Pleistocene hominids. *American Journal of Physical Anthropology* **99**, 119–133.

Beynon AD and Dean MC (1988) Distinct dental development pattern in early fossil hominids. *Nature* **335**, 509–514.

Beynon AD and Dean MC (1991) Hominid dental development. *Nature* **351**, 165.

Bogin B (1988) *Patterns of Human Growth.* Cambridge: Cambridge University Press.

Bogin B (1993) Why must I be a teenager at all? *New Scientist* **137**, 34–38.

Bogin B (1994a) Human learning, evolution of anthropological perspectives. In *The International Encyclopedia of Education*, 2nd edn, vol. 5, ed T. Husén and T.N. Postelthwaite, pp. 2681–2685. Oxford: Pergamon Press.

Bogin B (1994b) Adolescence in evolutionary perspective. *Acta Paediatrica Supplement* **406**, 29–35.

Bogin B (1995) Growth and development: recent evolutionary and biocultural research. In *Biological Anthropology: The State of the Science*, ed. N.T. Boaz and L.D. Wolfe, pp. 49–70. Bend, OR: International Institute for Human Evolutionary Research.

Bogin B (1997) Evolutionary hypotheses for human childhood. *Yearbook of Physical Anthropology* **40**, 63–89.

Bogin B and Smith H (1996) Evolution of the human life cycle. *American Journal of Human Biology* **8**, 703–716.

Bolk L (1926) On the problem of anthropogenesis. *Proceedings of the Section of Sciences, Koninklijke Akademie van Wetenschappen te Amsterdam* **29**, 465–475.

Bromage TG and Dean MC (1985) Re-evaluation of the age at death of immature fossil hominids. *Nature* **317**, 525–527.

Broom R and Robinson JT (1952) *Swartkrans Ape-Man.* Transvaal Museum Memoir 6.

Clark WE Le Gros (1950) Hominid characters of the australopithecine dentition. *Journal of the Royal Anthropological Institute of Great Britain and Ireland* **80**, 37–54.

Clark WE Le Gros (1967) *Man-Apes or Ape-Man?* New York: Holt, Rinehart and Winston.

Clements E and Zuckerman S (1953) The order of eruption of the permanent teeth in the *Hominoidea. American Journal of Physical Anthropology* **11**, 313–332.
Conroy GC and Vannier MW (1987) Dental development of the Taung skull from computerized tomography. *Nature* **329**, 625–627.
Coon CS (1962) *The Origin of Races.* New York: Knopf.
Dean MC (1985a) The eruption pattern of the permanent incisors and first permanent molars in *Australopithecus* (*Paranthropus*) *robustus. American Journal of Physical Anthropology* **67**, 251–257.
Dean MC (1985b) Variation in the developing root cone angle of the permanent teeth of modern man and certain fossil hominids. *American Journal of Physical Anthropology* **68**, 233–238.
Dean MC (1987a) Growth layers and incremental markings in hard tissues; a review of the literature and some preliminary observations about enamel structure in *Paranthropus boisei. Journal of Human Evolution* **16**, 157–172.
Dean MC (1987b) The dental developmental status of six East African juvenile fossil hominids. *Journal of Human Evolution* **16**, 197–213.
Dean MC (1989) The developing dentition and tooth structure in hominoids. *Folia Primatologica* **53**, 160–176.
Dean MC and Wood BA (1981) Developing pongid dentition and its use for ageing individual crania in comparative cross-sectional growth studies. *Folia Primatologica* **36**, 111–127.
Dean MC and Wood BA (1984) Phylogeny, neoteny and growth of the cranial base in hominoids. *Folia Primatologica* **43**, 157–180.
Dean MC, Stringer CB and Bromage TG (1986) Age at death of the Neandertal child from Devil's Tower, Gibraltar and the implications for studies of general growth and development in Neandertals. *American Journal of Physical Anthropology* **70**, 301–309.
Dienskc II (1986) A comparative approach to the question of why human infants develop so slowly. In *Primate Ontogeny, Cognition and Social Behaviour*, ed. J.G. Else and P.C. Lee, pp. 145–154. Cambridge: Cambridge University Press.
Feldesman MR (1992) Femur/stature ratio and estimates of stature in children. *American Journal of Physical Anthropology* **87**, 447–459.
Feldesman MR and Fountain RL (1996) Race specificity and the femur/stature ratio. *American Journal of Physical Anthropology* **100**, 207–224.
Feldesman MR, Kleckner JG and Lundy JK (1990) The femur/stature ratio and estimates of stature in mid and late-Pleistocene fossil hominids. *American Journal of Physical Anthropology* **83**, 359–372.
Friedlaender JS and Bailit HL (1969) Eruption times of the deciduous and permanent teeth of natives on Bougainville Island, Territory of New Guinea: a study of racial variation. *Human Biology* **41**, 51–65.
Frisancho AR (1990) *Anthropometric Standards for the Assessment of Growth and Nutritional Status.* Ann Arbor, MI: University of Michigan Press.
Gasser T, Müller H-G, Köhler W, Prader A, Largo R and Moninari L (1985). An analysis of the mid-growth and adolescent spurts of height based on acceleration. *Annals of Human Biology* **12**, 129–148.
Heim J-L (1982) *Les enfants de La Farrassie.* Paris: Masson.

Hellman M (1943) The phase of development concerned with erupting the permanent teeth. *American Journal of Orthodontics* **29**, 507–526.

Holliday TW and Ruff CB (1997) Ecogeographic patterning and stature prediction in fossil hominids: comment on M.R. Feldesman and R.L. Fountain, *American Journal of Physical Anthropology* (1996) **100**: 207–224. *American Journal of Physical Anthropology* **103**, 137–140.

Hrdlička A (1930) The skeletal remains of early man. *Smithsonian Miscellaneous Collections* **83**, 297–303.

Jaswal S (1983) Age and sequence of permanent tooth emergence among Khasis. *American Journal of Physical Anthropology* **621**, 77–186.

Johanson DC and White TD (1979) A systematic assessment of early African hominids. *Science* **202**, 321–330.

Kappelman J (1996) The evolution of body mass and relative brain size in fossil hominids. *Journal of Human Evolution* **30**, 243–276.

Klaatsch H and Hauser O (1909) *Homo mousteriensis Hauseri. Archiv für Anthropologie* **35**, 287–289.

Klein RG (1989) *The Human Career*. Chicago: University of Chicago Press.

Koski K and Garn SM (1957) Tooth eruption sequence in fossil and modern man. *American Journal of Physical Anthropology* **15**, 469–487.

Krogman WM and Iscan MY (1986) *The Human Skeleton in Forensic Medicine*. Springfield, IL: Charles C. Thomas.

Kuykendall KL and Conroy GC (1996) Permanent tooth calcification in chimpanzees (*Pan Troglodytes*): patterns and polymorphisms. *American Journal of Physical Anthropology* **99**, 159–174.

Laird AK (1967) Evolution of the human growth curve. *Growth* **31**, 345–355.

Largo RH, Gasser TH, Prader A, Stuetzle W and Huber PJ (1978) Analysis of the adolescent growth spurt using smoothing spline functions. *Annals of Human Biology* **5**, 421–434.

Legoux P (1966) *Détermination de l'Age dentaire du fossiles de la Lignée humaine*. Paris: Librarie Maloine.

Leigh SR (1996) Evolution of human growth spurts. *American Journal of Physical Anthropology* **101**, 455–474.

Leigh SR and Shea BT (1996) Ontogeny of body size variation in African apes. *American Journal of Physical Anthropology* **99**, 43–65.

Little MA and Johnson BR (1987) Mixed-longitudinal growth of nomadic Turkana pastoralists. *Human Biology* **59**, 695–707.

Little MA, Galvin K and Mugambi M (1983) Cross-sectional growth of nomadic Turkana pastoralists. *Human Biology* **55**, 811–830.

Mai LL, Gauld SC, Nelson AJ and Austin JK (1992) Evolutionary context of hominid body Mass Prediction Models. Paper presented to the 3rd International Congress in Human Paleontology, Jerusalem. [*l'Anthropologie*, in press].

Mann A (1968) The Paleodemography of *Australopithecus*. PhD thesis, University of California at Berkeley.

Mann A (1975) *Paleodemographic Aspects of the South African Australopithecines*. Pennsylvania: University of Pennsylvania Publications in Anthropology.

Mann A (1988) The nature of Taung dental maturation. *Nature* **333**, 123.
Mann A, Lampl M and Monge JM (1987) Maturational patterns in early hominids. *Nature* **328**, 673–675.
Mann A, Lampl M and Monge JM (1990) Patterns of ontogeny in human evolution: evidence from dental development. *Yearbook of Physical Anthropology* **33**, 111–150.
Mann A, Lampl M and Monge JM (1996) The evolution of childhood: dental evidence for the appearance of human maturation patterns. *American Journal of Physical Anthropology*, Supplement **21**, 156.
Maresh MM (1970) Measurements from roentgenograms. *In Human Growth and Development*, ed. R.W. McCammon, pp. 157–200. Springfield, IL: Charles C. Thomas.
Martin RD (1983) *Human Brain Evolution in an Ecological Context.* Fifty-second James Arthur lecture. New York: American Museum of Natural History.
Mayhall JT, Belier PL and Mayhall MF (1978) Canadian Eskimo permanent tooth emergence timing. *American Journal of Physical Anthropology* **49**, 211–16.
Meadows L and Jantz RL (1995) Allometric secular change in the long bones from the 1800s to the present. *Journal of Forensic Sciences* **40**, 762–767.
McHenry HM (1978) Fore- and hind-limb proportions in Plio-Pleistocene hominids. *American Journal of Physical Anthropology* **49**, 15–22.
McHenry HM (1991a) Sexual dimorphism in *Australopithecus afarensis. Journal of Human Evolution* **20**, 21–32.
McHenry HM (1991b) Femoral lengths and stature in Plio-Pleistocene hominids. *American Journal of Physical Anthropology* **85**, 148–158.
McHenry HM (1992) How big were early hominids? *Evolutionary Anthropology* **1**, 15–20.
McKern TW and Stewart TD (1957) *Skeletal Age Changes in Young American Males Analyzed from the Standpoint of Age Identification.* Environmental Protection Research Division Quartermaster Research and Development Center, and U.S. Army, Technical Report EP-45.
Merbs CF (1996) Spondylolysis of the sacrum in Alaskan and Canadian Inuit skeletons. *American Journal of Physical Anthropology* **101**, 357–367.
Merbs CF and Wilson WH (1962) *Anomalies and Pathologies of the Sadlermiut Eskimo Vertebral Column. Contributions to Anthropology 1960* Part I. National Museum of Canada Bulletin no. 180. Department of Northern Affairs and Natural Resources, Canada.
Minugh-Purvis N (1988) Patterns of craniofacial growth and development in Upper Pleistocene hominids. PhD dissertation, University of Pennsylvania.
Minugh-Purvis N (1993) Reexamination of the immature hominid maxilla from Tangier, Morocco. *American Journal of Physical Anthropology* **92**, 449–461.
Minugh-Purvis N (1995) Ontogenetic patterning through childhood and adolescence in the mandible of late Pleistocene *Homo sapiens. American Journal of Physical Anthropology* Special Supplement **20**, 155.
Molleson T (1995) Rate of ageing in the eighteenth century. In *Grave Reflections*, ed. S.R. Saunders and D.A. Herring, pp. 199–222. Toronto: Canadian Scholars' Press.

Moorrees CFA, Fanning EA and Hunt EE (1963) Age variation of formation stages for ten permanent teeth. *Journal of Dental Research* **42**, 1490–1502.

Nelson AJ, Gauld SC and Austin JK (1992) Models of fossil hominid body mass prediction using cranial and post-cranial bone thickness. Paper presented to the 3rd International Congress in Human Paleontology, Jerusalem. [*l'Anthropologie*, in press.]

Oakley KP, Campbell BG and Molleson TI (1971) *Catalogue of Fossil Hominids* Part II: *Europe*. London: Trustees of the British Museum (Natural History).

Owsley DW and Jantz RL (1983) Formation of the permanent dentition in Arikara Indians: timing differences that affect dental age assessments. *American Journal of Physical Anthropology* **61**, 467–471.

Pilbeam D and Gould SJ (1974) Size and scaling in human evolution. *Science* **186**, 892–901.

Robinson JT (1970) *The Dentition of the Australopithecinae*. Amsterdam: Swets and Zeitlinger.

Rosenberg KR (1992) The evolution of modern human childbirth. *Yearbook of Physical Anthropology* **35**, 89–124.

Ruff CB, Trinkaus E and Holliday TW (1997) Body mass and encephalization in Pleistocene Homo. *Nature* **387**, 173–176.

Saunders SR (1992) Subadult skeletons and growth related studies. In *Skeletal Biology of Past Peoples: Research Methods*, ed. S.R. Saunders and M.A. Katzenberg, pp. 1–20. New York: Wiley-Liss.

Saunders SR and Hoppa RD (1993) Growth deficit in survivors and non-survivors: biological mortality bias in subadult skeletal samples. *Yearbook of Physical Anthropology* **36**, 127–151.

Schultz AH (1960) Age changes in primate and their modification in man. In *Human Growth*. ed. J.M. Tanner, pp. 1–20. Oxford: Pergamon Press.

Simpson SW, Russell KF and Lovejoy CO (1996) Comparison of diaphyseal growth between the Libben population and the Hamann-Todd Chimpanzees sample. *American Journal of Physical Anthropology* **99**, 67–78.

Singh SP (1980) Eruption of permanent teeth in Gaddi Rajput males of Dhaula Dhar Range of Himalayas. *Zeitschrift für Morphologie und Anthropologie* **70**, 259–301.

Skinner MF (1978), Dental maturation, dental attrition and growth of the skull in fossil Hominidae. PhD thesis, University of Cambridge.

Smith BH (1986) Dental development in *Australopithecus* and early *Homo*. *Nature* **323**, 327–330.

Smith BH (1991) Dental development and the evolution of life history in Hominidae. *American Journal of Physical Anthropology* **86**, 157–174.

Smith BH (1993) The physiological age of KNM–WT 15 000. In *The Nariokotome* Homo erectus *Skeleton*, ed. A. Walker and R. Leakey, pp. 195–220. Cambridge, MA: Harvard University Press.

Smith BH and Tompkins RL (1995) Toward a life history of the Hominidae. *Annual Review of Anthropology* **24**, 257–279.

Stini WA (1972) Malnutrition, body size and proportion. *Ecology of Food and Nutrition* **1**, 121–126.

Stringer CB, Dean MC and Martin R (1990) A comparative study of cranial and dental development within a recent British sample and among Neandertals. In *Primate Life History and Evolution*, ed C.J. DeRousseau, pp. 115–152. New York: Wiley-Liss.

Szathmary EJE (1984) Human biology of the arctic. In *Handbook of North American Indians*. vol. 5, ed. D. Damas, pp. 64–71. Washington, DC: Smithsonian Institution.

Tanner JM (1962) *Growth at Adolescence*, 2nd edn. Oxford: Blackwell Scientific Publications.

Tanner JM (1990) *Foetus into Man*. Cambridge, MA: Harvard University Press.

Tanner JM and Whitehouse RH (1976) Clinical longitudinal standards for height, weight height velocity, weight velocity and the stages of puberty. *Archives of Disease in Childhood* **51**, 170–179.

Tanner JM, Whitehouse RH and Takaishi M (1966) Standards from birth to maturity for height, weight, height velocity, and weight velocity: British children 1965. *Archives of Disease in Childhood* **41**, 454–471, 613–635.

Thoma A (1963) The dentition of the Subaluyk Neandertal child. *Zeitschrift für Morphologie und Anthropologie* **54**, 127–150.

Thompson DW (1942) *On Growth and Form*, revised edn. Cambridge: Cambridge University Press.

Thompson JL (1995) Terrible teens: the use of adolescent morphology in the interpretation of Upper Pleistocene human evolution. *American Journal of Physical Anthropology Supplement* **20**, 210.

Thompson JL (1998) *Neandertal Growth and Development*. Cambridge Encyclopedia of Human Growth, ed. S.J. Ulijaszek, F.E. Johston and M.A. Preece, pp. 106–107. Cambridge: Cambridge University Press.

Thompson JL and Bilsborough A (1997) The current state of the Le Moustier 1 skull. *Acata Praehistorica et Archaeologica* **29**, 17–38.

Thompson JL and Nelson AJ (1997) Relative postcranial development of Neandertals. *Journal of Human Evolution* **32**, A23–24.

Tillier A-M (1986) Quelque aspects de l'ontogenèse du skelette cranien des Néandertaliens. *Anthropos* (Brno) **23**, 207–216.

Tobias PV (1991) *Olduvai Gorge* vol. IV. Homo habilis: *Skulls, Endocasts, and Teeth*. New York: Cambridge University Press.

Tompkins RL (1996a) Human population variability in relative dental development. *American Journal of Physical Anthropology* **99**, 79–102.

Tompkins RL (1996b) Relative dental development of Upper Pleistocene hominids compared to human population variation. *American Journal of Physical Anthropology* **99**, 103–118.

Tompkins RL and Trinkaus E (1987) La Ferrasie 6 and the development of Neandertal pubic morphology. *American Journal of Physical Anthropology* **73**, 233–239.

Trinkaus E (1976) The evolution of the hominid femoral diaphysis during the Upper Pleistocene in Europe and the Near East. *Zeitschrift für Morphologie und Anthropologie* **67**, 291–319.

Trinkaus E (1981) Neanderthal limb proportions and cold adaptation. In *Aspects of*

Human Evolution, ed. C.B. Stringer, pp. 187–224. London: Taylor and Francis.

Trinkaus E (1982) Evolutionary continuity among Archaid *Homo sapiens*. In *The Transition from Lower to Middle Palaeolithic and the Origin of Modern Man*, ed. A. Rohen, pp. 301–314. British Archaeological Report International Series no. 151.

Trinkaus E (1986) The Neandertals and modern human origins. *Annual Review of Anthropology* **15**, 193–218.

Trinkaus E and Tompkins RL (1990) The Neandertal life cycle: the possibility, probability, and perceptibility of contrasts with recent humans. In *Primate Life History and Evolution*, ed. C.J. de Rousseau, pp. 153–180. New York: Wiley-Liss.

Trotter M and Gleser GG (1952) Estimation of stature from long bones of American Whites and Negros. *American Journal of Physical Anthropology* **10**, 463–514.

Tupman GS (1962) A study of bone growth in normal children and its relationship to skeletal maturation. *Journal of Bone and Joint Surgery* **44B**, 42–67.

Ubelaker DH (1984) *Human Skeletal Remains*. Washington: Taraxacum.

Vlček E (1964) Einige in der ontogenese des modernen menschen untersuchte Neandertalmermale. *Zeitschrift für Morphologie und Anthropologie* **56**, 63–83.

Vlček E (1970) Étude comparative onto-phylogénétique de l'enfant du Pech-de-L'Azé par rapport à d'autres enfants Néandertaliens. In L'Enfant du Pech-de-L'Azé. *Institut de Paléontologie Humaine, Paris. Archives* **33**, 149–178.

Walker A and Leakey R (1993) *The Nariokotome* Homo erectus *Skeleton*. Cambridge, MA: Harvard University Press.

Watts ES (1986) The evolution of the human growth curve. In *Human Growth*, vol. 1, 2nd edn, ed. F. Falkner and J.M. Tanner, pp. 153–156. New York: Plenum Press.

Wolpoff MH (1978) The dental remains from Krapina. In *Krapinski Pracvojek i Evolucija Hominida*, ed. M. Malez, pp. 119–144. Zagreb: JAZU.

Wolpoff MH (1979) The Krapina dental remains. *American Journal of Physical Anthropology* **50**, 67–113.

Wolpoff MH (1996) *Human Evolution*. 1996–1997 edn. New York: McGraw-Hill.

5 *Hominoid tooth growth:*

using incremental lines in dentine as markers of growth in modern human and fossil primate teeth

CHRISTOPHER DEAN

Dentine structure and function

Dentine is a vital (i.e. a living) tissue that is extremely sensitive to temperature and to changes in osmotic pressure (e.g. in response to contact with sweet things). This is especially the case when it is exposed or cut. Enamel, on the other hand, has no nerves or cells of any description in it, either living or dead, and is, therefore, completely insensitive: enamel is essentially a secretory product that can neither repair nor remodel. Much of the protective feedback that prevents tooth damage during everyday use comes from the dentine but importantly also from the periodontal ligament that supports the tooth in its socket of alveolar bone. Dentine has two properties that make it ideal as a tooth tissue. It is elastic and can recoil under huge biting forces and it is able to resist fractures in all directions because there is no 'grain' or directionality to its structure in the way there is with, for example, wood or for that matter enamel. Unfortunately, these properties are attractive in the material world and are responsible for the ivory trade (ivory is simply elephant dentine). If you consider that billiard (pool) balls were once made of ivory, for obvious reasons, and that intricate ivory carvings made possible because of its structure were, and sadly still are, big business you can easily appreciate how special the physical properties of dentine are.

The odontoblasts that form dentine first lay down a predentine matrix. One function of predentine is to hold back the mineralisation process until the collagenous component of the matrix has matured into a dense feltwork of fibres. In the bulk of the dentine, the so-called circumpulpal dentine, collagen fibres all lie in a plane transverse to the cell sheet. When the collagenous portion of the predentine matrix has matured, which may be several days after the predentine is first secreted, there is degradation, and then removal, of the components of the non-collagenous predentine matrix that inhibit mineralisation at the predentine/dentine front. Subsequently, other components of the non-collagenous predentine matrix move to the mineralising front and promote mineralisation in a regular and orderly manner both within the collagen bundles and between them. Thus, dentine differs from enamel in that it is (1) a vital tissue, (2) is not as highly mineralised (about 70% by weight rather than 96%) and (3) it contains collagen. The odontoblasts have long cell processes that trail behind the cell as it passes from the future enamel or cement dentine junction towards the pulp chamber. Each odontoblast process lies in a dentine tubule that is approximately 2 μm in diameter and which therefore reveals the former direction of travel of the odontoblast during tooth formation. There are nerves in some dentine tubules but all of them contain tissue fluid that can be drawn towards the surface of the tooth or the pulp in response to changes in temperature and osmotic pressure. Throughout life a layer of dentine with no collagen in it but which is highly mineralised becomes deposited around the periphery of the tubule. This is called peritubular dentine. With age, peritubular dentine can occlude the whole of the tubule and dentine then appears sclerosed or translucent in section. The pulp chamber of a vital tooth is lined with odontoblasts that can retreat away from physical and harmful stimuli and secrete yet other forms of dentine called regular or irregular secondary dentine. Odontoblasts can continue to do this until the pulp chamber becomes obliterated. It follows that dentine differs from enamel in yet another respect; it remains able to react to wear, trauma and minor irritation throughout life.

Dentine is a complex tissue. Predentine secretion, the degradation and removal of some of its components and the modification of others, as well as the final mineralisation process itself may all occur in an incremental or rhythmic manner. It is likely that adult dentine still records both the movements of the odontoblast cell sheet and the circadian mineralisation process when viewed in ground or demineralised sections. If these markings in dentine can be securely identified then dentine is obviously a valuable source of information about tooth growth.

The nature and periodicity incremental markings in dentine

The use of enamel incremental markings in studies of human growth and in studies of comparative growth and dental development is now well established (FitzGerald 1998). The use of incremental markings in dentine, however, is less well established among studies of primates at least. Okada

(1943) has reviewed the experimental evidence for the daily nature of incremental lines in dentine. In some ways this evidence is better than that for enamel because dentine can be successfully labelled with a number of markers during growth of the teeth and examined after the event in demineralised histological sections and in ground sections. The first good experimental evidence for a circadian rhythm during dentine formation was obtained using labels of lead acetate administered at known times to a number of different species of animals (Okada 1943). Importantly, Okada (1943) also brought together previously published experimental evidence in his review that points to the mechanisms underlying these markings; he concluded that circadian shifts in acid–base equilibrium alternately produce lines that are less well mineralised (when pH falls) or more highly mineralised (when pH rises) and which appear white or dark blue respectively in demineralised sections stained with haematoxylin (Okada *et al.* 1939, 1940; Okada and Mimura 1940, 1941; Fujita 1943).

Shinoda (1984) reconfirmed these findings and established that the dark bands in demineralised sections of dentine, which are more highly mineralised, form in the active period in several experimental animals. Animals with no strong rhythm of activity and inactivity (e.g. hamsters, as opposed to rats and chipmunks which are nocturnal and diurnal, respectively, in their activity patterns) do not show such strongly alternating incremental markings in dentine (Shinoda 1984). Shinoda (1984) also showed that altering the lighting or feeding regime in experimental animals over a prolonged period of time can result in the appearance of intra- or ultradian markings in dentine every 12 hours. The experimental work of Shinoda (1984) has also revealed the circadian nature of plasma calcium, inorganic phosphate and alkaline phosphatase levels at the times incremental lines are formed in dentine. Okada (1943) provided additional evidence for the hypothesis that shifts in acid–base equilibrium caused markings in dentine: he monitored the blood pH of pregnant rabbits in the last days up until the time of parturition. As the rabbits became more acidotic toward term, so incremental lines in dentine became less distinct, but immediately after birth, and as blood pH rose to normal values, so then the strong alternating dark and light lines returned. This physiological demonstration of how normal body function is reflected in dentine structure is powerful evidence for the fact that dentine is a sensitive indicator for some body functions. Stress brought on for any number of reasons, parturition lines, hypoplastic lesions in dentine and enamel due to exanthematous fevers or dysentery are all recorded in growing dental tissues in a reliable chronological manner (Bowman 1991; Hillson 1997). It is learning to read the 'clock' of past events that is difficult after teeth have been sectioned and prepared for histology.

Yilmaz and colleagues (1977) also demonstrated experimentally that short period (von Ebner's) lines in pig dentine are circadian increments but did not report on the spacing between the lines in these pigs nor on the site at which lines were measured within the pig teeth. Ohtsuka and Shinoda (1995) have gone on to establish the time of first appearance after birth of circadian markings in rat dentine and tied this in with the appearance of other circadian rhythms in the body.

In addition to all this, Miani and Miani (1971) presented experimental evidence for a circadian rhythm in the rate of advancement of the mineralising dentine front in dog dentine. Maximum advancement of the mineralising dentine front occurred at the end of the active period, early evening (which the authors note happens to correspond to minimum adrenal cortex activity). Minimum advancement occurred at the end of the inactive dark phase (6 a.m.).

Despite this convincing evidence, Kawasaki and colleagues (1980) concluded that the calcospheritic lines visible in human dentine were 12 hour increments of dentine mineralisation and not therefore circadian in nature. If incremental markings in dentine are to be regarded with confidence as daily markings in growth studies, this finding needs to be examined more carefully. Kawasaki *et al.* (1980) first calculated the average daily rate of enamel formation in ground sections of teeth by measuring cross-striations. They then used this average rate to calculate the rate of dentine mineralisation occurring at the same time. The rates for dentine formation were not, therefore, directly derived experimental data for dentine alone. Rather, they were data reliant upon the accurate cross-matching of daily enamel cross-striations with lines in the dentine formed at the same time. Nonetheless, the authors concluded that rates of human dentine formation are of the order of 4 μm in the cusps of premolars. This estimate is in direct agreement with results from measurements of the spacings of short period lines made in the axial plane of ground sections of human canines (Fig. 5.1), if one assumes them to be circadian increments (Dean 1998).

However, Kawasaki and co-workers (1980) took their argument further and examined the spacing of incremental markings in demineralised sections of human dentine that had been stained with silver. They interpreted the closely spaced calcospheritic short period in lines in demineralised and silver-stained sections of premolar teeth to be 12 hourly increments of dentine formation solely because the spacing of short period lines in these sections was only 1.7–2.0 μm apart (i.e roughly half the 4 μm cuspal daily rate). They reasoned that two lines must be formed in a single day, in order to form 4 μm of dentine every day. Moreover, as 10 short period lines were visible between more prominent and widely spaced long period markings

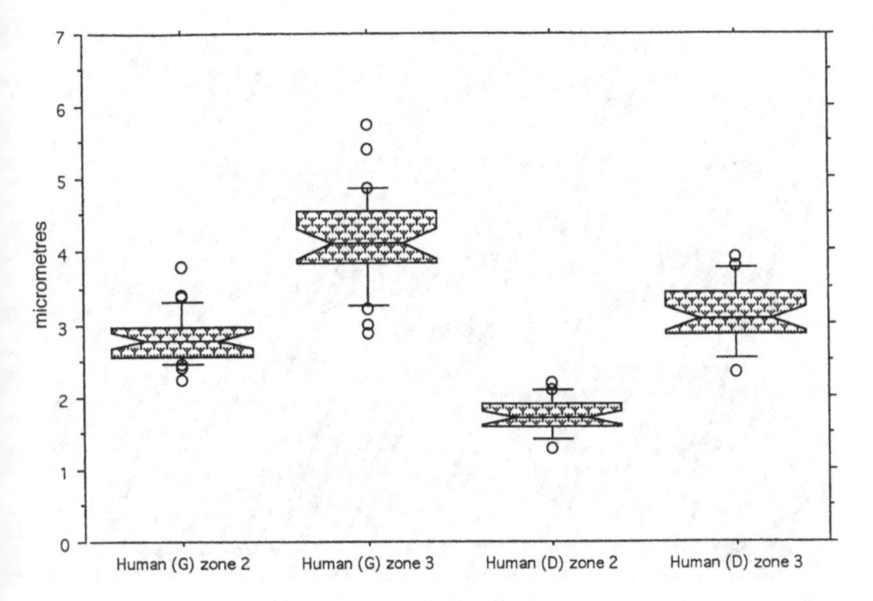

Fig. 5.1. Box plots illustrating the distribution of measurements of the spacing between daily lines in the dentine of two human teeth, one a ground section (G) and the other a demineralised section (D) stained with silver. Measurements were made 100 μm from the granular layer of Tomes, where lines are spaced close together (zone 2), and also in the axial plane of the cusp, where lines are spaced maximally apart (zone 3). In the demineralised section the spacing of the lines is significantly less than in the ground section in each region.

in some of these demineralised silver-stained sections, the authors concluded it is likely there is also a 5 day rhythm during human dentine formation. This widely quoted conclusion about long period lines in human dentine also requires some attention.

Measurements of the spacing of short period lines in ground sections of human teeth reflect rates of dentine formation well (Fig. 5.1). Similar measurements in demineralised and silver-block-stained human teeth, however, may not, since they are spaced much closer together. This suggests there is considerable shrinkage during preparation of demineralised sections of dentine (Brain 1966) and that the spacings measured from these sections cannot therefore be used as a reliable indicator for rates of dentine mineralisation. Furthermore, while a long period 5 day rhythm in dentine (and enamel) formation has been reported for macaques (Bowman 1991) it is generally accepted that humans most often show a long period rhythm with a modal value of between 7 and 10 days for a given population of individuals (Okada 1943; Beynon 1992; FitzGerald 1995, 1998). Figure 5.2 shows two sections of human teeth, one demineralised the other not, each

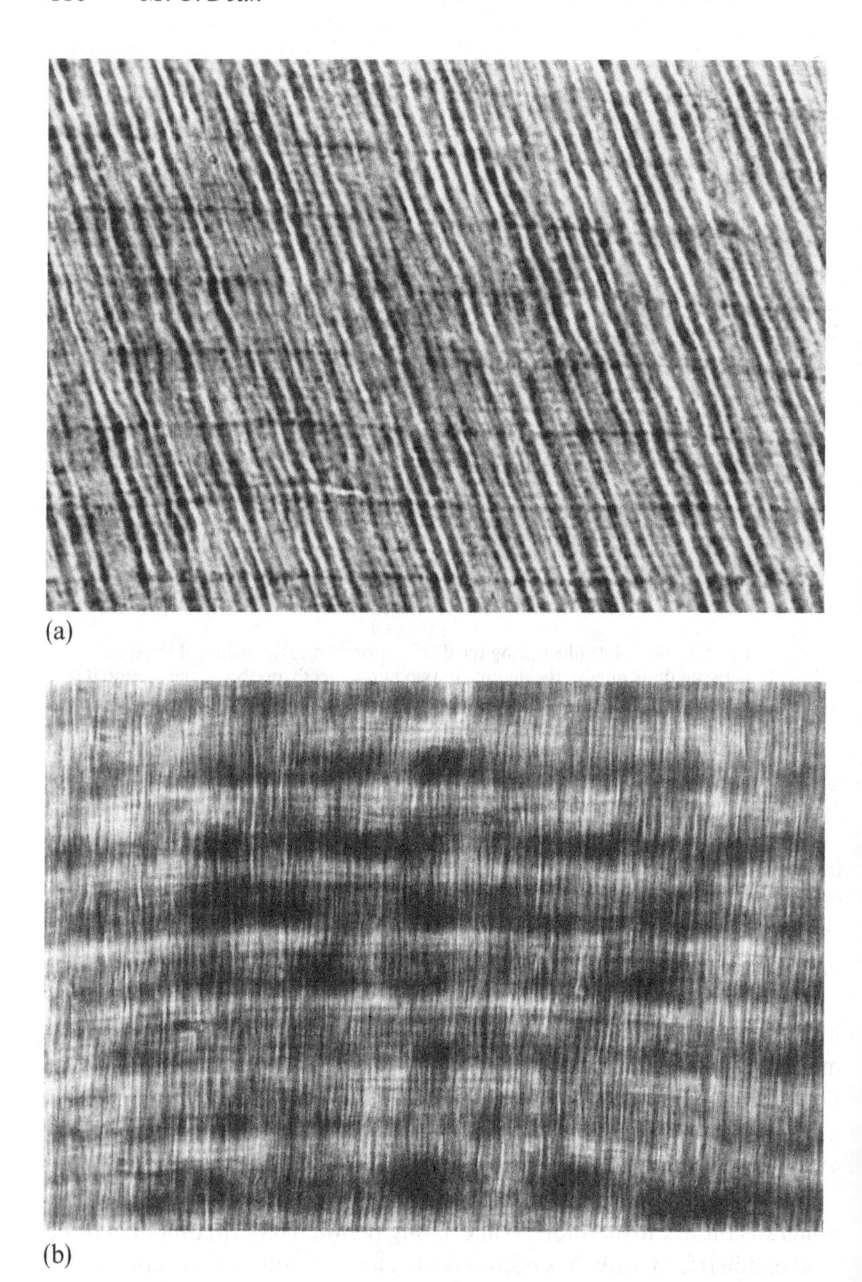

Fig. 5.2. Long- and short-period incremental lines in the dentine of human teeth. One micrograph (a) is of a demineralised section and the other (b) of a ground section. (Magnification 265 ×.)

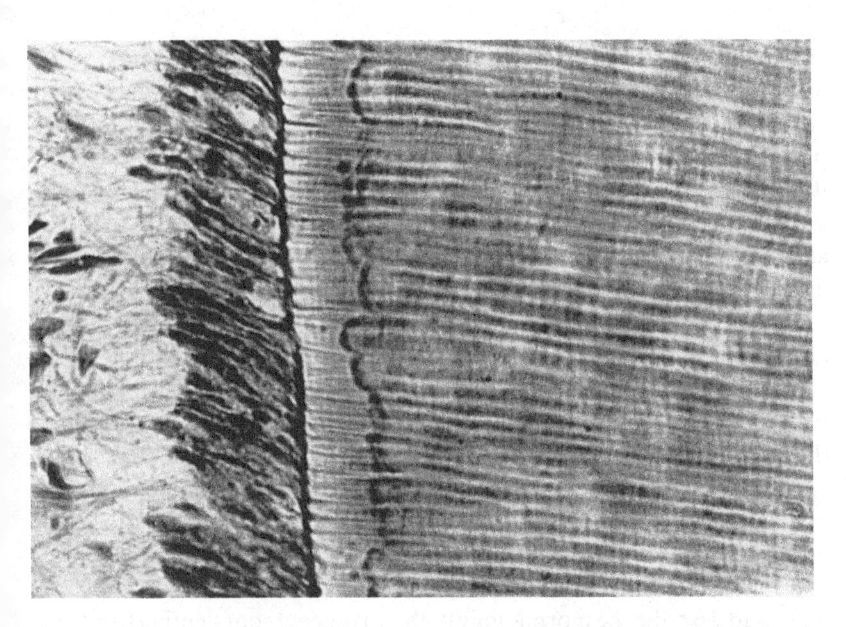

Fig. 5.3. A demineralised section of human dentine and predentine showing incremental markings in the predentine as well as in the dentine. (Magnification 265 ×; field width 220 μm.)

with seven to eight short period markings between long period lines and not five. It seems far more likely that there is as much variation in this long period rhythm in human dentine as there is in human enamel and equally likely that the same long period periodicity exists in both enamel and dentine from the same tooth and even the same mouth (Dean *et al.* 1993a; Dean 1995; Dean and Scandrett 1995, 1996).

On balance it appears that the proposal for a regular 12 hour periodicity in dentine mineralisation may not be the most obvious explanation for observations on the spacing of short period lines in demineralised sections of human dentine. Nonetheless, it is clear that ultradian or infradian markings in dentine and enamel do exist and are a well-documented phenomenon (Rosenberg and Simmons 1980; FitzGerald 1995; Ohtsuka and Shinoda 1995). Since it remains possible that silver-stained sections of human dentine reveal a different incremental phenomenon to those visible in ground sections they are probably to be avoided in any studies of primate tooth growth until they are better understood. A point worth noting is that incremental markings are visible in predentine as well as in dentine that has mineralised (Fig. 5.3) in silver-stained demineralised sections of human dentine. It is not enough to simply presume that mineralisation is the only process that occurs in a circadian manner. Predentine secretion and the breakdown and removal of some components of the non-collagenous predentine matrix in order to allow the onset of mineralisation may each occur in a rhythmic manner. Furthermore, the predentine may be secreted several days before dentine at the same position it mineralises but some components of predentine are known to be transported directly to the mineralising front as soon as they are secreted. This all points to a very complicated process indeed and a picture of incremental lines that may be easily prone to misinterpretation.

The incremental lines in human and non-human primate dentine do in fact appear to mirror the gradual increase in the rate of dentine mineralisation that must occur as tooth formation proceeds from the root surface towards the circumpulpal dentine and onward to the pulp. Importantly, they also reflect the manner in which dentine is known to mineralise and can therefore be confidently associated with the mineralisation process and not any of the others discussed above. Shellis (1983) has described the normal structural arrangement of calcospherites in human dentine. Boyde and Jones (1983) and Jones and Boyde (1984) used back-scattered electron images to describe the morphology of the advancing root dentine front and demonstrated how calcospherites change in shape and size in different regions of the dentine, becoming larger and flatter as the mineralising front proceeds from the outer root dentine. Whittaker and Kneale (1979) also described the transition from 'dome-shaped' calcospherites to those with a more flattened 'cogwheel' appearance (because of the way dentine tubules notch their periphery) as dentine formation progresses. These observations explain the shift in appearance from concentric to laminar short period lines seen in histological sections of dentine (Fig. 5.4).

Furseth (1974), Boyde and Jones (1983) and Jones and Boyde (1984) also demonstrated that the granular layer of Tomes itself is formed as minute calcospherites of unequal size fail to coalesce completely, perhaps because of a high rate of initiation of new mineralising centres in this region. It is important to note that closely spaced lines in the small calcospherites suggest slow rates of mineralisation within them but that the rate of advancement of the mineralising front results from two things: the number of new mineralising centres that form ahead of the mineralised dentine surface, and how far ahead of it these centres form. Boyde and Jones (1983) indicated that this situation of a slow rate of mineralisation within calcospherites but a fast advancement of the mineralising front as a whole (Kawasaki 1975) may be conducive to the unmineralised 'interglobular' spaces that persist and which give the 'granular' appearance to the granular layer of Tomes. The somewhat irregular manner in the way dentine forms close to the root surface means that it would be wise to be cautious about

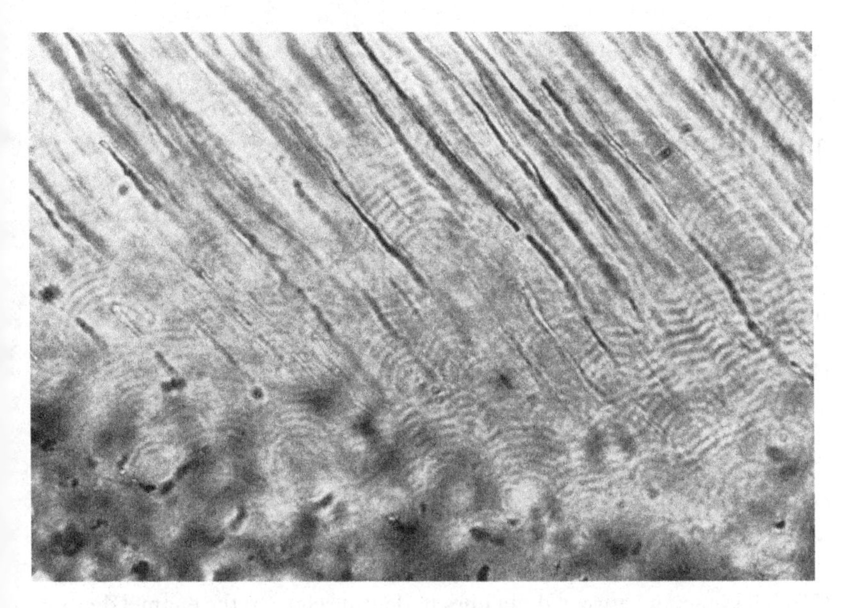

Fig. 5.4. Calcospherites in dentine approximately 100 μm from the granular layer of Tomes, showing the transition from concentric daily lines to a more laminar form of daily line that is more widely spaced. (Magnification 265 ×; field width 220 μm.)

the way rates of root growth are estimated using various techniques (see below) that make use of incremental markings as measures of linear daily dentine apposition.

How can incremental markings in dentine be identified in primate tooth sections?

Identifying any kind of marking in either enamel or dentine requires experience and care. Dentine is notorious for being a difficult tissue to make use of in primate growth studies but for some reason fossil dentine shows incremental markings that on more occasions are clearer than in the dentine of recent primates. Several authors have observed that incremental markings in dentine can be enhanced in one of several ways. Ground sections can be made anorganic, or can be demineralised (sections of carious teeth or of teeth that have become partially denatured post mortem may be both demineralised and wholly or partially anorganic). Markings can be enhanced by heating them sufficiently to drive off CO_2 from carbonate-rich regions, by staining demineralised blocks or sections with silver, or by treating fractured blocks of dentine with NaOCl and then etching them with EDTA (Andresen 1899; von Ebner 1902; Mummery 1914, 1924; Kawasaki *et al.* 1980; Jones and Boyde 1984; Posner and Tannenbaum 1984; Kodaka and Higashi 1995).

In addition to preparing sections of teeth in one of these ways as may be appropriate (and it may be that fossil teeth by virtue of their anorganic nature are preprepared in an ideal way) the following eight criteria need to be carefully considered when one is identifying lines in dentine as circadian increments:

(1) Markings in dentine should show a calcospheritic pattern (Boyde and Jones 1983) close to the granular layer of Tomes in the root and gradually become more laminar in their contour.
(2) They should appear as a continuous series of evenly spaced lines.
(3) They should follow the contours of the growing tooth crown and root.
(4) They should be maximally spaced in the axial plane of the tallest cusp where rates of dentine formation are fastest.
(5) The spacing between daily lines in dentine close to the enamel dentine junction should match that predicted from the geometry of the enamel forming at the same time.
(6) The number of short period daily increments in enamel and dentine growing at the same time (between accentuated markings that occur in both enamel and dentine) should be equal in number.
(7) When visible, the number of daily lines between long period markings in dentine should be the same as that for cross-striations counted between adjacent striae of Retzius in enamel of the same individual (Dean 1995).
(8) The spacing of dentine increments in a given part of the tooth crown or root should be equal, or close, to values for the rate of dentine formation determined in experimental studies of humans and non-human primates.

Using incremental markings in dentine to estimate the rates of root growth in primates

Some evidence exists (Dean 1995) to suggest that long period lines are in dentine (sometimes called Andresen lines after the person who first described them; Andresen 1898) crop out at the root surface as periradicular bands. But because these bands lie beneath cementum and because they are often quote close together and not as prominent as the perikymata on enamel they are difficult to see. Theoretically, it should be possible to count these bands and calculate their number for a given length of tooth root. If the periodicity of the bands is known, either directly from histological examination of the dentine, or from a knowledge of the periodicity of the striae of Retzius in enamel from the same tooth, then the rate of root elongation is equal to the length of root divided by the time taken to form that root. The same technique would be possible if long period dentine lines could be seen in histological sections at the root surface. However, the microanatomy of dentine and the way the surface dentine forms in tooth roots is probably the reason these lines are not clearly observed in this position.

Given that it is not possible to calculate the rate of root length in growing teeth in either of these ways then three things must be measured in order to reconstruct the geometry of the growing tooth root and to estimate the rate of 'extension' of the root margin. (1) The daily rate at which cells produce matrix, as represented by the spacing between daily incremental lines. (2) The direction of cell movement, as represented by the direction of the dentine tubules. (3) The number of mature secretory cells active at any one time (their rate of differentiation), as represented by the length and angulation of an incremental line making up one side of a triangle.

Shellis (1984) has expressed the extension rate of teeth at the cement–dentine junction (CDJ) in the root mathematically. In the equation

$$c = d\left[\left(\frac{\sin I}{\tan D}\right) - \cos I\right],$$

c is the extension rate, d the daily rate of dentine secretion, angle I is the angle the dentine tubules make with the root surface and angle D is the angle between an incremental or accentuated line and the root surface. These variables are illustrated with respect to the root dentine shown in Fig. 5.5. The equation defines how measurements of these variables can be used to estimate the rate of tooth root extension.

This approach has been used to estimate the time taken to grow tooth roots in the early Miocene fossil hominoid *Proconsul* (Beynon *et al.* 1998). No other way exists of reconstructing the time it took to form roots in this early fossil hominoid. This was only made possible by the exquisitely preserved incremental lines in the dentine of the crowns and roots visible in ground sections made from 13 teeth belonging to three individuals of *Proconsul heseloni* and *P. nyanzae*. By calculating the rates of root exten-

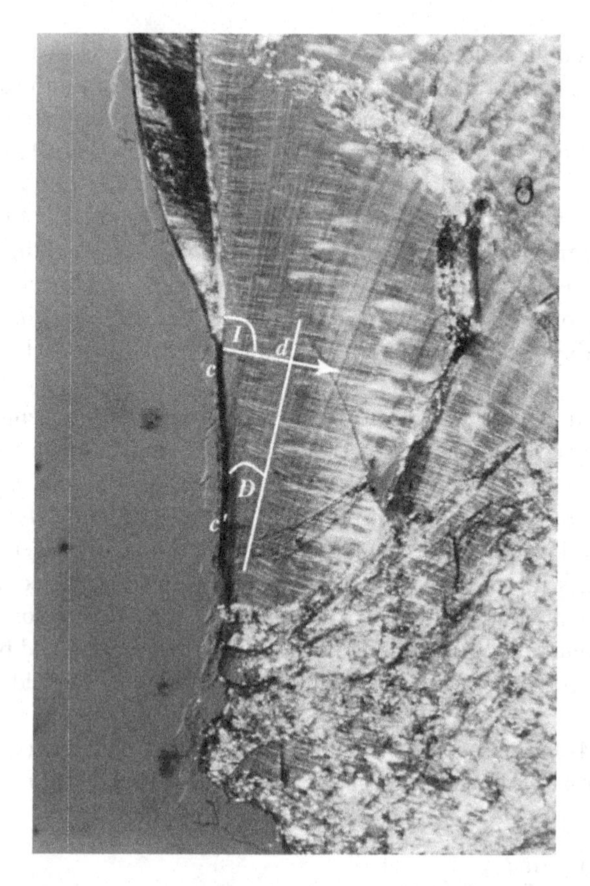

Fig. 5.5. The root dentine of a tooth with the variables used to estimate the extension rate at the root surface superimposed over the tubules and incremental lines. See text for details.

sion along tooth roots at several locations and then by dividing the length of the tooth root by the rate of root extension it was possible to estimate root formation period for both permanent and deciduous teeth belonging to these fossils. Extension rates in the cervical third of permanent tooth roots of *P. heseloni* were on average 6.5 μm per day. In the apical third of the root they were on average 14.5 μm per day and close to the apex 21.5 μm per day. A 7 μm or 8 μm long root probably took around 2.5 years to form (Beynon *et al.* 1998). Root extension rates in a deciduous tooth, which one would expect to form much faster, were estimated at 35 μm per day.

Using daily incremental lines in dentine to estimate crown formation times

Estimates of crown formation times are important in studies of modern humans and of fossil primates. The time to grow the enamel cap is an important yardstick for comparing growth between different species. The time to form enamel can be estimated using incremental markings in enamel but it can also be estimated using the dentine. Dean and colleagues (1993b) estimated the time to grow the enamel cap of a permanent canine belonging to a juvenile specimen of *Paranthropus robustus* (SK 63) from Swartkrans, South Africa, as between 3.18 and 3.48 years. Extra time had to be added to this estimate to account for lost tooth tissue at the buccal cervix post mortem. Cuspal enamel that was sectioned obliquely posed another problem in the calculation. Unpublished information on the dentine of this canine, however, allowed us to calculate crown formation time and the whole of the undamaged part of root dentine formation in another way.

Figure 5.6 is a micrograph of dentine in SK 63 taken in the midaxial plane of the tooth. Faint markings in the dentine fulfil some of the criteria defined above and resemble daily incremental lines. Many of them are clear enough for the spacing between them to be measured. The average spacing for 12 groups of 6 consecutive lines measured at different places in the axial plane is 2.6 µm (1 standard deviation = 0.31). This translates into what seems like a very low rate of dentine formation for the cusp of a tall tooth type, but nonetheless it is possible to estimate a crown formation time from this. Since the distance from the first-formed dentine in the cusp to the last-formed dentine in the axial plane at death is 3600 µm along the direction of the central dentine tubules this suggests about 3.8 years of canine tooth growth in SK 63 at the time of death. The period of time between birth and initial mineralisation of the canine was probably between 0.25 and 0.75 years, which means an age at death for this juvenile estimated using dentine rather than enamel is roughly between 4 and 4.5 years of age. The estimates using enamel were 3.45–4.23 years of age. In this situation, where the markings in the dentine are less reliable than those in the enamel (in the sense that too few can be measured along the whole length of the axis of the cusp for comfort) it is astonishing that the estimates made in these completely different ways are so close. The age at death using dentine increments supports an age at death at the top end of the age range estimated using enamel.

This same approach has been used to calculate crown formation times for the *Proconsul heseloni* and *P. nyanzae* specimens described by Beynon

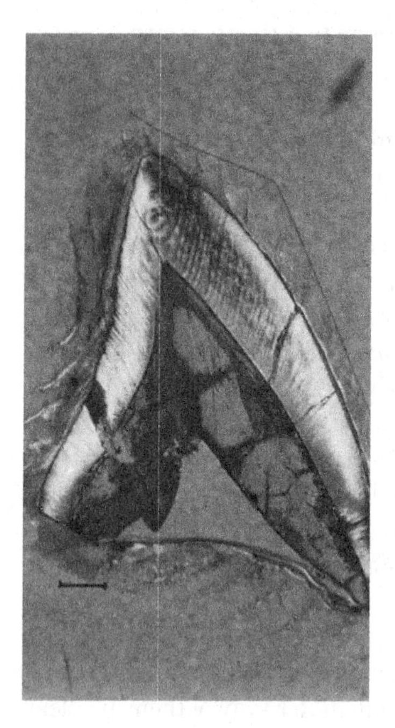

Fig. 5.6. An outline of the tooth crown of SK 63 together with a micrograph of daily incremental markings in the axial plane of the crown. Scale bars represent: *left*, 1 mm; *right*, 50 μm. (Magnification of the micrograph 275 ×; field height 220 μm.)

and colleagues (1998). Estimates based on daily lines in dentine match those made using daily cross-striations in enamel very closely indeed. This all supports an internal consistency in tooth formation and provides some security that dental development can be reconstructed in this way.

Conclusions

Dentine is a very complicated tissue. The fact that it records a faithful sequence of physiological events and that it is not turned over during growth, together with the fact that it can be labelled experimentally, makes it a very useful tissue for studies of human and non-human primate growth and development. At present, studies that have made use of dentine in reconstructing dental development have done so only in conjunction with enamel from the same tooth. As better data become available for ranges of rates of dentine formation in the teeth of humans and non-human primates, so dentine will come to be more useful in its own right. In the long term, dentine is likely to prove even more useful than enamel, since a longer period of development is recorded within it than in the enamel of any given tooth. The timing of stress events due to, for example, emotional or seasonal changes, or to physiological upsets as well as to first parturition might all in the future be retrieved from fossil and recent teeth when the things that affect dentine formation are better understood (Bowman 1991; Macho *et al.* 1996; Hillson 1996).

Acknowledgements

I am grateful to the editors for asking me to contribute to this volume. Research included in this review was made possible by grants from The Leverhulme Trust and The Royal Society. I am grateful to the Office of the President and to the Governor and Trustees of the Kenya National Museum, Kenya, and to Dr Meave Leakey for permission to work on valuable fossil material in the Department of Palaeontology. I thank Francis Thackeray and the Director of the Transvaal Museum, Pretoria, South Africa, for permission to work with SK63.

References

Andresen V (1899) Die Querstreifung des Dentins. *Deutsche Monatsschrift für Zahnheilkunde. Sechzehnter Jahrgang* **xxxviii**, 386–389.

Beynon AD (1992) Circaseptan rhythms in enamel development in modern humans and Plio-Pleistocene hominids. In *Structure, Function and Evolution of Teeth*, ed. P. Smith and E. Tchernov, pp. 295–309. London and Tel Aviv: Freund Publishing House Ltd.

Beynon AD, Dean MC, Leakey MG, Reid DJ and Walker A (1998) Comparative dental development and microstructure in *Proconsul* teeth from Rusinga Island, Kenya. *Journal of Human Evolution* **35**, 163–209.

Bowman JE (1991) Life history, growth and dental development in young primates: a study using captive rhesus macaques. PhD thesis, University of Cambridge.

Boyde A and Jones SJ (1983) Backscattered electron imaging of dental tissues. *Anatomy and Embryology* **168**, 211–226.

Brain EB (1966) *The Preparation of Decalcified Sections*, pp. 151–156. Springfield, IL: Charles C. Thomas.

Dean MC (1995) The nature and periodicity of incremental lines in primate dentine and their relationship to periradicular bands in OH 16 (*Homo habilis*). In *Aspects of Dental Biology: Paleontology, Anthropology and Evolution*, ed. J. Moggi-Cecchi, pp. 239–265. Florence: International Institute for the Study of Man.

Dean MC (1998) Comparative observations on the spacing of short-period (von Ebner's) lines in dentine. *Archives of Oral Biology* **43**, 1009–1021.

Dean MC and Scandrett AE (1995) Rates of dentine mineralisation in permanent human teeth. *International Journal of Osteoarchaeology* **5**, 349–358.

Dean MC and Scandrett AE (1996) The relation between enamel cross striations and long-period incremental markings in dentine in human teeth. *Archives of Oral Biology* **41**, 233–241.

Dean MC, Beynon AD, Thackeray JF and Macho GA (1993a) Histological reconstruction of dental development and age at death of a juvenile *Paranthropus robustus* specimen, SK 63, from Swartkrans, South Africa. *American Journal of Physical Anthropology* **91**, 401–419.

Dean MC, Beynon AD, Reid DJ and Whittaker DK (1993b) A longitudinal study of tooth growth in a single individual based on long and short period incremental markings in dentine and enamel. *International Journal of Osteoarchaeology* **3**, 249–264.

Ebner V von (1902) Histologie der Zähne mit Einschluss der Histogenese. In *Handbuch der Zahnheilkunde*, ed. J. Scheff, pp. 243–299. Wien: A. Holder.

FitzGerald CM (1995) Tooth crown formation and the variation of enamel microstructural growth markers in modern humans. PhD thesis, University of Cambridge.

FitzGerald CM (1998) Do enamel microstructures have regular time dependency? Conclusions from the literature and a large-scale study. *Journal of Human Evolution* **35**, 371–386.

Fujita T (1943) Über die Entstehung der Interglobularbezirke im Dentin. *Japanese Journal of Medical Science*, Part 1 Anat. **11**, 1–17.

Furseth R (1974) The structure of peripheral root dentine in young human premolars. *Scandinavian Journal of Dental Research* **82**, 557–561.

Hillson S (1996) *Dental Anthropology*. Cambridge: Cambridge University Press.

Jones SJ and Boyde A (1984) Ultrastructure of dentine and dentinogenesis. In *Dentine and Dentinogenesis*, vol. 1, ed. A. Linde, pp. 81–134. Boca Raton, FL: CRC Press Inc.

Kobayashi K (1984) Scanning electron microscopic studies on spherical calcification treated with HCL collagenase method. *Shiwa Gakuho* **84**, 1077–1104. [In Japanese with an English abstract.]

Kawasaki K, Tanaka S and Ishikawa T (1980) On the daily incremental lines in human dentine. *Archives of Oral Biology* **24**, 939–943.

Kodaka T and Higashi S (1995) Scanning electron microscopy of spherical and linear laminate structure in human dentine etched with EDTA after treatment with sodium hypochlorite. *Japanese Journal of Oral Biology* **37**, 80–85.

Macho GA, Reid DR, Leakey MG, Jablonski NG and Beynon AD (1996) Climatic effects in dental development of *Theropithecus oswaldi* from Koobi Fora and Olorgesailie. *Journal of Human Evolution* **30**, 57–70.

Miani A and Miani C (1971) Circadian advancement rhythm of the calcification front in dog dentine. *Minerva Stomatology* **20**, 169–178.

Mummery JH (1914) On the process of calcification in enamel and dentine. *Philosophical Transactions of the Royal Society, series B* **205**, 95–113.

Mummery JH (1924) *The Microscopic and General Anatomy of the Teeth, Human and Comparative.* Oxford: Oxford Medical Publications, Oxford University Press.

Ohtsuka M and Shinoda H (1995) Ontogeny of circadian dentinogenesis in the rat incisor. *Archives of Oral Biology* **40**, 481–485.

Okada M (1943) Hard tissues of animal body. Highly interesting details of Nippon studies in periodic patterns of hard tissues are described. *Shanghai Evening Post*, Medical Edition of September 1943, pp. 15–31.

Okada M and Mimura T (1940) Zur Physiologie und Pharmakologie der Hartgewebe. III. Über die Genese der rhythmischen Streifenbildung der harten Zahngewebe. Proceedings of the Japanese Pharmacological Society, 14th Meeting, Japan. *Journal of Medical Science IV, Pharmacology* **13**, 92–95.

Okada M and Mimura T (1941) Zur Physiologie und Pharmakologie der Hartegewbe. VII. Über den zeitlichen Verlauf der Schwangerschaft und Entbindung gesehen von der Streifenfigur im Dentin des mutterlichens Kaninchens. Proceedings of the Japanese Pharmacological Society, 15th Meeting, Japan. *Journal of Medical Science IV, Pharmacology* **14**, 7–10.

Okada M, Mimura T, Ishida T and Matsumoto S (1939) The hematoxylin stainability of decalcified dentin and the calcification. *Proceedings of the Japanese Academy* **35**, 42–46.

Okada M, Mimura T and Fuse S (1940) Zur Physiologie und Pharmakologie der Hartegewbe. VI. Eine Methode der pharmakologischen Untersuchung durch die Anwendung von Streifenfiguren im kaninchendentin. Proceedings of the Japanese Pharmacological Society, 14th Meeting, Japan. *Journal of Medical Science IV, Pharmacology*, **13**, 99–101.

Posner AS and Tannenbaum PJ (1984) The mineral phase of dentine. In *Dentine and Dentinogenesis*, vol. 2, ed. A. Linde, pp. 17–36. Boca Raton, FL: CRC Press Inc.

Rosenberg GD and Simmons DJ (1980) Rhythmic dentinogenesis in the rabbit incisor; circadian, ultradian and infradian periods. *Calcified Tissue International* **32**, 29–44.

Shellis RP (1983) Structural organisation of calcospherites in normal and rachitic human dentine. *Archives of Oral Biology* **28**, 85–95.

Shellis RP (1984) Variations in growth of the enamel crown in human teeth and a possible relationship between growth and enamel structure. *Archives of Oral Biology* **29**, 697–705.

Shinoda H (1984) Faithful records of biological rhythms in dental hard tissues. *Chemistry Today* **162**, 34–40 [in Japanese].

Whittaker DK and Kneale MJ (1979) The dentine–predentine interface in human teeth; a scanning electron microscope study. *British Dental Journal* **146**, 43–46.

Yilmaz S, Newman HN and Poole DFG (1977) Diurnal periodicity of von Ebner growth lines in pig dentine. *Archives of Oral Biology* **22**, 511–513.

6 *New approaches to the quantitative analysis of craniofacial growth and variation*

PAUL O'HIGGINS AND UNA STRAND VIDARSDOTTIR

Why compare growth allometry amongst human populations?

In this chapter we outline some recent developments in the analysis of morphological variation. The focus is on the human facial skeleton but the methods and approaches described can be applied to any situation in which variations in form are to be studied. In particular this chapter presents an account of some recent developments in statistical and graphical approaches to the study of landmark data. Collectively the class of methods that we will use forms a part of the toolkit of 'geometric morphometrics' (Marcus *et al.* 1996).

In order to focus our description of the implementation of geometric morphometric techniques it is illustrated by an example study in which morphological variation is examined between the faces of two groups of people: Aleutians and Alaskans (Inupiaq Eskimos). This study aims to assess the degree and nature of any differences in cranial morphology between adults and to investigate the extent to which such differences can be attributed to differences in patterns of cranial growth.

The example study illustrates an approach that might be applied more widely. Many fossil crania are known and much discussion of human origins hinges on this material. If, however, crania are to form a focus of studies of human evolution and variation in the past then it is important that adult morphology is interpreted from a developmental perspective. This is because adult morphology arises through developmental processes and, in consequence, variations between crania arise through variations in these processes. This leads us to seek an understanding of differences amongst adults in terms of differences in ontogeny in the expectation that adaptive and evolutionary transformations might, in turn, be understood in terms of pattern and process.

Differences between distinct populations might arise simply through truncation or extension (in time or rate; Shea 1983, 1986) of a common growth pattern. Alternatively differences might be due to the evolution of distinctive patterns of growth. Such distinctiveness might be a feature of the whole of the growth period or it may be confined to one stage of growth. A further possibility is that differences in early development lead to the early establishment of morphological distinctiveness and that this persists into adulthood despite common growth patterns. It is also possible that some combination of differences in early developmental patterning, growth patterning, growth timing and growth rate accounts for differences between adult forms. Knowledge of ontogeny can be expected to provide insights into variation and adaptation through unravelling of the morphogenetic basis of evolutionary adaptation.

Cranial growth

The craniofacial skeleton is made up of distinct skeletal elements. Each develops and grows under the influence of diverse local and systemic factors. Many individual bones can, in turn, be divided into subunits, each of which is potentially subject to different influences during growth. Despite this, the cranium remains a functional whole during growth, and this is achieved through co-ordinated growth and remodelling of individual bones.

Parts of the cranium develop through ossification of cartilaginous models. Three pairs of cartilages contribute to the cranial base: the prechordal and hypophyseal cartilages, mainly derived from the neural crest; and the parachordal cartilages, derived from the occipital sclerotomes and the first cervical sclerotome (Sperber 1989). Endochondral ossification is preceded by hyaline cartilage prototype models of the future bone and is characteristic of the bones of the cranial base. In contrast, intramembranous ossification is characteristic of the bones in the cranial vault, most of the facial bones and the mandible. It takes place in tissues of neural crest origin that form sheet-like osteogenic membranes (Sperber 1989; Moore and Persaud 1993). Sometimes secondary cartilage will appear and later ossify endochondrally in membranous bones, such as the mandible.

Growth involves not only increases in size of the individual elements of the cranium during development but also changes in the spatial relationships and shapes of those elements (Enlow 1975). In humans and mammals generally, growth contributes significantly to final adult morphology, since the craniofacial skeleton undergoes changes in shape as well as size. Thus adult human crania differ considerably in form from those of neonates because different parts of the skull enlarge at different rates and in different directions. In this chapter we use the term growth allometry to refer to the changes in shape consequent upon differential growth.

Craniofacial skeletal growth consists of three principal processes: conversions of cartilage to bone, sutural deposition and periosteal remodelling (Thilander 1995). The postnatal conversions of cartilage at the cranial base are largely confined to the spheno-occipital synchondrosis, which persists into the late teens, or early adulthood. Although this synchondrosis allows for some linear growth in the cranial base, its most important function is thought to be to adjust the flexure of the cranial base (Thilander 1995). The nasal septum appears (Moss 1964) to grow secondary to displacement of the midfacial bones. It is not, as previously thought, an active participator in the displacement of the midfacial bones during development (Scott 1953, 1956). Bone deposited at sutural edges contributes to growth allometry in that differential deposition at sutures with different spatial orientations results in transformation of cranial form. Sutural deposition is presently thought to occur in response to mechanical stimuli in the sutural membranes (Enlow and Hans 1996).

Growth of cartilages and deposition at sutures leads to the relative displacement of skeletal elements during growth. Functional alignment is maintained in part through co-ordination of these processes and in part through remodelling of existing bone. Bone remodelling is directed towards coordinated resorption and formation. It is regulated by systemic factors such as hormones that control osteoblastic and osteoclastic activity and by the local mechanical, hormonal and vascular environment. Remodelling of the bone surface contributes, together with sutural growth, to the normal development of the sizes and shapes of the bones of the face and vault. This ontogenetic remodelling process is termed 'bone growth remodelling' (Bromage 1986). It is a process that acts to a large extent as a compensatory mechanism, maintaining proper bone alignment, function and proportionate growth during bone displacement (Enlow 1975). The surface distribution of bone growth remodelling processes is therefore considered to be an important indicator of craniofacial growth as a whole (Enlow 1975; Bromage 1986). It is currently hypothesised that growth remodelling acts as a compensatory mechanism to maintain proper bone alignment during displacement (Enlow 1968, 1975). Consequently it has been suggested that the spatial distribution, direction and rate of surface remodelling activity should serve as an indication of the pattern of displacement (Enlow 1975; Bromage 1986).

Cartilaginous growth, sutural deposition and cortical remodelling are regulated and coordinated to ensure functional integrity during cranial growth. Each region of the skull is subject to its own particular mix of genetic and epigenetic influences during growth and the way in which these are regulated is not fully understood. Moss and his colleagues (Moss 1964; Moss and Salentijn 1969a,b) have proposed a widely accepted model of regulation through 'functional matrices'. Under this model the growth of the skeletal elements is considered secondary to, and guided by, the growth of the functional matrices. These matrices are considered to be of two basic types: periosteal and capsular. Periosteal matrices are ones in which growth is influenced by local effects such as the forces generated by muscles acting on the skeleton through the periosteum. Capsular matrices are ones in which skeletal elements forming a capsule are influenced by their contents.

The neurocranium has thus been considered a capsular functional matrix (Moss and Salentijn 1969a; McLachlan 1994) containing the brain, the leptomeninges and the cerebrospinal fluid. The expanding brain displaces the bones of the neurocranium outward, causing tension in the sutural membranes which in turn respond by depositing bone at the sutural edges (Enlow and Hans 1996). The brain develops very rapidly in early childhood, especially in the first year but its growth is completed long before most other parts of the human body. Thus the neurocranium follows a similar growth course, although the spheno-occipital synchondrosis will keep growing into adulthood to accommodate the posterior expansion of the maxilla as space is made for the molars and growing nasopharynx. Once neurocranial expansion slows down, sutural growth becomes negligible and remodelling becomes the most important factor in further growth and shape modification. This remodelling is mostly influenced by the masticatory muscles, either directly through periosteal functional matrices in the area of attachment or indirectly as loading of bones causes bending or torsion stresses. Further influence comes from the changing form (shape and size) of the growing elements of the anterior cranial base and the facial skeleton.

The facial skeleton is made up of numerous bones joined by sutures. Growth at the sutural margins is believed to be secondary to bone displacements influenced by capsular functional matrices such as the orbital and nasal capsules. Different parts of individual bones may be influenced by different functional matrices. The maxilla, for example, is influenced by most of the functional matrices acting on the facial skeleton: orbital, nasal, basal, pneumatic and alveolar. In turn, the orbital unit responds to the growing eyeball, the alveolar to the development of the teeth etc. Additionally, the masticatory muscles and other periosteal matrices influence bone surface remodelling.

Growth at facial sutures ceases on average at 17 years of age, which is 2 years earlier than growth of the mandibular condyle and total body height (Björk 1966). In addition developmental skeletal changes in the face slow markedly in girls not long after puberty, but in boys not until late adolescence (Enlow and Hans 1996), this difference is related to the development of facial sexual dimorphism in the teenage period.

Analysis of growth

Thorough understanding of the differences between adult crania therefore depends on study of growth changes in size and shape. Quantitative analyses of cranial growth are, however, not straightforward. Many methodological difficulties are presented in dealing with complex variations in size and shape such as are found between growing and adult crania.

For example, a common approach to the study of growth is through cephalometric radiography and comparisons of superimposed radiographic tracings at different ages. There is, however, no biologically or statistically 'correct' method of superimposition and each researcher has to use as a reference the landmarks or planes that seem most appropriate to the study at hand (Broadbent 1996). The registration method inevitably introduces problems in interpretation in that all landmarks will appear to move away from the registration plane. This makes different studies difficult to compare. One solution (Björk 1968) is to position metallic implants in the developing craniofacial skeleton that can be followed longitudinally on radiographs. Their movements indicate directions and magnitudes of bone displacement and changes in bone orientation. This method has proved valuable in craniofacial research but is limited by the areas in which implants can be safely positioned (especially in the facial skeleton where developing teeth can shift them) and the ethical constraints in repeated radiography.

An alternative is to dispense with data relating to geometric relations of bony points and to focus instead on the distances between such points. Using such measurements it is possible readily to compare growth in lengths but more difficult to thoroughly appreciate the full three dimensionality of growth processes. Recent advances in methodology do, however, show some promise in dealing with the issues of registration and geometry. The application of these methods is the focus of this chapter and we illustrate them through a study of differences in adult cranial morphology and growth allometry between two closely related human populations.

The study populations: Alaskans and Aleutians

The groups being compared in this analysis came from two distinct, but closely related archaeological populations: Aleutians from the Kagomil and Shiprock Islands and Inupiaq Eskimos from northwest Alaska. The two populations are thought to have arisen from a common Proto-Eskimo-Aleut Group that split *ca.* 3000 years ago (Heathcote 1986). Despite their close relatedness the two populations have quite distinct craniofacial morphologies (Fig. 6.1), the Aleut skull being shorter, broader and lower than that of the Inupiaq in addition to possessing less pronounced brow-ridges, a low sloping forehead and no sagittal keel (Hrdlička 1945).

The example geometric morphometric study presented here aims to test two hypothesis.

H1: The first is relatively straightforward; that the adult crania of these populations do not differ in morphology. This is addressed through an examination of the degree of difference between adult means.

H2: The second hypothesis depends on the first being falsified; that the differences between adult populations arise through different growth allometries. This will be tested by modelling shape changes with growth in size (growth allometry) and with increasing dental age in each and comparing these models.

The skeletal material studied consists of 35 crania from the Kagomil and Shiprock Islands in the Aleutians (24 subadults and 11 adults from the National Museum of Natural History, Smithsonian Institution, Washington DC), collected by A. Hrdlička in the late 1930s (Hrdlička 1945); and 43 crania from northwest Alaska (9 adults from the Natural History Museum, London, and 34 subadults from the American Museum of Natural History, New York). The adult northwest Alaskan crania are mostly donations to the Oxford skeletal collection, now housed at the Natural History Museum, London; the subadult Alaskan crania are all from Point Hope and Point Barrow in northwest Alaska. They were collected in 1932 by J.A. Ford (Ford 1959) and in the 1940s by H. Larsen and F. Rainey (Rainey 1971). The cultural affinities of the skeletons were determined by accompanying grave goods as well as general physical appearance of the skeletons.

Crania were selected so as to represent the broadest possible age range between 1 year and adulthood. In total 11 of the Aleutian and 9 of the Alaskan crania were full adult. Sampling at the youngest ages proved difficult because of the paucity of well-preserved very young crania, hence the youngest specimen has a dental age of 1 year. The individuals were aged

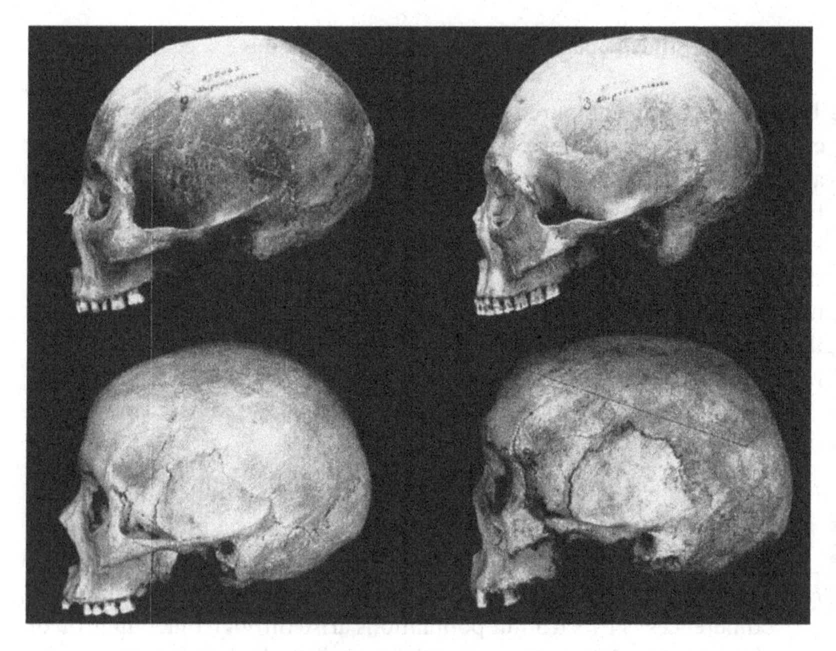

Fig. 6.1. Examples of Alaskan and Aleutian crania. Left, females; right, males; top, Aleutians; bottom, Alaskans. Not to scale.

according to their dental eruption patterns, using the revised standard of Ubelaker (1989) for 'non-white' populations. This standard is compiled from 17 other populations, the standards for deciduous dentition being based on American Caucasians and those for adult dentition on Amerindians. Where possible, because of loss of alveolar bone, crown and root formation stages were used to refine estimates based on eruptions. It is recognised that such data lead to relatively crude age estimates, but this is a practical limitation that would have required considerable resources in terms of radiology and dental histology to overcome and with dubious end benefits in terms of the conclusions of this study.

Quantification of morphology

The issue of measurement is frequently considered to be a rather straightforward matter: simply take calipers to specimens and record relevant lengths. In this chapter, we take the opportunity to consider measurement more deeply. First, it is necessary to define some basic terms: form, shape and size.

Unless they are identical, sets of measurements describing objects or figures will differ in their absolute scale, in their proportions and, if the measurements are taken with reference to the surroundings, in location and rotation (reflection also comes into play in some situations). We use the term 'form' to refer to the spatial organisation of an object independent of its location ('translation' is the term used for differences due to location) and rotation. Form itself we subdivide into two components 'size', which is a measure of scale of the form and 'shape' refers to aspects of form independent of scale. The term 'registration' is used to refer to the way in which objects are translated, rotated and scaled with respect to each other.

These definitions lead us to seek a biologically sensible quantitative representation of the spatial organisation of each cranium. Landmarks form the basis of most morphometric analyses. There are, however, numerous practical issues surrounding their identification and philosophical issues surrounding their nature. Principal amongst the latter is the issue of equivalence from specimen to specimen.

In biology a special type of equivalence forms the basis of many studies, homology (Hall (1994) provides a recent review). Homology in evolutionary studies relates to the matching of parts between organisms according to common evolutionary origin. In developmental studies, however, 'homology' is used in a different sense to refer to the matching of structures through ontogenetic time. This matching is not necessarily physical, since local growth phenomena (e.g. bony remodelling, shifting muscle insertions) may result in replacement of material between different ages such that structures that appear equivalent in terms of their local relations need not necessarily reflect the location of homologous material. Wagner (1994) has recently noted that, despite material replacement, structural identity is maintained. This maintenance of identity requires the action of 'morphostatic' mechanisms and, although structures may not be equivalent in the sense of material, they may be equivalent in terms of the continuity of morphostatic mechanisms. Developmental equivalence between landmarks may therefore be considered to equate to homology in the sense of van Valen (1982): 'correspondence caused by continuity of information' – a homology of the processes giving rise to structure.

In morphometric studies of growth we are faced with the task of representing the spatial relations of developmentally homologous parts in a suitable way. The classic approach is through the use of landmarks defining the limits or meeting points of structures (e.g. Martin 1928; Trevor 1950). Landmarks are samplings of the map of homologies between specimens (Bookstein 1991) and the density with which landmarks can be sited in regions of a specimen is dependent on the resolution with which the homology map can be discerned. This definition of the homology map depends, in turn, on purely biological rather than mathematical or geometric criteria. The identification of landmarks on the homology map may, however, depend on geometric features. The practical difficulties in identifying landmarks are recognised in a commonly quoted taxonomy of landmarks that is designed to encourage critical appraisal (Bookstein 1991; Marcus *et al.* 1996). Below we summarise and modify it slightly.

Type I landmarks are those whose homology from case to case is supported by the strongest (local) evidence (meeting of structures or tissues; local unusual histology etc.).

Type II landmarks are those in which claimed homology from case to case is supported by geometric, not histological evidence (tooth tip etc.). Type II landmarks include landmarks that are not homologous in a developmental or evolutionary sense but are equivalent functionally, such as wing tips.

Type III landmarks have at least one deficient coordinate (which means that they can be reliably located to an outline or surface but not to a very specific location, e.g. tip of a rounded bump).

In terms of the homology map therefore we can be most confident about landmarks of type I and least about landmarks of type III. This should not necessarily preclude the use of all types of landmark but it should lead us to expect greater (possibly directional) variation due to error alone in data based on type III rather than on type I landmarks when interpreting the results of analyses.

Landmarks in the study populations

In our example study of Aleutians and Alaskans we quantify the morphology of each face using 26 landmarks per half face. Most of these are of type I, some of type II and a few of type III. Their name and type is listed below (definitions in Martin 1928; Trevor 1950; where no definition exists we give brief details): alare, III; alveolare/infradentale superius, II; bregma, I; dacryon, II; frontomalare orbitale, II; frontomalare temporale, II; frontotemporale, III; glabella, III; infraorbital foramen upper boarder, III; jugale, II; maxillofrontale, I; nasion I; nasospinale, II; orbitale, III; point at which the palatine-maxillary suture crosses the midline (pmx), I; staphylion, II; stephanion, II; most superior point on rim of the orbit, III; midpoint supraorbital torus, III; the external alveolus at the distal margin of the canine (deciduous or permanent), II; the external alveolus at the distal margin of the most posterior tooth in the tooth row, II; the external alveolus at the distal margin of the second incisor (deciduous or permanent), II; zygomaxillare, II; zygoorbitale, II; zygotemporale inferior, II; zygotemporale superior, II.

Landmark coordinates were taken in no particular registration using a Polhemus 3 Space Isotrak II digitiser (Polhemus Incorporated, 1 Hercules Drive, PO Box 560, Colchester, VT 05446, USA). This operates electromagnetically through detection of the location and orientation of a coal within a pointing stylus relative to three reference coils. Tests of accuracy using a cube of known dimensions indicate that measurements of coordinates are accurate to within approximately 0.5 mm, although this figure varies according to ambient electromagnetic conditions. All data were gathered by one of us (U.S.V.).

Analysis and modelling of form transformations in growth and evolution

Once landmark data are gathered, the task of analysis can begin. In this study, adult differences and growth allometry are to be addressed. In particular we wish to investigate changes in cranial shape with increasing size and age during growth. It is therefore necessary to partition size from shape, and the calculation of a size measure and appropriate scaling of forms are called for. Following scaling, analyses can be directed to the study of covariances between shape, age and size. Interpretation can then proceed through examination of significance statistics and, importantly, through visualisation of the analytical results.

Size

Interlandmark distances, like the coordinates of landmarks are dependent on both size and shape; form = size + shape. When the focus of interest is growth it seems sensible to partition form into size and shape and to examine the relationship between these, but this presents several difficulties.

Size is not a straightforward quantity and there are difficulties in discussing size independently of shape in most circumstances. One difficulty arises because 'size' is often loosely defined. Sneath and Sokal (1973) ask 'which is bigger, a snake or a turtle?'. The term 'size' in this instance might relate to the differences in scale over whole objects. A suitable size measure is one that relates to the magnitude of many dimensions such as their sum or their mean.

In analyses of form (size + shape) based on landmark coordinates such as we undertake in this chapter a mathematically natural size measure is centroid size, the summed deviation of landmarks from the mean of all landmarks (centroid). Size measures may be chosen because they are appropriate, given the hypothesis at hand. There are no absolutely 'correct' choices in every circumstance, yet different choices may lead to different conclusions. If shape variations amongst specimens are fairly small with respect to size differences, the differences in result through different choices of size measure will also be small. As such, the same biological conclusions will be reached through different approaches.

The relationship between size and age in our study populations

In our example study we choose centroid size as an appropriate measure of scale. Centroid size is sensible biologically, since it takes account of the overall spread of landmarks and so of scale in a general sense. Of interest is the relationship between scale and estimated (dental) age in each of the Alaskan and Aleutian samples.

Figure 6.2 is a plot of deviation from the mean centroid size (cm; vertical axis) against estimated age (years; horizontal axis) for the two populations. No attempt was made to estimate the ages of specimens beyond the possession of fully erupted and occluded permanent dentition and fused spheno-occipital synchondrosis; as such, adults are simply allocated a nominal age of 21 years or more. In consequence they show a range of variation in centroid size but are represented in a vertical scatter above the 21 year mark on the horizontal axis. The scatters are such that Aleutian adults appear to show a wider spread and marginally larger mean centroid size. A t-test indicates, however, that the apparent difference between means is not significant ($p = 0.1$) although the variance ratio indicates that Aleutian adults are significantly more variable in centroid size than are the Alaskans ($p = 0.015$).

Amongst the subadult specimens there is a highly significant correlation between centroid size and age ($r = 0.92$, $p < 0.001$). There is no apparent difference in this scaling relationship between Alaskans and Aleutians below the estimated age of 10. The rate of increase in size with age appears to diminish after the age of 10 years in the Alaskan sample. We lack sufficient data from Aleutians in their teens to determine whether this is the same in both populations. The increased size variability amongst adult

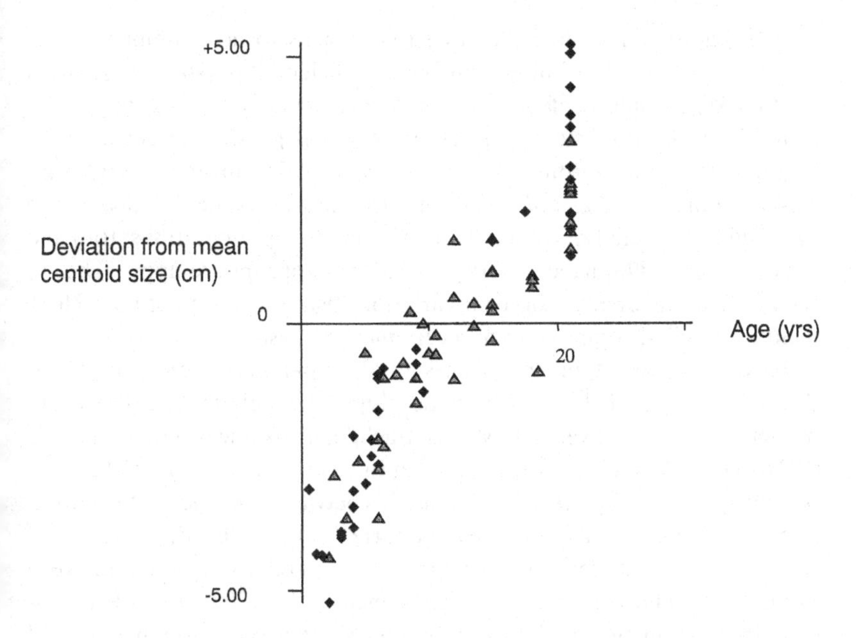

Fig. 6.2. Plot of dental age (horizontal axis) vs. deviation from mean centroid size (vertical axis). Triangles, Alaskans; diamonds, Aleutians; r between age and size = 0.92, p = 0.001.

Aleutians may be due to differences in later growth but this also remains an open question because of lack of data.

These findings lead us to consider whether any differences in shape exist between adult Aleutians and Alaskans. If there is any evidence of differences, are these due to differences in growth allometry or are they present at birth and simply continued into adulthood?

Shape

In this chapter we focus on the analysis of form variations using landmark co-ordinates and use statistical tools from geometric morphometrics (Bookstein 1991; Marcus *et al.* 1996). Classically, however, landmarks form the basis of analysis of variations in form through the taking of interlandmark distances ('ilds'). This is because ilds are independent of location and rotation of the forms under comparison and they are very easy to acquire using calipers. Furthermore ilds in themselves often conform to the biological notion of a 'character' or feature of interest.

Multivariate morphometric methods (Sneath and Sokal 1973; Mardia *et al.* 1979) allow relationships amongst specimens to be examined on the basis of several ilds simultaneously. If sufficient ilds ($k(k-1)/2$, for k landmarks) are taken or fewer ilds are taken in a systematic way (e.g. in the form of a truss; Bookstein *et al.* 1985), then it is possible to generate the original landmark coordinates from the matrix of ilds through multidimensional scaling (Mardia *et al.* 1979). In turn, this allows the visualisation of the results of an analysis in terms of co-ordinate representations. Rao and Suryawanshi (1996) have recently considered appropriate approaches to the multivariate analysis of form variations using such sets of ilds. These include principal components and canonical analysis.

An alternative set of approaches to the analysis of form variations through the use of ilds has been developed by Lele (1993). These approaches are collectively known as Euclidean distance matrix analysis (EDMA). EDMA allows form variation to be examined through the comparisons of ratios of pairs of equivalent ilds between specimens. It results in large matrices of ild ratios that can be turned both to the identification of landmarks which appear to differ significantly in relative location between forms and to analyses of growth (Richtsmeier *et al.* 1993). This identification depends on the careful examination of often very large matrices of form differences, growth differences etc. and visualisations of such differences can be achieved using multidimensional scaling to generate landmark coordinates of interesting forms.

In examining ilds rather than co-ordinates, issues concerning registration are to some degree sidestepped; however, visualisation and interpretation of results are somewhat more difficult and issues arise relating to the estimation of means, scaling and the morphometric space in which statistical inference is to be undertaken. Lele (1993) argued strongly in favour of EDMA and against registration based approaches such as those used in this study. Many other statisticians and biometricians place confidence in registration-based approaches (see e.g. Bookstein 1978, 1987; Marcus *et al.* 1996; Dryden and Mardia 1998).

In this chapter we focus on methods for the direct analysis of landmark co-ordinates because these directly address geometry, are the best understood at present and form the focus of much current interest (e.g. Marcus *et al.* 1996; Dryden and Mardia 1998).

Geometric morphometrics

Approaches to analysis based on landmarks are fundamentally different from those using ilds in that differences in co-ordinate values due to location and orientation alone (registration) need to be factored out of the comparison. We aim to preserve geometric information throughout the analysis and the class of approaches known as geometric morphometrics or statistical shape analysis is appropriate for this purpose (Rohlf and Bookstein 1990; Marcus *et al.* 1996; Dryden and Mardia 1998).

The task of describing relative landmark movements has proved intractable until recently. In the space of the original specimens (the real world) all landmarks will appear to 'move away' from the reference points chosen for the superimposition and so different registrations will appear to indicate different patterns of growth. In terms of multivariate statistical analysis of registered co-ordinate data, the particular patterns of variation represented by a particular principal component or canonical axis will be entirely dependent on the way in which operational taxonomic units (OTUs) have been registered with respect to each other.

Thus, the perceived displacement of any particular landmark from one shape or another depends upon the way in which OTUs are scaled, reflected, rotated and translated with respect to each other. Different registrations will produce different impressions of the shape transformations and regions close to the registration points will appear to change less than those more distant. These difficulties are most significant when the shapes under comparison are very different and unimportant when they are very similar. The important issue is therefore not one of choice of registration method but rather of the magnitude of differences in shape. When variations are small, the effects of registration method are also small. Dryden and Mardia (1998, p. 287) give a tentative suggestion that 'if the data lie within full Procrustes distance of about 0.2 of an average shape then methods give very similar conclusions'.

Procrustes registration

Given small variations in shape we seek a registration method that is sensible in terms of biology and well understood statistically. Unless we have an *a priori* basis for selecting a particular fixed baseline, and in our example facial data we do not, it is reasonable to register forms on the basis of a 'best fit' of all landmarks. The methods of Procustes analysis (reviewed by Rohlf and Slice 1990; Dryden and Mardia 1998) register forms by translating, rotating, reflecting and scaling forms with respect to each other to maximise fit.

In Fig. 6.3a we illustrate the results of Procrustes registration of the co-ordinate data from adult Aleutian and Alaskan crania included in the example study. It is difficult to appreciate the full three-dimensional geometry of the resulting landmark clusters but it is clear from Fig. 6.3a that the scatters of registered landmarks are fairly small. Figure 6.3b illustrates a wireframe model drawn between mean coordinates. It gives a clearer impression of the three-dimensional nature of our data and it is drawn such that approximate boundaries of the palate, maxilla, frontal and zygomatic bones are indicated.

Registered sets of coordinates can be interpreted visually but it is usually desirable to undertake statistical analyses and model shape variability in an abstract 'shape space'. The shape space for Procrustes registered data is non-linear and statistical analysis needs to account for this.

Statistical analysis of Procrustes registered data

When figures (described by k landmarks in m dimensions) are scaled (centroid size = 1) and registered to remove translational and rotational differences by generalised least squares superimposition (generalised Procrustes analysis; Gower 1975; Rohlf and Slice 1990; Goodall 1991) they can be represented as points in a shape space which is of $km - m - m(m-1)/2 - 1$ dimensions ($= km - 7$ when $m = 3$; $= km - 4$ when $m = 2$). This dimensionality arises because location (m dimensions), rotation ($m(m-1)/2$ dimensions) and scale (1 dimension) differences have been removed.

This space was first described by Kendall (1984) and it is commonly referred to as Kendall's shape space. We have already noted that the relative locations of points representing specimens in this space are more or less independent of registration if variations are small. Additionally and importantly from a statistical perspective, isotropic distributions of landmarks about the mean results in an isotropic distribution of points representing specimens in the shape space. Kendall's shape space is, however, non-Euclidean (non-linear). For the most simple shapes, populations of triangles, the space can be visualised as being spherical but for more than three landmarks the space is much more complex, being high dimensional (Le and Kendall 1993).

Since the shape space is non-linear, great care is needed in carrying out statistical analyses. One approach that is particularly appealing, since it naturally allows the study of multivariate allometry, is to carry out principal components analysis (PCA) in the tangent space to Kendall's shape space (Dryden and Mardia 1993; Kent 1994). For triangles we take the scatter of points on the spherical shape space representing variation within our sample and project it into a Euclidean tangent plane in exactly the same way as a cartographer might project a map from a globe onto a flat sheet of paper. The co-ordinates of the points representing specimens are no longer given in terms of the sphere but rather as co-ordinates in the plane. As long as the projection has not resulted in excess distortion (as might occur if the projection encompasses a large proportion of the sphere) we can carry out useful analyses in this plane. For higher dimensions the tangent plane to the shape sphere can be imagined as a tangent space of $km - m - m(m-1)/2 - 1$ dimensions.

Procrustes tangent co-ordinates can be estimated using the Procrustes tangent space projection given by Dryden and Mardia (1993). This projection results in a $(k-1)m$ vector of tangent space shape co-ordinates with respect to the mean for each specimen. Both of these vectors of tangent space co-ordinates are of rank $km - m - m(m-1)/2 - 1$. Principal components analysis can be carried out using tangent space co-ordinates to extract $km - m - m(m-1)/2 - 1$ eigenvectors; the principal components of variation of shape. In the case of a growth study we expect that the first few principal components will serve as an adequate model of allometry. Note that since Procrustes analysis involves scaling to centroid size the variations we examine through PCA are shape rather than form variations. If we wish to examine the relationship between size and shape (allometry) we can do this by examining plots and correlations of principal components (PC) scores vs. centroid size for the significant principal components.

Visualisation of patterns of shape variation

Variation in the shape space

Since the PCs are mutually orthogonal they each represent statistically independent modes of variation in the sample. Further interpretation of the PCs depends in part on visualisation of the shape variation represented by each. A graphical representation of shape variation along each axis can be achieved by reconstructing hypothetical specimens with scores of 0 on all PCs except the PC of interest. By inspecting a range of scores on this PC it is possible to visualise the variability along it through series of reconstructed forms or as an animation.

Transformation grids

An alternative strategy for comparing co-ordinate representations of form is to represent differences in a single diagram as a deformation that smoothly rearranges the configuration of landmarks as a whole. The best known representation of such a deformation is in the form of a ‘Cartesian transformation grid' (Thompson 1917) in which differences in morphology are described through distortions of a regular grid.

An appropriate approach to drawing transformation grids uses mathematical functions, known as thin plate splines (TPS: Bookstein 1989; Marcus *et al.* 1996; Dryden and Mardia 1998). The grids derived from TPS indicate how the space (or a regular Cartesian grid) in the vicinity of a reference figure might be deformed into that surrounding a target such that landmarks in the reference map exactly into those of the target. The thin plate spline ensures that this deformation involves minimum bending; and it is chosen for this purpose, since this seems a sensible minimisation criterion. The statistical and graphic models of shape transformations which result from these approaches are readily interpretable and highly visual (e.g. Bookstein 1978, 1989; O'Higgins and Dryden 1992, 1993; Marcus *et al.* 1996). The thin plate spline is not the only possible choice of method for drawing a grid but the fact that it minimises 'bending energy' is intuitively appealing. Besides producing a transformation grid thin plate splines can be extended to examine the affine and non-affine components of shape difference and to explore variation at different scales (localised variations vs. global) amongst OTUs. These refinements are beyond the scope of this chapter but full accounts have been given by Marcus and colleagues (1996) and Dryden and Mardia (1998).

Geometric morphometric analysis of Aleutian and Alaskan craniofacial variation

We illustrate the approaches described above by comparing adult morphology and postnatal growth in the facial skeletons of Aleutians and Alaskans. The first hypothesis (H1) we test is that adults of the two populations do not differ in facial morphology.

Differences between adult Aleutian and Alaskan samples

Figure 6.3a presents the results of Procrustes analysis of adult Aleutians and Alaskans. In this diagram, each point represents the location of a landmark on a specimen. The scatters of points indicate variability at each landmark and this appears to be small. The mean of the landmark configurations is presented in Fig. 6.3b, landmarks are joined by lines indicating approximate boundaries of the frontal, zygomatic, maxilla and palate.

Principal components analysis of the tangent coordinates results in the

(a)

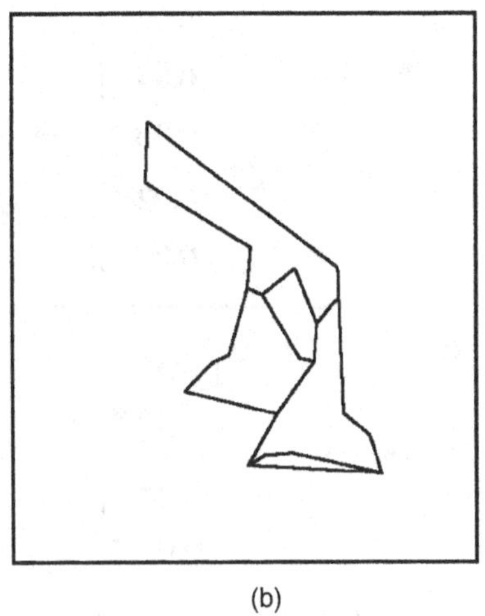

(b)

Fig. 6.3. Left frame (a), scatter of points representing adult facial shapes after generalised Procrustes registration. Right frame (b), the mean of all adult faces with landmarks connected by a wire frame approximately outlining facial bones.

scatter of specimens on PCs I and II presented in Fig. 6.4. In this plot, the Alaskans are completely separated from Aleutians such that Alaskans occupy the upper right-hand side of the plot and Aleutians, the lower left. This leads us to consider that the differences between adults might be significant, although small sample size in relation to the rank of the shape space does not allow conventional statistical testing of these differences (such as might be carried out using Hotelling's T^2). An alternative approach is through a permutation test (Good 1993) in which the true difference between adult means is compared with the range of differences between the means of many randomly permuted samples drawn from the same data. In total we have only 20 specimens, so the number of permutations we can draw is rather small; however, different runs of the permutation test over 100 permutations indicate that the adult means are significantly different at a level of $p < 0.01$. H1 is falsified.

Having established that the adult means are different we turn to an examination of the nature of these differences. In Fig. 6.5 configurations of landmarks in lateral view are represented as rendered images constructed through triangulations of landmarks. Figure 6.5a represents the Alaskan mean and Fig. 6.5c the Aleutian mean. In Fig. 6.5b the difference between these means in represented by the transformation grid calculated using a three-dimensional thin plate spline. This is drawn in a single plane just to the right of the midline since it is in this plane and in this region that the

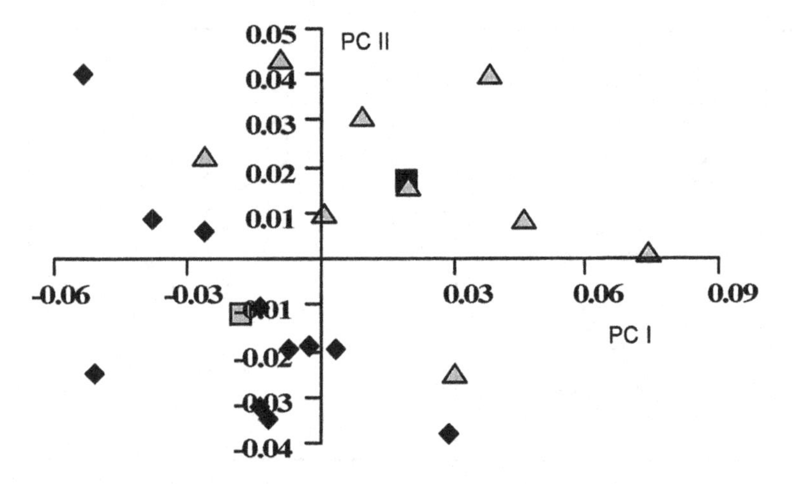

Fig. 6.4. Principal components analysis of adult data after generalised Procrustes registration. Triangles, adult Alaskans; black rectangle, mean adult Alaskan; diamonds, adult Aleutians; grey rectangle, mean adult Aleutian. Horizontal scale, PC I 24% total variance; vertical scale, PC II 16% total variance.

differences appear greatest in magnitude. The grid was regular rectangular over the Alaskan mean (the reference shape) and its deformation to the Aleutian mean (target shape) is multiplied by a factor of 2 for ease of interpretation. It can be seen that in this plane, the principal difference between Aleutian and Alaskan means consists of a relative midfacial and nasal projection in Aleutians with respect to Alaskans. The transformation grid showed little distortion in the coronal and transverse planes other than that already noted in the vertical. We conclude therefore that Alaskan and Aleutian adult faces differ significantly and that these differences consist principally of a more prominent prognathism of the midface and projection of the nasal region in Aleutians with respect to Alaskans.

The ontogenetic basis of differences between Alaskans and Aleutians

Despite a relatively short period of isolation of 3000 years or so the adults of these two populations present different facial morphologies. It is of interest to investigate the ontogenetic basis of these differences since the results will reflect on evolutionary and adaptive mechanisms in the face. Several possible explanations exist for the differences we encounter between adults;

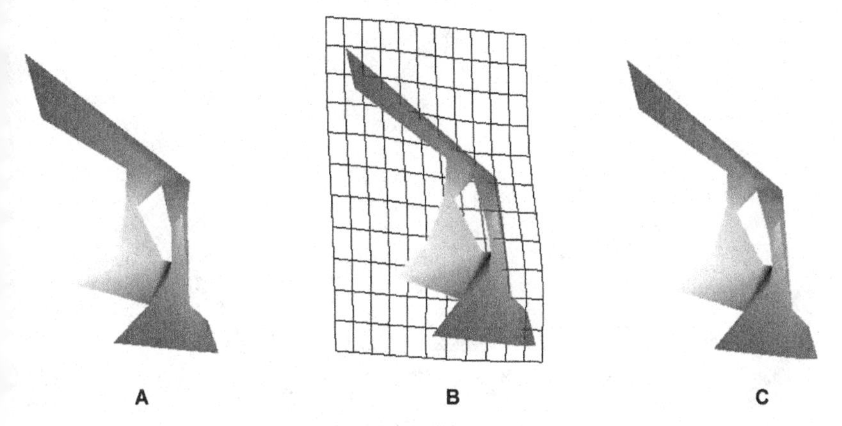

Fig. 6.5. The contrast between Alaskan and Aleutian adult means. (a) Alaskan mean; (c) Aleutian mean; (b) Cartesian transformation grid illustrating the difference between Alaskan (reference) and Aleutian (target) means. The deformation between Alaskan and Aleutian is emphasised for the purposes of drawing the grid by multiplying the transformation by a factor of 2. Permutation tests give a significance of $p < 0.01$ for this difference in means.

Differences may be present at birth and these may persist into adulthood, postnatal growth allometries being identical between the populations.
Populations may be identical at birth and diverge through different growth allometries.
Populations may be identical at birth and diverge through relative extension/truncation of a common growth trajectory.
Some combination of the above might operate in concert.

This study sets out to test the hypothesis (H2) that different growth allometries exist between the populations. If this is falsified we can examine the nature of differences between populations at birth and the extent to which growth is relatively extended or truncated between the populations.

We begin the study of growth changes in facial shape by undertaking a Procrustes analysis of the whole data set: infants, juveniles and adults of both populations. The resulting deviations of co-ordinates from the Procrustes mean are then submitted to PCA. The first PC from this analysis accounts for 47.5% of the total variance, the second for 6.2% and the third 5.8%. The first 20 PCs together account for >90% of the total shape variance and there are ($km - 7 =$) 71 non-zero eigenvectors in total. This means that the first, second and third PCs in combination account for 60% of the total shape variance and so can be used to examine the major features of shape variability in the 71-dimensional shape space. Examin-

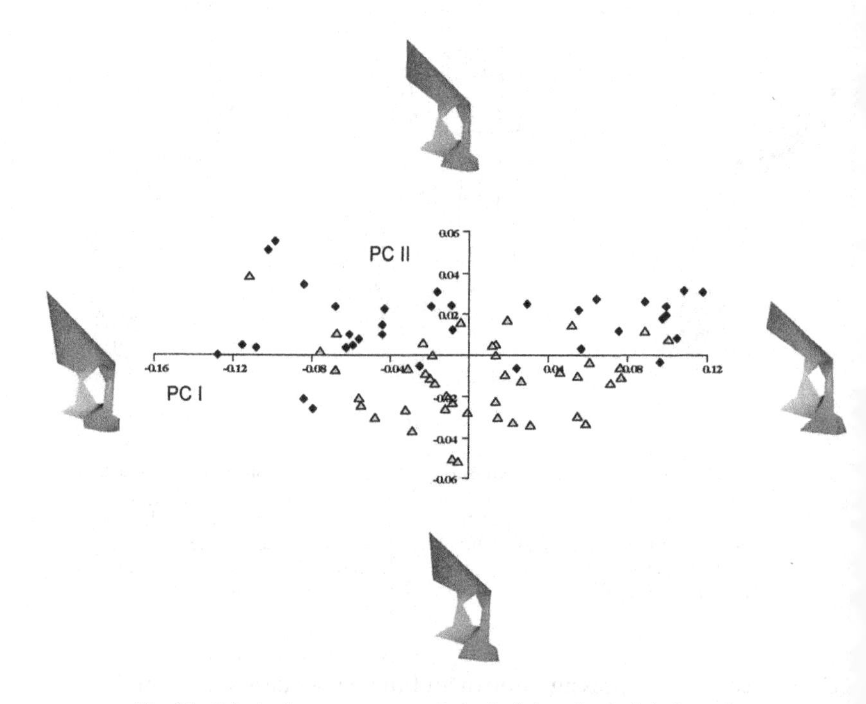

Fig. 6.6. Principal components analysis of adult and subadult data after generalised Procrustes registration. PC I (horizontal axis) vs. PC II (vertical axis). Triangles, Alaskans; diamonds, Aleutians. Small inset figures illustrate shapes represented by scores of ± 0.12 on PC I and ± 0.05 on PC II.

ation of PCs of order >3 showed no biologically interesting variation and so we focus on the first three only.

Figure 6.6 presents a plot of the first two PCs. At the extremes of each PC is drawn a rendered reconstruction of the mean configuration after transformation along it. The first PC does not separate Alaskans from Aleutians, rather the reconstructions at its extremes and the ordering of specimens suggests that it represents size-related shape variation during growth (ontogenetic allometry). The reconstructions indicate that it represents a mode of variation in which specimens with low scores are orthognathic, with relatively large orbits, and those at the other extreme are prognathic, with relatively small orbits.

In order to investigate more deeply the biological correlates of variation along the first PC we examine the relationship between scores on this PC and centroid size. Figure 6.7 shows a plot of PC I against centroid size (units are cm) and it seems from this that the two are strongly related. The correlation between scores on PC I and size is 0.91 and this is significant at the $p < 0.001$ level. PC I can therefore be said to represent a mode of facial

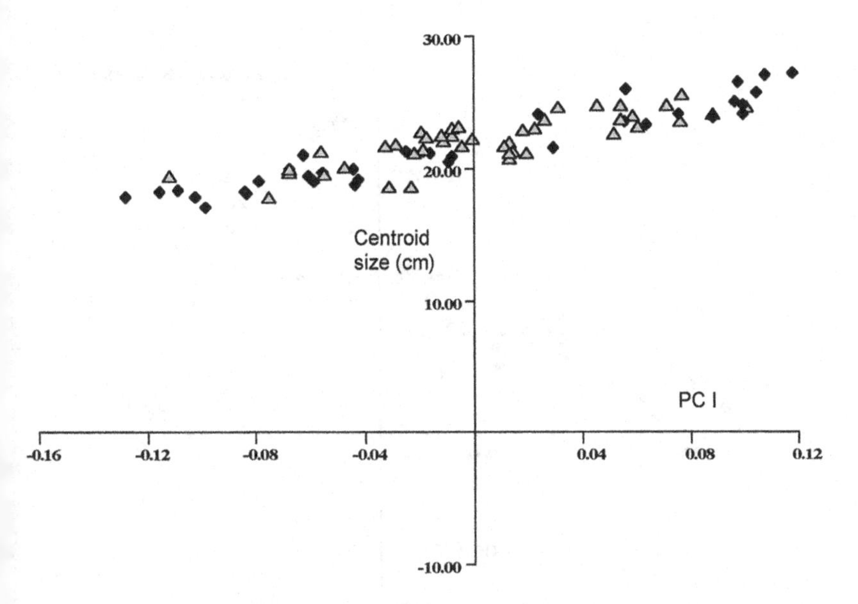

Fig. 6.7. PC I (horizontal axis) vs. centroid size (vertical axis) from the analysis of adult and subadult crania. (See Fig. 6.6 for key.) $r = 0.91$, $p < 0.001$.

shape variation that is strongly related to cranial size variation during growth. As such it is a good model of ontogenetic allometry. It is noteworthy that this relationship appears identical in both populations: there is no evidence of differences in the mean or gradient. No other PC shows evidence of a size-related shape change so we conclude that the allometric relationship observed on PC I is identical for both populations.

One subtle difference is, however, possible in that some adult Aleutians have higher scores on PC I than Alaskans, suggesting that Aleutians may extend the common allometry into larger size ranges. The difference in mean adult scores on PC I is, however, not quite significant ($p = 0.055$).

Of related interest is the relationship between relative dental age and scores on PC I. These are plotted in Fig. 6.8, the vertical axis represents mean dental age in years and the horizontal represents scores on PC I. This plot indicates a relationship between age and the mode of shape variation represented by the first PC. The correlation between these variables, $r = 0.90$, is highly significant, $p < 0.001$. This is not surprising, given that we have already noted a highly significant relationship between age and size. It is noteworthy, however, that age is less strongly correlated with shape than is size. This might be expected, since size and shape are biologically interwoven through growth phenomena while age is simply the temporal axis within which the biological processes occur.

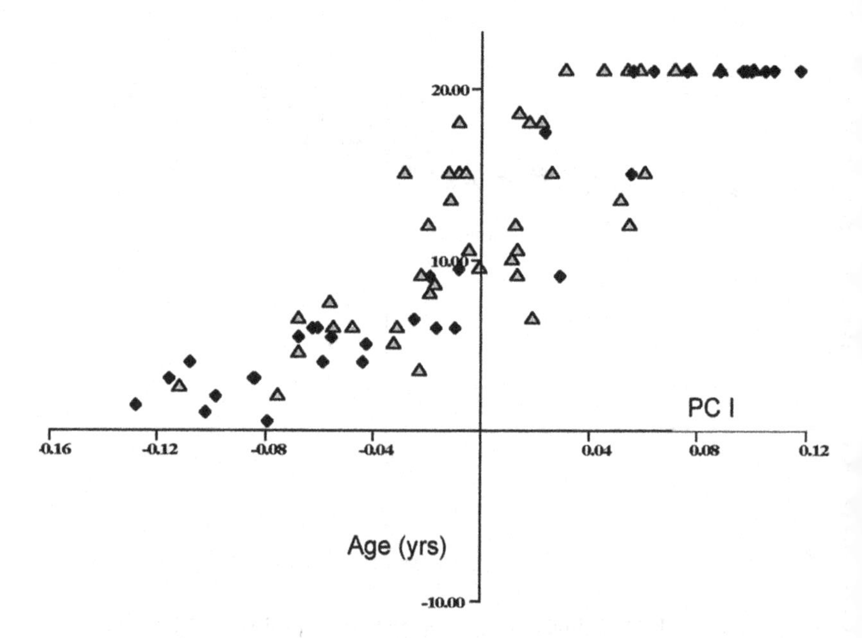

Fig. 6.8. PC I (horizontal axis) vs. relative dental age in years (vertical axis) from the analysis of adult and subadult crania (adults all nominally allocated age 21). (See Fig. 6.6 for key.) $r = 0.90$, $p < 0.001$.

We conclude that the data fail to falsify the hypothesis (H2) that different growth allometries exist between the populations. There is, however, a suggestion that Aleutians extend this common allometry (hypermorphosis) beyond Alaskans but this just fails to achieve statistical significance.

Having demonstrated a common ontogenetic allometry for our study populations it is of interest to examine its nature. We examine the geometric aspects of allometric growth by drawing (Fig. 6.9) transformation grids between small/young (PC I score -0.12; reference shape) and large/old (PC I score $+0.12$; target shape) specimens.

In Fig. 6.9a a Cartesian transformation grid is drawn in a coronal plane such that it passes through the lateral maxilla and just within the orbit of a smooth-rendered representation of the transformed mean at PC I score 0.12 (mean of oldest and largest specimens). The grid is expanded (principally laterally but also vertically) and curved upwards over the zygomatic part of the maxilla. This indicates that during growth this region undergoes substantial relative increase in size and relative lateral displacement.

In Fig. 6.9b the target form is shown in lateral view and a grid is drawn passing through the middle of the orbit. Its deformation confirms that relative vertical expansion of the maxilla is a feature of growth. The grid is

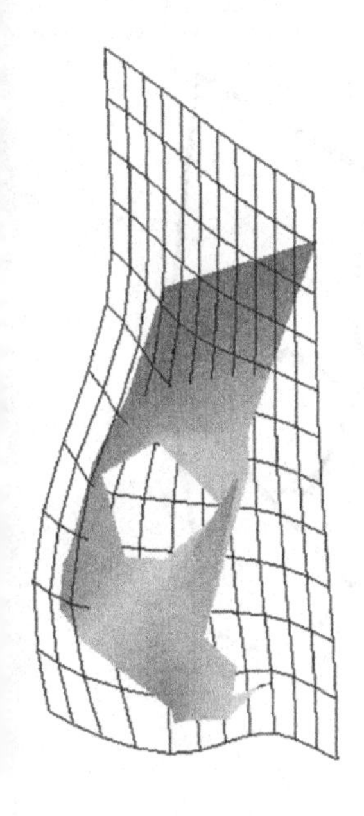

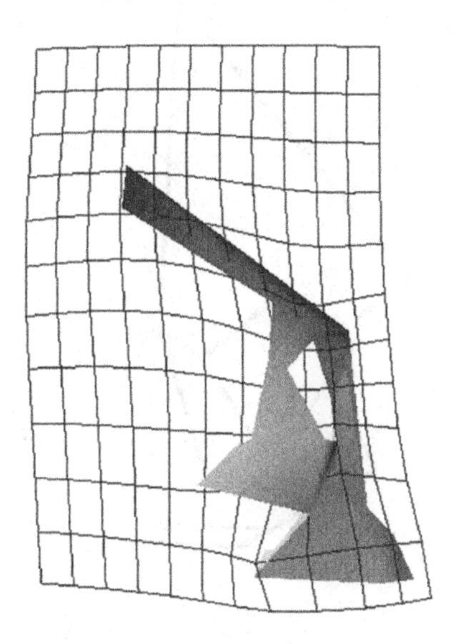

Fig. 6.9. Cartesian transformation grids illustrating the shape changes during growth of both populations (PC I score − 0.12 to + 0.12 from the analysis of adult and subadult data after generalised Procrustes registration). Frontal view, A; lateral view, B. Note the relative lateral expansion of the zygomatic region and the relative vertical and horizontal expansion of the maxilla during growth.

also expanded horizontally along the alveolar part of the maxilla, indicating that a feature of growth is relative lengthening of the tooth-bearing part of the maxilla. This is confirmed by more pronounced similar deformations observed in the region of the maxillary alveolus in more medially sited grids (not drawn because of space limitations). This maxillary alveolar expansion accommodates the dentition and results in a moderate increase in maxillary prognathism and greater relative posterior positioning of the posterior limit of the maxilla.

A wireframe model delimiting the principal bones of the face is drawn in Fig. 6.10. Figure 6.10a represents the 'small' or 'young' reference (PC I score −0.12) and 9.10b, the 'large' or 'old' target (PC I score +0.12). The principal difference between these figures is that the face of the older individuals is relatively much larger than the younger. This can be seen to be due, in the main, to relative maxillary expansion in all directions, resulting in more lateral siting of an expanded zygomatic. The zygomatico-maxillary suture is relatively increased in length during growth. The palate is increased and the orbit and frontal are decreased in relative size.

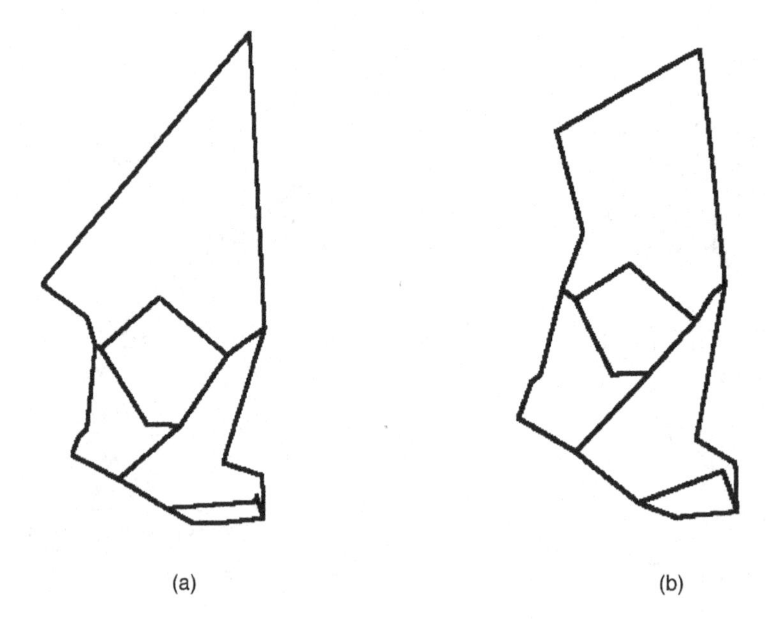

Fig. 6.10. Proportional changes in the skeletal elements of the face during growth illustrated by wireframe model approximately delimiting bone boundaries. (a) Shape represented by score of − 0.12 on PC I; (b) shape represented by score of + 0.12 on PC I.

These analyses lead us to conclude that there is no evidence of different growth allometries between the Alaskans and Aleutians; we have failed to falsify our hypothesis, H2. This common allometry might be extended (hypermorphosis) in Aleutians but our statistical findings are equivocal so we set this possibility aside. In the main, allometric growth in both populations features a relative lateral, vertical and horizontal expansion of the face, especially of the maxilla. The result is that the frontal and orbital regions show a relative decrease in size.

The findings indicate that the principal distinctions in facial morphology between Alaskans and Aleutians are present very early and probably at birth. In our PCA we therefore expect to be able to differentiate Alaskans and Aleutians irrespective of age. Since the first PC represents an allometric growth vector we expect these differences to be present on higher-order PCs (PC II and above).

PC II is plotted against PC I in Fig. 6.6 where it separates Alaskans from Aleutians to some degree. The small-rendered reconstructions drawn at the limits of PC II indicate the transformed mean at the extremes of this component. They show differences that are small in comparison to those on PC I. This finding is consistent with the eigenvalue for PC II, which

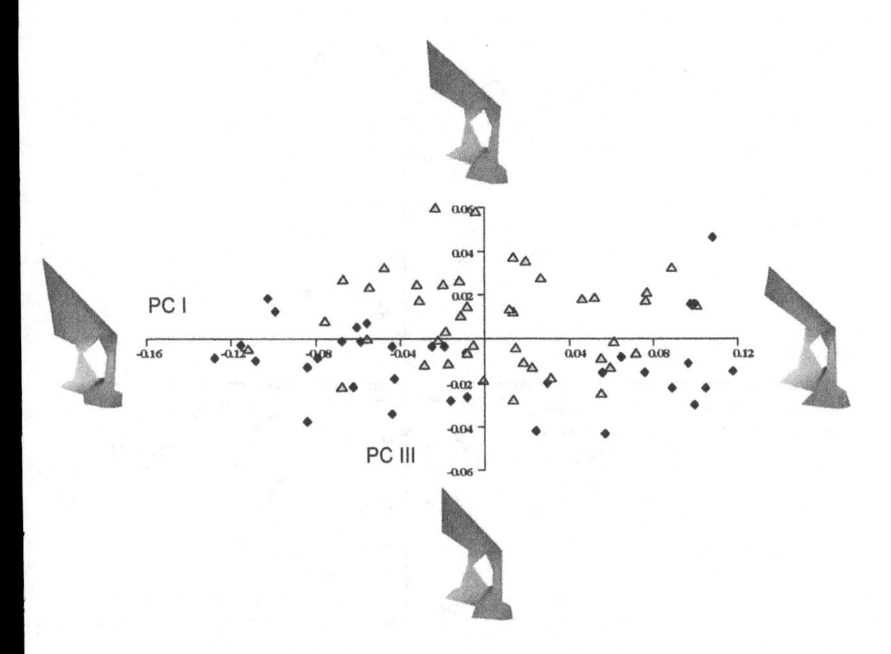

Fig. 6.11. Principal components analysis of adult and subadult data after generalised Procrustes registration. PC I (horizontal axis) vs. PC III (vertical axis). Triangles, Alaskans; diamonds, Aleutians. Small inset figures illustrate shapes represented by scores of $\pm$ 0.12 on PC I and $\pm$ 0.05 on PC III; 5.8% total variance.

indicates that it accounts for only 6.2% of the total variance while PC I accounts for 47.5%. PC II also contrasts with PC I in partially differentiating Alaskans (low scores) from Aleutians (large scores). This differentiation is, however, incomplete in that populations overlap.

Further differences between Alaskans and Aleutians are represented by PC III (Fig. 6.11). The transformed means reconstructed at the extremes of this axis again indicate that differences on this PC are small relative to those occurring during growth. This is consistent with the relative values of the eigenvectors for these components: PC III 5.8% of total variation, PC I 47.5%. PC III, like PC II offers some separation, with overlap, between Alaskans (high scores) and Aleutians (low scores).

No other PC shows clear evidence of separating Alaskans and Aleutians. The contrast between populations is clear when PCs II and III are plotted (Fig. 6.12). Triangles representing Alaskans occupy the upper left of this diagram while diamonds representing Aleutians occupy the lower right, and there is little overlap. The one exception is the Alaskan at PC II $\approx$ 0.035 PC III $\approx$ 0.005, which sits squarely with Aleutians. This is probably explained by distortions in this very young, fragile, specimen. The

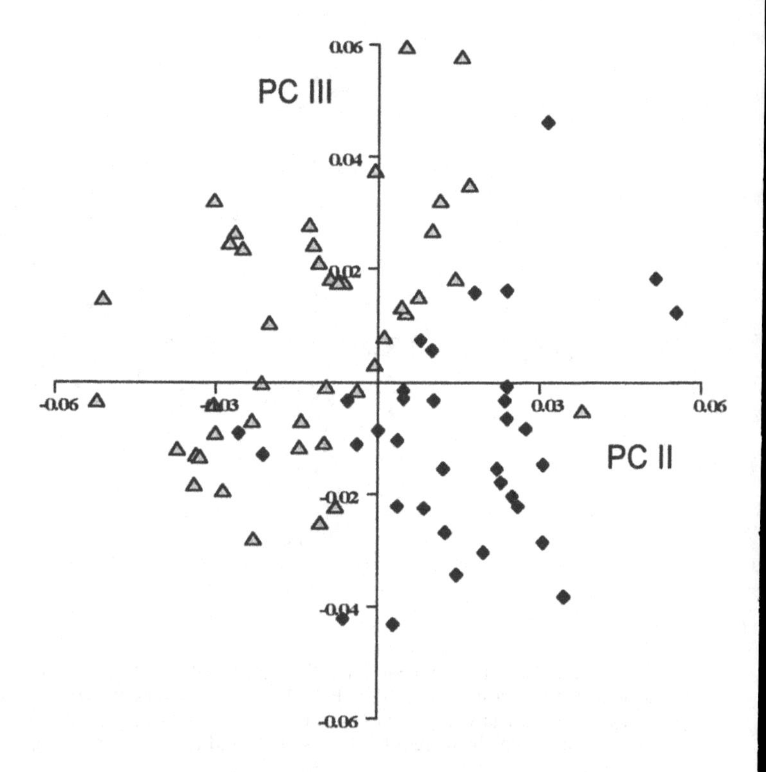

Fig. 6.12. Principal components analysis of adult and subadult data after generalised Procrustes registration. PC II (horizontal axis) vs. PC III (vertical axis). Triangles, Alaskans; diamonds, Aleutians.

differences evident between the populations on PCs II and III are independent of the growth allometry modelled by PC I.

The plot of PCs II and III is reminiscent of that of the first two PCs from the analysis of adults (Fig. 6.4) except that scores of specimens are reflected with respect to the horizontal axis. In both, Alaskans and Aleutians occupy opposite semicircles of the scatter and are nearly completely separated. It is of interest, therefore, to compare this age-independent difference in facial shape with the difference found between adults in the earlier analysis. In Fig. 6.13 the difference at all ages is visualised by comparing rendered transformed means. That in Fig. 6.13a represents the mean Alaskan on PCs II and III and that in Fig. 6.13c, the mean Aleutian. In Fig. 6.13b a transformation grid between these two reconstructions is drawn in a plane which just passes to the right of the midline, since it is in this plane and in this region that the differences are greatest in magnitude. This grid is

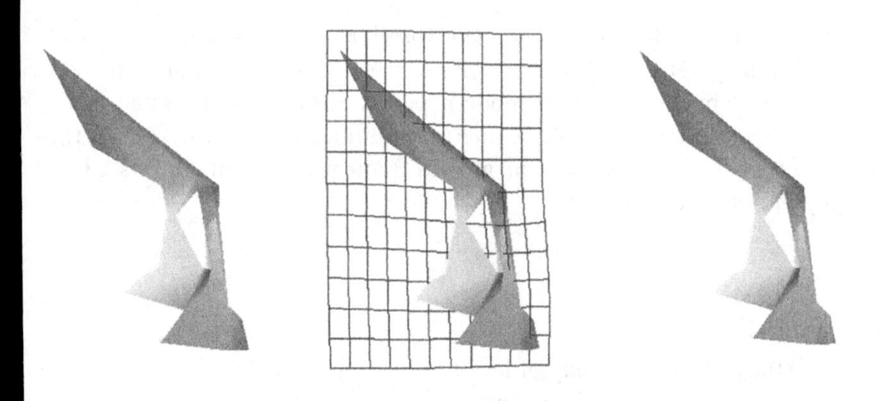

Fig. 6.13. Differences between Alaskan and Aleutians on PCs II and III in the principal components analysis of adult and subadult data after generalised Procrustes registration. (a) Representation of mean Alaskan on PCs II and III; (b) Mean Aleutian on PCs II and III; (c) Cartesian transformation grid illustrating the difference between Alaskan (reference) and Aleutian (target) means on PCs II and III. The deformation between Alaskan and Aleutian is emphasised for the purposes of drawing the grid by multiplying the transformation by a factor of 2.

multiplied by a factor of 2 to aid interpretation. It indicates that the age-independent difference between Alaskans and Aleutians is small but consists in the main of a greater anterior projection of the midface and nasal region in Aleutians with respect to Alaskans. The transformation grid showed little distortion in the coronal and transverse planes.

Figure 6.13 is very similar to Fig. 6.5, in which adult means alone are compared. The differences between adult population means are very similar to the age-independent difference found between populations at all ages. The degree of difference between adults is, however, greater and the difference includes some relative upward bending of the grid to the bottom left (Fig. 6.5b vs. Fig. 6.13b). This is probably explained by the omission of small differences represented by PCs IV–LXXI in the reproduction of Fig. 6.13b. The overwhelming similarity between Fig. 6.5b and 6.13b, together with the similarities found in all other planes examined, lead us to conclude that the differences in shape found between adults are largely present postnatally and are continued into adulthood.

Thus Alaskans and Aleutians share a common growth allometry that serves to preserve into adulthood differences in facial morphology present at birth. This finding contrasts with the differences found amongst other populations in as yet unpublished studies being undertaken by one of us

(U.S.V.). It should not be assumed, therefore, that differences in facial morphology between modern human populations are generally fully expressed at birth or that all modern populations share the same facial growth allometry. It will be of interest in future analyses to examine the extent to which differences in growth allometries might be related to population divergence.

Discussion

The study of the ontogeny of facial form we present in this chapter has served to demonstrate the potential of the techniques of geometric morphometrics in the study of three-dimensional growth changes. These methods result in highly visual representations of shape differences and allometric growth models. Additionally conventional statistical analyses are possible in the tangent space.

Our findings are, first, that the adult crania of Alaskans and Aleutians differ in morphology; H1 is falsified. Secondly, that the populations share a common facial growth allometry; H2 is falsified. This leads to the third finding that a substantial part of the differences in shape present between adults can be observed between even the youngest individuals.

Differences in midfacial prognathism and nasal projection are present by the end of the first postnatal year (we have no earlier data) and are continued into adulthood. Growth contributes little to further accentuation of these differences. Our data suggest that Aleutians extend the common growth allometry relative to Alaskans but this finding just fails to achieve statistical significance. More data are needed to confirm or deny this.

It seems likely that the influences on postnatal facial growth (see Introduction) are identical between these closely related populations. However, before the first postnatal year, a fundamental difference in facial morphology is established. The nature of this difference is such that it does not modify subsequent growth allometry between populations and this implies that growth trajectories are to some degree independent of, or can compensate for, initial form. The limits of this independence between growth allometry and form at the end of the first postnatal year need to be explored through growth studies of more divergent populations and different species. The findings of such studies will cast further light on the mechanisms regulating growth allometry.

The finding of a difference in facial morphology between these populations that is independent of age opens up the possibility of generating an age-independent discriminant function to enable forensic identification of subadult material. This is an interesting finding that we intend to pursue with respect to other human populations, since the practical application to forensics is of great potential value.

This example study has served to indicate how facial growth might be readily compared using tools from the geometric morphometric toolkit. These tools are relatively new and advances in statistical understanding and computer graphics will inevitably open up exciting new possibilities in the future. An important new horizon lies in combining these tools with modern imaging modalities in order to allow analysis of images from computed tomography (CT) (Spoor and Zonneveld 1995; Spoor 1997) and magnetic resonance imaging (MRI). It should be soon possible to undertake, with reasonable effort, studies where internal morphology as well as external contributes to models of growth and where the three-dimensional visualisations and animations are of the level of quality we have come to expect from modern CT and MRI.

In particular, the methods we have outlined here offer considerable potential in understanding the ontogenetic basis of morphological variation and its relationship to evolutionary adaptation and divergence. Insights into the means by which modern taxa become different through growth might well prove of value in the interpretation of evolutionary divergence. Future studies will indicate the extent to which such growth variations are useful in understanding adaptation and phylogeny. Thus, interpretation of the significance of growth variations in the past depends to a great degree on knowledge of variations in the present.

These technologies are, of course, applicable in studies of the morphology of any anatomical region. In applying such technologies, however, it is important that a keen eye is kept on the biological issues at hand. It is easy to fall into the trap of producing visually appealing analyses devoid of biological hypothesis testing simply because the technology is there. The difficulty for anthropologists of the future will lie not in technical issues of analysis but rather in devising testable hypotheses of biological merit.

Acknowledgements

This work is entirely dependent on the considerable programming efforts of Nicholas Jones, who made it possible to carry out the analyses and generate images in a relatively straightforward way. Professor Christoper Stringer of the Natural History Museum, London, has been pivotal in enabling and supporting U.S.V. in these studies. The programming and the studies described in this chapter could not have been undertaken without the support of our statistical colleagues – Professor Kanti Mardia, Dr Ian Dryden and Professor John Kent of the Department of Statistics, University of Leeds. Others have also contributed valuable advice. These include Dr Fred Bookstein (Michigan), Dr Les Marcus (New York) Professor Jim Rohlf (New York) and Dr Subhash Lele. Professor Christoper Dean and Dr Fred Spoor have been supportive and stimulating in discussions leading to this work.

We must also thank the curators of specimens used in this study: Dr Ian Tattersall, the American Museum of Natural History, New York; Dr David Hung, National Museum of Natural History, Smithsonian Institution, Washington DC; Professor Chris Stringer, The Natural History Museum, London. U.S.V.'s role in this work was supported by the University of London, the UCL Graduate School and the CVCP.

References

Björk A (1966) Sutural growth of the upper face, studied by the implant method. *Acta Odontologica Scandinavica* **24**, 109–127.

Björk A (1968) The use of metallic implants in the study of facial growth in children: method and application. *American Journal of Physical Anthropology* **29**, 243–254.

Bookstein FL (1978) *The Measurement of Biological Shape and Shape Change.* Lecture Notes in Biomathematics no. **24**. New York: Springer-Verlag.

Bookstein FL (1987) On the cephalometrics of skeletal change. *American Journal of Orthodontics* **83**, 177–182.

Bookstein FL (1989) Principal warps: thin-plate splines and the decomposition of deformations. IEEE. *Transactions in Pattern Analysis Machine Intelligence* **11**, 567–585.

Bookstein FL (1991) *Morphometric Tools for Landmark Data: Geometry and Biology.* Cambridge: Cambridge University Press.

Bookstein FL, Chernoff B, Elder R, Humphries J, Smith G and Strauss R (1985) *Morphometrics in Evolutionary Biology. The Geometry of Size and Shape Change with Examples from Fishes.* Philadelphia: Academy of Natural Sciences of Philadelphia.

Broadbent BH (1996) Cephalometrics. In *Essentials of Facial Growth,* ed. D.H. Enlow and M.G. Hans, pp. 241–264. Philadelphia: WB Saunders Co.

Bromage TG (1986) A comparative scanning electron microscope study of facial growth and remodelling in early hominids. PhD thesis, University of Toronto.

Dryden IL and Mardia KV (1993) Multivariate shape analysis. *Sankya* **55(A)**, 460–480.

Dryden IL and Mardia KV (1998). *Statistical Shape Analysis.* London: John Wiley and Sons.

Enlow DH (1968) *The Human Face: An Account of the Postnatal Growth and Development of the Craniofacial Skeleton.* New York: Harper and Row Publ. Inc.

Enlow DH (1975) *Handbook of Facial Growth.* Toronto: WB Saunders Co.

Enlow DH and Hans MG (1996) *Essentials of Facial Growth.* WB Saunders Co.

Ford JA (1959) Eskimo prehistory in the vicinity of Point Barrow, Alaska. *Anthropological Papers of the American Museum of Natural History* **47**(1), New York.

Good P (1993) *Permutation Tests: A Practical Guide to Resampling Methods for Testing Hypotheses.* New York: Springer-Verlag

Goodall CR (1991) Procrustes methods and the statistical analysis of shape (with discussion), *Journal of the Royal Statistical Society* B. **53**, 285–350.

Gower JC (1975) Generalised Procrustes analysis. *Psychometrika* **40**, 33–50.

Hall BK (ed.) (1994) *Homology: The Hierarchical Basis Of Comparative Biology.* San Diego: Academic Press.

Heathcote GM (1986) *Exploratory Human Craniometry of Recent Eskaleutian Regional Groups from the Western Arctic and Sub-Arctic of North America: A New Approach to Population Historical Reconstruction.* BAR International Series 301, Oxford.

Hrdlička A (1945) *The Aleutian and Commander Islands.* Philadelphia: The Wistar Institute of Anatomy and Biology.

Kendall DG (1984) Shape manifolds, Procrustean metrics and complex projective spaces. *Bulletin of the London Mathematical Society* **16**, 81–121.

Kent JT (1994) The complex Bingham distribution and shape analysis. *Journal of the Royal Statistical Society* B **56**, 285–299.

Le H and Kendall DG (1993) The Riemannian structure of Euclidean shape spaces: a novel environment for statistics. *Annals of Statistics* **21**, 1225–1271.

Lele S (1993) Euclidean distance matrix analysis: estimation of mean form and form difference. *Mathematical Geology* **25**, 573–602.

Marcus LF, Corti M, Loy A, Naylor GJP and Slice D (eds.) (1996) *Advances in Morphometrics.* Nato ASI Series. New York: Plenum Press.

Mardia KV, Kent JT and Bibby JM (1979) *Multivariate Analysis.* London: Academic Press.

Martin R (1928) *Lehrbuch der Anthropologie,* vols. 1–3, 2nd edn. Jena: Gustav Fisher.

McLachlan J (1994) *Medical Embryology.* Wokingham: Addison-Wesley

Moore KL and Persaud TVN (1993) *The Developing Human: Clinically Oriented Embryology.* 5th edn. Philadelphia: SB Saunders Co.

Moss ML (1964) Vertical growth of the human face. *American Journal of Orthodontics* **50**, 359–376.

Moss ML and Salentijn L (1969a) The capsular matrix. *American Journal of Orthodontics* **56**, 474–490.

Moss ML and Salentijn L (1969b) The primary role of functional matrices in facial growth. *American Journal of Orthodontics* **55**, 566–577.

O'Higgins P and Dryden IL (1992) Studies of craniofacial growth and development. *Perspectives in Human Biology* **2**/ *Archaeology in Oceania* **27**, 95–104.

O'Higgins P and Dryden IL (1993) Sexual dimorphism in hominoids: further studies of craniofacial shape differences in *Pan*, *Gorilla* and *Pongo*. *Journal of Human Evolution* **24**, 183–205.

Rainey F (1971) *The Ipiutak Culture Excavations at Pt. Hope, Alaska*. Addison-Wesley Modular Publications, no. 8.

Rao CR and Suryawanshi S (1996) Statistical analysis of objects based on landmark data. *Proceedings of the National Academy of Sciences, USA* **93**, 12132–12136.

Richstmeier JT, Cheverud JM, Dahaney SE, Corner BD and Lele S (1993) Sexual dimorphism in the of ontogeny in the crab-eating macaque (*Macaca fascicularis*). *Journal of Human Evolution* **25**, 1–30.

Rohlf F and Bookstein FL (1990) *Proceedings of the Michigan Morphometrics Workshop*. Special Publication no. 2, University of Michigan Museum of Zoology, Ann Arbor.

Rohlf F and Slice DE (1990) Extensions of the Procrustes method for the optimal superimposition of landmarks. *Systematic Zoology* **39**, 40–59.

Scott JH (1953) The cartilage of the nasal septum. *British Dental Journal* **95**, 37–43.

Scott JH (1956) Growth at facial sutures. *American Journal of Orthodontics* **42**, 381–387.

Shea BT (1983) Allometry and heterochrony in the African apes. *American Journal of Physical Anthropology* **62**, 275–289.

Shea BT (1986) Ontogenetic approaches to sexual dimorphism in the anthropoids. *Human Evolution* **1**, 97–110.

Sneath PHA and Sokal RR (1973) *Numerical Taxonomy*. San Francisco: W.H. Freeman & Co.

Sperber GH (1989) *Craniofacial Embryology*, 4th edn. London: Wright.

Spoor CF (1997) Basicranial architecture and relative brain size of Sts 5 (*Australopithecus africanus*) and other Plio-Pleistocene hominids. *South African Journal of Science* **93**, 182–187.

Spoor CF and Zonneveld FW (1995) Morphometry of the primate bony labyrinth: a new method based on high resolution computed tomography. *Journal of Anatomy* **186**, 271–286.

Thilander B (1995) Basic mechanisms in craniofacial growth. *Acta Odontologica Scandinavica* **53**, 144–151.

Thompson D'AW (1917) *On Growth and Form*. Cambridge: Cambridge University Press.

Trevor J (1950) Osteometry. In *Chamber's Encyclopaedia*, vol. I, pp. 458–462. London: George Newnes.

Ubelaker DH (1989) *Human Skeletal Remains, Excavation, Analysis, Interpretation*, 2nd edn. Washington DC: Taxacum.

van Valen L (1982) Homology and causes. *Journal of Morphology* **173**, 305–312.

Wagner GP (1994). Homology and the mechanisms of development. In *Homology: The Hierarchical Basis of Comparative Biology*, ed. B.K. Hall, pp. 274–301. San Diego: Academic Press.

7 *Invisible insults during growth and development:*

contemporary theories and past populations

SARAH E. KING AND STANLEY J. ULIJASZEK

Introduction

It is important to understand patterns of human physical growth and development when attempting to infer health among past populations. Despite some problems inherent in archaeological material, and in the standardisation of techniques (Saunders 1992; Saunders and Hoppa 1993; Saunders *et al.* 1993), comparisons of growth profiles between different archaeological groups and/or between archaeological and modern populations can provide information on the timing and extent of growth faltering of one population relative to another, or relative to European or North American norms. Interpretations of these patterns, however, require a detailed knowledge of the environmental factors that can influence growth.

Human growth is an outcome of complex interactions between genes and the environment (Bouchard 1991), of which nutrition and infection are the most important components (Ulijaszek and Strickland 1993; Ulijaszek 1997). Study of the relationships between nutrition and specific disease entities (Tomkins and Watson 1989), disease–immune system relationships (Brandtzaeg 1992), adaptive immunity (Fearon and Locksley 1996) and the genetics of the resistance to infection (Good 1994; Steel and Whitehead 1994) have improved the understanding of how these major environmental factors impact on growth (e.g. Schelp 1994; Ulijaszek 1997). In addition, it has been suggested that immunological stress from continuous exposure to infectious disease agents may retard growth and development independently of undernutrition (Solomons 1993). This chapter briefly surveys archaeological studies involving the inference of growth patterns from skeletal material, presents an overview of current literature on environmental factors which can influence growth, and considers previous interpretations of growth faltering from archaeological contexts in light of this literature.

Interpretations of poor growth in the archaeological record

Growth studies have been made on a number of ecologically, temporally and spatially diverse archaeological groups. Some studies are descriptive (e.g. Johnston 1962; Armelagos *et al.* 1972; y'Edynak 1976; Stloukal and Hanáková 1978; Hummert and Van Gerven 1983; Jantz and Owsley 1984a), whereas others are more analytical and attempt to support or refute hypotheses regarding health and disease states (e.g. Jantz and Owsley 1984b; Owsley and Jantz 1985). A number of growth studies also address methodological issues (e.g. Merchant and Ubelaker 1977; Sundick 1978; Lovejoy *et al.* 1990; Hoppa 1992; Saunders *et al.* 1993) often concerning standards for comparison. Some studies are entirely methodological in focus and are concerned with issues such as the estimation of diaphyseal length from fragmentary remains (e.g. Hoppa and Gruspier 1996) or the standardisation of measures of long bone growth (e.g. Goode *et al.* 1993; Sciulli 1994).

When provided, interpretations of growth faltering from archaeological contexts vary in complexity, the simplest involving an acknowledgement that genetic and environmental factors affect patterns of growth (e.g. Johnston 1962; Merchant and Ubelaker 1977). More detailed interpretations predominantly include explanations involving undernutrition and/or infection, and ecological factors that may have affected them (e.g. Cook 1984; Goodman *et al.* 1984; Jantz and Owsley 1984b; Mensforth 1985; Owsley and Jantz 1985; Lovejoy *et al.* 1990; Saunders *et al.* 1993; Miles and Bulman 1994, 1995).

Studies inferring nutritional state from growth patterns

In Cook's (1984) study of prehistoric North Americans from west central Illinois, growth patterns were examined, along with other types of skeletal evidence, to assess the impact of intensified food production on health. Femur lengths of juveniles (birth to 6 years of age) were compared between foragers (Middle Woodland period), transitional maize agriculturalists (late Late Woodland period) and intensive maize agriculturalists (Mississippian period). The late Late Woodland population was observed to have shorter femurs for dental age than groups from earlier or succeeding time periods. Higher frequencies of stress indicators (cribra orbitalia and circular caries) in the stunted juveniles supported an inference that growth retardation was the result of nutritional deficiencies. Cook suggested that the introduction of a low protein maize diet during the late Late Woodland was associated with a decline in child health, but acknowledged that other factors, such as population pressure on scarce resources, may also have been responsible for poor childhood health during this period.

A study of skeletal groups from Dickson Mounds, Illinois (*ca.* AD 950–1300) (Lallo 1973; Goodman *et al.* 1984) revealed growth disruption in children aged 2 to 5 years from the Mississippian period (AD 1200–1300) relative to children from earlier periods (AD 950–1200). The Mississippian period has been characterised by the intensification of maize agriculture along with increased population density and sedentism, and the extension and intensification of trade (Goodman *et al.* 1984). Although Goodman and co-workers concluded that increased levels of stress were a result of factors involving nutrition and infection, growth faltering in the young children was considered to be indicative of nutritional stress. The increased adoption of a low protein maize diet, especially during weaning, was implicated as a cause of poor nutrition and growth faltering.

Studies inferring the effects of infection from growth patterns

Mensforth (1985) interpreted growth faltering predominantly as a result of infection. He compared patterns of tibia growth in a Late Woodland sample from Ohio (AD 800–1100) and a Late Archaic sample from Kentucky (2655–3922 BC). Between the ages of 6 months and 4 years, the Woodland sample demonstrated reduced rates of growth in comparison to the Archaic sample. The author's suggestion that high levels of infectious disease during the first year of life were probably the cause of early growth retardation in the Woodland sample was supported with palaeopathological evidence; a significantly greater frequency of periosteal lesions was observed in the Woodland subadults. Mensforth argued that both groups had access to adequate resources and suggested that the differences in growth were a consequence of diseases associated with the denser and more sedentary horticultural Late Woodland population in comparison with the seasonally mobile Late Archaic hunter–gatherers.

Similarly, in a study of growth velocity in a Libben Late Woodland population, Lovejoy and colleagues (1990) observed depressed growth during the first 3 years of life when compared to modern growth references. The authors acknowledged that nutritional factors could have caused growth faltering, but suggested that from previous archaeological studies the Libben diet was quantitatively and qualitatively adequate. Lovejoy and co-workers then examined the incidence of periosteal new bone formation and found that 50% of individuals under 5 years of age had active or postinfective bone changes. The authors concluded that growth retardation in that age group was most likely a result of high levels of systemic infectious disease.

Miles and Bulman (1994, 1995) examined the growth curves of a skeletal sample from the island of Ensay, Scotland (AD 1500–1850) and found them to be below those from modern samples. As there was no 'skeletal evidence of malnutrition' the authors suggested that factors other than undernutrition contributed to growth retardation: hard work and cold, damp, smoky, dark and overcrowded houses. An interpretation weighted on infection was also supported by the assumption that these island peoples ate fish as a staple, and were unlikely to have suffered from malnutrition.

Studies inferring the influence of nutrition and infection from growth patterns

Jantz and Owsley (1984b) presented a study in which growth faltering was attributed to both undernutrition and infection. Long bone growth variation was compared among 10 Arikara skeletal groups from South Dakota from different historical time periods, ranging from AD 1600 to 1832. The prehistoric and protohistoric periods were characterised by climatic changes and the introduction of the horse; the historic period was characterised by European-introduced epidemic diseases and social disruption. Impacts on health and disease in each of these periods were expected to be reflected in patterns of long bone growth. The predictions of the authors were supported by the skeletal evidence: growth status in the protohistoric period was improved in relation to the prehistoric period, whereas the historic period was characterised by growth failure except in 'early childhood'. Jantz and Owsley (1984b) attributed this growth failure to high levels of morbidity and undernutrition.

Interpretations of undernutrition and infection have also extended to the mothers of stunted children. Owsley and Jantz (1985) examined perinatal growth among the Arikara of South Dakota from the early (AD 1600–1733) and late (AD 1760–1835) postcontact period. Smaller skeletons from the late postcontact period were deemed a result of maternal malnutrition and illness during pregnancy, causing higher frequencies of preterm infants and of infants that were small for gestational age.

In a more recent paper, Saunders and co-workers (1993) examined growth profiles of a subadult skeletal sample from a historic church cemetery in Belleville, Ontario. The cemetery was predominately used by settlers of British descent from 1821 to 1874. This sample followed modern growth patterns except for those under the age of 2 years. The authors suggested that the slightly lower growth rates in this age group may have been a result of undernutrition and infection, or otherwise a result of poor maternal health and prenatal growth.

We have seen that one approach to the interpretation of growth faltering is to isolate undernutrition or infection as the dominant factor. Evidence for contemporary populations suggests that this may be overly simplistic, given that undernutrition and infection are interrelated, and can influence growth in complex ways (Pagezy and Hauspie 1989; Smith *et al.* 1991; Ulijaszek 1997).

Undernutrition–infection interactions and their influence on growth among contemporary populations

In contemporary populations, interactions between undernutrition and infection lead to the well-known patterns of growth faltering from about the age of supplementation, and this is common in developing countries (Waterlow 1988; Neumann and Harrison 1994). Although children in developing countries often have lower mean birth weights, their weight velocities are similar to, and sometimes higher than, those of Western children in the first 3 months of life (Offringa and Boersma 1987). Subsequently, growth faltering relative to European reference values takes place. This is largely due to the combined stresses of low nutrient intakes and exposure to infectious agents associated with the introduction of foods other than breast milk and the weaning process. Most women are unable to produce sufficient breast milk to sustain a European pattern of infant growth beyond 6 months post partum, and dietary supplementation of the infant will often begin around or before that time (Nabarro 1984). Delayed supplementation may lead to growth faltering and undernutrition, leaving an infant more susceptible to infectious disease, while earlier dietary supplementation may provide adequate nutrient intake, but concomitantly introduce potential diarrhoeal agents to a child (Rowland *et al.* 1988). Such interactions are well documented for populations in developing countries (Tomkins 1986; Tomkins and Watson 1989; Ulijaszek 1997).

In addition, the impact of generalised malnutrition on infection (Bairagi *et al.* 1987; Tomkins and Watson 1989; Chowdhury *et al.* 1990) and immunocompetence (e.g. Chandra 1988; Hoffman-Goetz 1988; Crevel and Saul 1992) is well known, as is the influence of infection on the impairment of nutritional status (Tomkins 1981; Briend 1990, 1998). While easy to describe generally, the combined influence of undernutrition and infection on growth and development is complex, varying with disease ecology, the age of a child, and the type and pattern of infant and young child feeding. In addition, the duration and severity of infection and subsequently repeated infections influences the extent to which disease plays a more dominant role in growth faltering than undernutrition, while cultural patterns of infection management influence the duration and sometimes the severity of infection.

Undernutrition–infection interactions

Undernutrition–infection interactions can be initiated in either of two ways. The first involves poor nutritional status leading to impaired immunocompetence and reduced resistance to infection (Chandra 1994), while the second involves an exposure to infectious disease, which leads to a combination of the following: anorexia (Schelp 1994), malabsorption (Briend 1990), elevated basal metabolic rate as a consequence of fever (Duggan and Milner 1986), and protein catabolism (Tomkins *et al.* 1983) to fuel (in part or total) the production of acute phase proteins needed in the immune response (Grimble 1992). Once initiated, the interactions between undernutrition and infection become increasingly complex.

The diseases associated with nutrition and immune function are numerous, and impact on growth and development in different ways. The specific interactions between undernutrition and infectious disease of importance for growth and development (Table 7.1) are: (1) diseases known to affect nutritional status; (2) disease susceptibilities known to be affected by nutritional status; (3) nutritional factors known to cause immunoparesis; and (4) diseases known to cause immunoparesis (impairment of immune function). Of the diseases shown, all except leprosy and trypano-somiasis are population density dependent.

The balance between possible immune system depression and adaptation as a consequence of one bout of infection is important in determining the immune system response to subsequent exposure to infectious agents and disease experience, and the extent, if any, of anorexia, fever, and malabsorption during infectious episodes (Tomkins 1986), any of which can affect the nutritional status of the child. Specific micronutrient deficiencies can also influence immune status and responsiveness (Beisel 1982; Ulijaszek 1990), and the success of adaptive immunity in responding to possible infectious agents. Micronutrient deficiencies that impair immune function include those of vitamin A, pyridoxine, folic acid, zinc and iron

Table 7.1. *Nutrition–infection processes associated with growth faltering of children*

Diseases known to affect nutritional status
Diarrhoea
Upper respiratory tract infections
Pneumonia
Measles
Malaria
Intestinal parasites
Diseases known to be influenced by nutritional status
Tuberculosis
Diarrhoea
Cholera
Leprosy
Pertussis
Respiratory infections
Measles
Pneumocystis carinii pneumonia
Intestinal parasites
Trypanosomiasis
Malaria
Nutritional deficiencies clearly associated with immunoparesis
Energy
Protein
Vitamin A
Pyridoxine
Folic acid
Iron
Zinc
Diseases known to cause immunoparesis
Acquired immunodeficiency syndrome
Measles
Leprosy
Malaria

(Schelp 1994). Disease ecology varies greatly across the contemporary world, making it difficult to move beyond generalities in the description of nutrition–infection cycles outside local and regional situations. Growth faltering is also culturally mediated, in that patterns of disease management and sickness behaviour can influence the incidence, severity and duration of infections and their effects on nutritional status (Tomkins, 1986). Such patterns may include the withholding of food when a child has diarrhoea, which can contribute to impaired nutritional status directly.

The growth faltering associated with the interaction of undernutrition and infection may continue for months or years, depending on the severity of the disease environment, and the abundance and quality of the nutritional environment. In most populations, the process of growth faltering is complete by the age of 2 years, after which a shorter, stunted child may follow a parallel trajectory to European growth references (Eveleth and Tanner 1990). This period of departure from the growth references derived from measures of European populations can be regarded as an adaptation to the disease and nutritional environment. This adaptation is usually associated with high mortality rates (van Lerberghe 1989).

Nutrition, infection and immunity

One aspect of growth and development that mediates the overall growth response to undernutrition and infection is the development of the immune system. B lymphocytes of the newborn child are functionally immature (Hayward 1986), and this is reflected in the low levels, relative to adult values, of circulating immunoglobulins G, A and M (IgG, IgA and IgM). While newborn infants are highly susceptible to most infectious agents in the environment because of this immaturity, they are largely protected from pathogens in the birth canal by the IgG antibody from the mother (Baker and Kasper 1976). After birth, breastfeeding infants obtain maternal IgA, which is protective against a broad spectrum of infectious agents (Ogra *et al.* 1979). While exclusively breastfeeding, infants are largely shielded from the infectious disease environment. Furthermore, growth factors and their binding proteins found in breast milk have trophic effects in an infant's intestine, as well as effects on the autocrine regulation of intestinal growth (Donovan and Odle 1994). Although not essential for gut growth, insulin-like growth factor I (IGF-I) may enhance the development of intestinal lactase (Houle *et al.* 1996) and also be important under pathological conditions in which endogenous IGF-I may be compromised (Burrin 1997), including during the course of undernutrition–infection interactions.

Maternal lactational performance

Under optimal breastfeeding, infants who have suffered intrauterine growth retardation and are small for dates as a consequence, can show considerable catch-up growth in the first 3 months of postnatal life, to the extent that the earlier deficit disappears. In general, milk nutrient levels are well protected against maternal undernutrition, with low body mass index in mothers having little influence on breast milk output at 3 months post partum (Prentice and Prentice 1995). However, there is limited evidence to suggest that severe malnutrition may impair breast milk quantity after this time (Chavez and Martinez 1980). There is little evidence that maternal illness, other than mastitis (Prentice *et al.* 1985), influences breast milk output in any way other than behaviourally, inasmuch as a breastfeeding child might be weaned earlier than usual. In the Gambia, the normal milk of mastitis sufferers relative to non-sufferers has been shown to be low in the immunoproteins IgA, C3 and lactoferrin, but not IgG, IgM, C3, lysozyme or the secretory component (Prentice *et al.* 1985). It is not clear whether undernutrition at the time of weaning leads to maturational delays in the child's immune system development, as it does to more generalised growth faltering. Nor is it clear whether IgM B lymphocyte maturation and class switching to IgA synthesis are influenced by nutritional status. However, protein-energy malnutrition (PEM) influences most aspects of the immune system, including B lymphocyte development and response to antigenic challenge (Tomkins 1986). Thus, nutrition–infection interactions are particularly potent in influencing growth and development, especially in contexts where infections leading to diarrhoeal and respiratory disease are common among children.

Total serum white blood cell level is a gross measure of the immunological components B and T lymphocytes, and leukocytes. Elevated levels of this measure, sometimes in the absence of apparent clinical infection, have been identified as immunological stress by Solomons (1993), who postulated that this could have a role in growth faltering independently of nutritional status. This author drew a parallel between the elevated immune response of some children undergoing growth faltering in developing countries, with the elevated levels of immunological markers seen in battery-reared poultry, which do not routinely receive antibiotics in their feed. In the case of poultry, inclusion of antibiotics with feed has a growth-promoting effect irrespective of environmental quality, by reducing the immune response to infectious agents in the generally poor environment experienced under battery-farming conditions. The extension of this principle to the study of human growth faltering has been termed the 'dirty chick hypothesis'; it has been suggested that a similar mechanism might operate among human populations, particularly in areas of high and stable disease prevalence.

Repeated exposure to infectious agents and an associated continuous activation of the acute phase response of the immune system (Solomons 1993) is catabolic, stimulating gluconeogenesis (Grimble 1992). This process generates the amino acids required to produce the acute phase proteins needed for the recognition of pathogenic agents by antibodies, B and T cells. However, the source of these amino acids, in the absence of plentiful intake of protein (often common when a child has anorexia associated with infection) is from the body, and contributes to impairment of nutritional status. The cytokines interleukin-1, interleukin-6, and tumor necrosis factor α are involved in the acute phase response (Beutler and Cerami 1988; Dinarello 1998), but also mediate bone growth (Price *et al.* 1994). Elevated levels of interleukin-1 and interleukin-6 stimulate bone resorption (Gowen and Mundy 1986; Ishimi *et al.* 1990), while elevated tumor necrosis factor α production can inhibit growth hormone release by pituitary cells (Watson and Cronin 1989). In combination, their effect on growth inhibition is likely to be powerful. While it is usual for peak immune response to last but a few days in the course of infection, constant exposure to infectious agents can result in elevated production of cytokines across long periods. Thus, an elevated immune response, even in the absence of overt clinical signs of infection, if prolonged, may impair child growth independently of undernutrition. This is most likely to be the case once some adaptive immunity has developed, and where exposure to infectious agents does not necessarily lead to overt disease.

Breastfeeding, weaning and growth

Dietary supplementation of infants is culturally mediated (Stuart-Macadam and Dettwyler 1995) and often takes place when breast milk is insufficient to meet the dietary energy needs of the child. Theories about appropriate age for weaning based on biological characteristics such as gestation length, size of mother, or multiple of birth weight do not hold true for humans (Dettwyler 1995), and reasons for supplementation vary greatly across the world. Dietary supplementation of infants may have been a behavioural adaptation that became culturally fixed, assuming it showed greater infant survivorship among those practising supplementation. This may well have been the case among hunter–gatherer populations, where acute infectious diseases were uncommon. However, after intensification of food-obtaining practices and the rise of human population density-dependent diseases that came with it, supplementary feeding probably acquired a negative loading, since food became a novel source of infection. Among contemporary populations of the developing world, supplementary food may act as a vehicle for exposure to pathogenic agents, as well as supplying nutrition. For example, in the Gambia, Barrell and Kolley (1982)

Table 7.2. *Cow's milk as a potential vehicle of diarrhoeal infection in the Gambia*

	Presence in 10^{-5} to 1 ml	
	Coliforms	*E. coli*
From cow	10^{-5}	10^{-4}
Filtered through cloth	10^{-5}	10^{-3}
After standing at ambient temperature		
6 h	10^{-5}	10^{-3}
12 h	10^{-2}	10^{-2}
24 h	10^{-2}	10^{-2}

From Barrell and Kolley 1982.

identified the high potential of cow's milk, used as a supplementary food from the age of 4 months onwards, as a vehicle for diarrhoeal infection (Table 7.2). Filtering milk through a cloth to remove particles accumulated during milking, and allowing milk to stand for more than 6 hours results in exponential increases in coliforms and *Escherichia coli*, both of which are markers for potential diarrhoeal agents.

Breastfeeding supplies both nutrition and immunological protection to the infant; cessation of breastfeeding both increases pathogenic exposure by way of food and removes the maternal antibody contribution to the infants' immune system.

During the first 3 days of lactation, concentrations of leukocytes and immunoglobulins G, A and M in breast milk are high, these values falling rapidly by 15 days post partum, but remaining present at residual levels until at least 15 months post partum (Lentner 1981). However, these values are based on women in industrialised countries, and higher values may occur in breast milk of women experiencing exposure to infection. If so then it is possible to speculate that the maternal antibody response to infection may also confer additional passive immunity to the infant.

Given the relative immaturity of the immune system in infancy, diarrhoeal and respiratory infections of varying kinds are the most common at this stage of life, in developing countries. Figures 7.1 and 7.2 give the prevalence of diarrhoeal morbidity from a large study in Imo State, Nigeria. Clearly, prevalence is both high and forms the largest proportion of all illness in children below 1 year of age. Figure 7.3 shows the prevalence of acute respiratory infections in four countries in Sub-Saharan Africa. In three of the four countries, rates are highest in early childhood, but do not peak within the first year of life. Thus diarrhoeal and acute respiratory

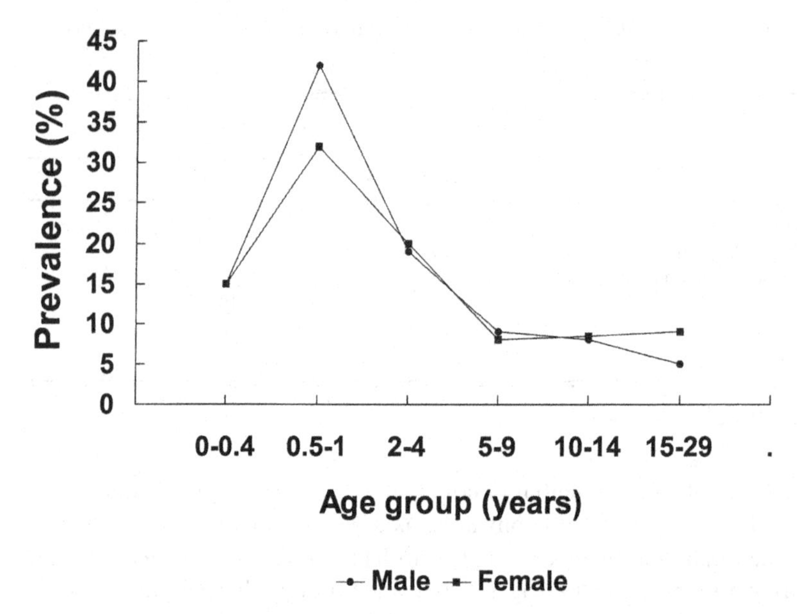

Fig. 7.1. Prevalence of diarrhoeal morbidity across an 8 day period, Imo State, Nigeria, 1983. (Adapted from Kirkwood 1991a.)

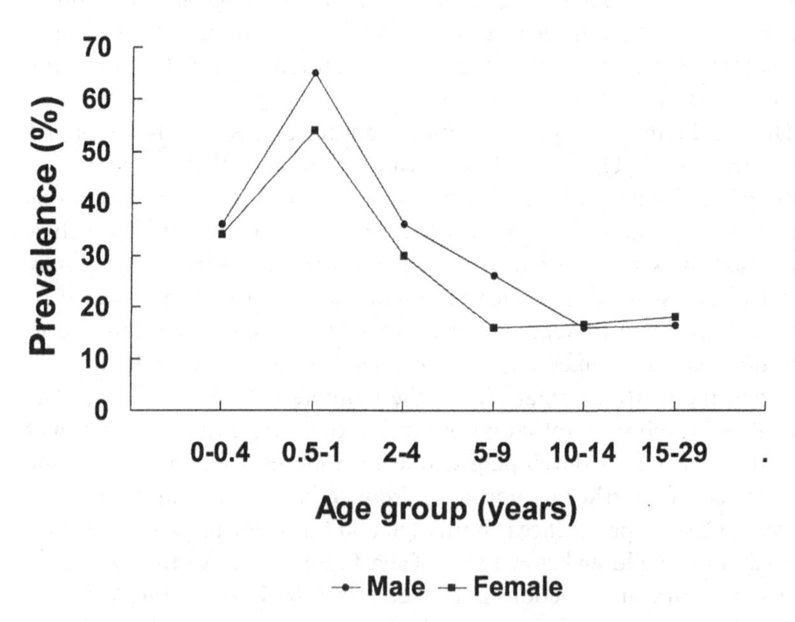

Fig. 7.2. Diarrhoea as a percentage of all illness episodes, Imo State, Nigeria, 1983. (Adapted from Kirkwood 1991a.)

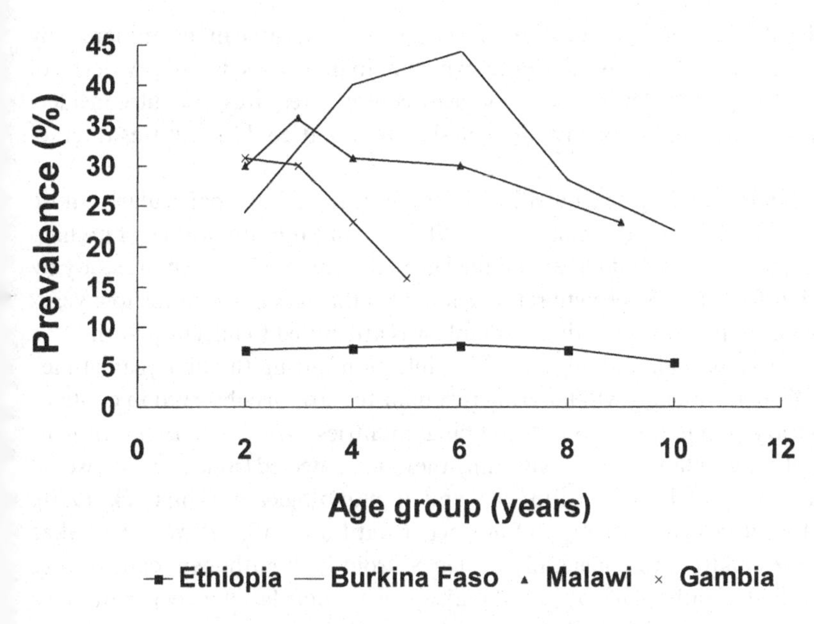

Fig. 7.3. Daily prevalence of acute respiratory infections in four countries of Sub-Saharan Africa. Ethiopia, Freij and Wall (1977); Burkina Faso, Lang *et al.* (1986); Malawi, Lindskog (1987); the Gambia, Rowland *et al.* (1988). (Adapted from Kirkwood 1991b.)

infections come to play an important role in the growth faltering process at different, but overlapping, times in the young child. The effects of different infections on growth are a function of relative exposure, adaptive immunity, illness management, and duration of infection. In the Gambia, Rowland *et al.* (1977) found that malaria had a greater influence on weight gain than diarrhoea by day of infection, but that diarrhoea had a greater overall influence on growth because of its greater prevalence.

Nutrition–infection interactions applied to archaeological contexts

Studies of growth from contemporary populations have revealed the importance of nutrition–infection interactions, and these can be applied to archaeological populations at different levels of analysis. In a study by Jantz and Owsley (1984b), general patterns of growth faltering were attributed to high levels of morbidity and undernutrition. This interpretation is probably correct, as growth faltering is a result of the interactions between the two. The amount and extent of growth disruption, however, depends on the severity of the environment and the quantity and quality of the diet. Although nutrition–infection interactions can be initiated by either poor nutritional status or exposure to infectious disease, evidence of growth disruption in archaeological contexts requires careful consideration of a number of environmental factors that could affect this interaction.

Understanding the nutrition–infection interactions, particularly those occurring during the first years of life, may aid interpretations of archaeological studies that have examined growth profiles by age. In the study by Mensforth (1985), growth retardation from the ages of 6 months to 4 years of age in the Late Woodland sample was attributed to infection during the weaning period, and 'high levels of infection during the first year of life'. This pattern of growth faltering is similar to patterns observed in contemporary populations from developing countries (Waterlow 1988). In general, while infants are breastfeeding they are protected from a broad disease environment both nutritionally and immunologically (Ulijaszek 1990). Thus, if growth faltering did not occur until 6 months, it was likely that these children were generally protected from high pathogen loads during the first months of life by breast milk. Growth retardation between the ages of 6 months and 4 years suggests continued exposure to an environment that affected nutrition–infection interactions. A growth trajectory that parallels the Archaic sample (and North American and European norms) after the age of 4 years suggests an adaptation to the disease and nutritional environment, associated with high mortality rates. Exposure to infection during supplementation could have initiated the nutrition–infection cycle in this population, particularly if increased sedentism and population density occurred.

In contrast to Mensforth's interpretation, Goodman and co-workers (1984) suggested that increased use of maize as a supplementary food during weaning may have resulted in a decreased growth velocity and an increase in growth disruption in children of 2 to 5 years of age at Dickson Mounds. Supplementation of food must have occurred much earlier than 2 years of age. In contemporary populations, breast milk may not be sufficient to sustain well-nourished patterns of growth after 6 months of age (Whitehead and Paul 1984). If the children did not experience growth faltering until much later, the supplementary food provided may have been adequate, and infection loads associated with supplementation may not have been significant. Although the adoption of a poor protein quality maize diet may have affected growth faltering, it is likely that other factors, such as the reported increase in population and sedentism during the Middle Mississippian (Goodman *et al.* 1984), may have had an impact on nutrition–infection interactions after 2 years of age.

In the study by Saunders *et al.* (1993), slight growth faltering in the St Thomas' Church sample was observed in children below the age of 2 years. The authors' explanations included undernutrition and infection, and poor maternal health during pregnancy. In contemporary populations, maternal nutrition has been observed to be associated with infant length at birth (see e.g. Neumann and Harrison 1994). However, children that are small for gestational age can show rates of growth that are higher than norms for the first 3 months, and often catch up to their expected levels, especially if breastfed. Early supplementation in the St Thomas' sample could have resulted in an inadequate dietary intake as well as the introduction of pathogens to the child. Conversely, it is likely that the small foetal and preterm infants observed in the Arikara may have been a result of maternal malnutrition and illness as suggested by Owsley and Jantz (1985).

A number of the studies presented above have incorporated skeletal evidence such as indicators of stress (e.g. cribra orbitalia, circular caries, dental enamel hypoplasia) to support nutritional interpretations of stunting (e.g. Cook 1984). The 'absence of malnutrition' on skeletons has also been used to support, in association with other types of archaeological evidence, a non-dietary interpretation of growth faltering (e.g. Mensforth 1985; Miles and Bulman 1994). Since growth faltering has been shown to be most commonly the outcome of nutrition–infection interactions among contemporary populations, the presence of stress indicators could represent a compromised nutritional state, but may not be used as unequivocal evidence for growth faltering due to malnutrition. Conversely, the absence of stress indicators should not be used as positive evidence for adequate nutrition.

Other studies have included palaeopathological evidence of infection (periosteal lesions) to support interpretations of growth faltering (e.g. Mensforth 1985; Lovejoy *et al.* 1990). However, the prevalence of acute phase response activation due to immunological challenge by endemic disease in children from developing countries may be an important contributing factor to growth faltering, even in the absence of any overt signs or symptoms of disease (Solomons *et al.* 1993). Moreover, repeated diarrhoeal and acute respiratory infections are also important to growth faltering and are unlikely to cause skeletal lesions. Thus, insults affecting growth may be invisible in palaeopathological skeletal evidence.

Interpretations based on positive or negative associations between stress indicators, infection and growth faltering are questioned on the basis of current understanding of nutrition–infection interactions. For example, Ribot and Roberts (1996) found no consistent relationship between the number of osteoarchaeological stress indicators (including dental enamel hypoplasia, porotic hyperostosis, cribra orbitalia, subperiosteal new bone formation and Harris lines) and growth faltering in subadults from an early Mediaeval site at Raunds, Northamptonshire, and a late Mediaeval site in Chichester, Sussex. Furthermore, Mays (1995) did not observe statistically significant relationships between Harris lines and femoral length in a group of juveniles from Wharram Percy, England. While this lack of association might be attributable to small sample size, it might also be due to a more complex relationship not represented by the two variables being compared. Thus, growth faltering can occur as a result of environmental insults that are paleopathologically invisible.

Discussion and conclusions

Before growth profiles of past populations can be interpreted, analysts must be aware of the methodological issues that influence or bias the profiles. For instance, Saunders and colleagues (1993) have investigated how different methods of age determination can affect reported timings of growth faltering, and hence the interpretation of the growth pattern. When poor growth is observed in archaeological contexts, the complexity of nutrition–infection relationships make any interpretation of growth faltering challenging. The age and immune status of the child, disease ecology, cultural patterns of feeding and disease management are a few of the factors that affect the extent of growth faltering. In addition, impaired growth may be a result of invisible insults, such as acute infections, or chronic low level immunostimulation (Solomons *et al.* 1993). Thus, archaeological evidence, such as sanitation and hygiene should be carefully considered in analyses. In addition, the types of disease associated with various population sizes and settlement patterns need consideration.

Evidence of population size and sedentism are important to interpretations of disease ecology, and pathogen loads. Early foraging groups with a population size below 300 people would not have been large enough to sustain and fix acute infections such as diarrhoea (including cholera and typhoid), respiratory tract infections and measles. If introduced, acute infections with brief and rapid stages of infection would have run their course in susceptible individuals and then died out (Inhorn and Brown 1998). Sedentism, the clearing of land for agriculture and animal husbandry, increased the potential for human contact with both human and animal faeces. This provided ideal conditions for the transmission of helminthic and protozoal parasites, with substantially increased infection rates (Cohen 1989), exposure to acute infections, and the onset of infectious disease profiles now common in the less-industrialised world.

From the brief survey of archaeological studies presented above, it appears that previous interpretations of growth faltering have been influenced in part by the types of archaeological evidence available and the use (and the prevailing interpretations of) palaeopathological indicators of stress. The succeeding overview of the growth literature has demonstrated how interpretations of archaeological studies may need some reconsideration. Although archaeological evidence may provide a convincingly weighted interpretation of growth faltering, the biological evidence suggests that these interpretations cannot be simplistic or narrowly focused. Ecological factors that could affect nutrition–infection interactions should also be included in analyses.

Interpretations of growth faltering within defined age ranges, particularly under the age of 5 years, may not be easily dichotomised, given current paradigms on nutrition–infection interactions during breastfeeding and weaning. Growth status at birth, within the first 6 months, and during weaning represent different outcomes of nutrition–infection interactions. Although the interactions may be initiated by either undernutrition or infection, subsequent growth faltering is most likely to be the outcome of both factors. With an understanding of some of these interactions, however, it may be possible to make some inferences on environmental or cultural grounds. For instance, the immunological and nutritional consequences of breastfeeding make interpretations of growth faltering below 6 months different from those made above that age.

Finally, studies that incorporate the presence or absence of other skeletal indicators of stress in weight interpretations may need more careful consideration in light of the undernutrition–infection interaction model. The biological evidence presented suggests that growth faltering can be a result of insults that are archaeologically invisible. Although the presence of different types of stress indicator may support an association with poor growth, such evidence does not represent causation.

References

Armelagos G, Mielke J, Owen K, Gerven DV, Dewey J and Mahler P (1972) Bone growth and development in prehistoric populations from Sudanese Nubia. *Journal of Human Evolution* **1**, 89–119.

Bairagi R, Chowdhury MK, Kim YJ, Curlin GT and Gray RH (1987) The association between malnutrition and diarrhoea in rural Bangladesh. *International Journal of Epidemiology* **16**, 477–481.

Baker CJ and Kasper DL (1976) Correlation of maternal antibody deficiency with susceptibility to neonatal group B streptococcal infection. *New England*

Journal of Medicine **294**, 753–756.

Barrell RAE and Kolley SSMI (1982) Cow's milk as a potential vehicle of diarrhoeal disease pathogens in a West African village. *Journal of Tropical Pediatrics* **28**, 48–52.

Beisel WR (1982) Single nutrients and immunity. *American Journal of Clinical Nutrition* **35**, Supplement, 417–468.

Beutler B and Cerami A (1988) Tumor necrosis, cachexia, shock, and inflammation: a common mediator. *Annual Reviews in Biochemistry* **57**, 505–518.

Bouchard C (1991) Genetic aspects of anthropometric dimensions relevant to assessment of nutritional status. In *Anthropometric Assessment of Nutritional Status*, ed. J.H. Himes, pp. 213–231. New York: Wiley-Liss, Inc.

Brandtzaeg P (1992) Humoral immune response patterns of human mucosae: induction and relation to bacterial respiratory tract infections. *Journal of Infectious Diseases* **165**, Supplement 1, S167–S176.

Briend A (1990) Is diarrhoea a major cause of malnutrition among the under-fives in developing countries? A review of available evidence. *European Journal of Clinical Nutrition* **44**, 611–628.

Briend A (1998) Infection. In *The Cambridge Encyclopedia of Human Growth and Development*, ed. S.J. Ulijaszek, F.E. Johnston and M.A. Preece, pp. 334–336. Cambridge: Cambridge University Press.

Burrin DG (1997) Is milk-borne insulin-like growth factor-I essential for neonatal development? *Journal of Nutrition* **34**, Supplement, 975S–979S.

Chandra RK (1988) Nutritional regulation of immunity: an introduction. In *Nutrition and Immunology*, ed. R.K. Chandra, pp. 1–8. New York: Alan R. Liss.

Chandra RK (1994) Nutrition, immune function and infectious disease. In *Nutrition in a Sustainable Environment*, Proceedings of the XV International Congress of Nutrition, IUNS, Adelaide, ed. M.L. Wahlqvist, A.S. Truswell, R. Smith and P.J. Nestel, pp. 486–487. London: Smith-Gordon.

Chavez A and Martinez C (1980) Effects of maternal nutrition and dietary supplementation. In *Maternal Nutrition During Pregnancy and Lactation*, ed. H. Aebi and R. Whitehead, pp. 274–284. Bern: Huber.

Chowdhury MK, Gupta VM, Bairagi R and Bhattacharya BN (1990) Does malnutrition predispose to diarrhoea during childhood? Evidence from a longitudinal study in Matlab, Bangladesh. *European Journal of Clinical Nutrition* **44**, 515–525.

Cohen MN (1989) *Health and the Rise of Civilization*. New Haven, CT: Yale University Press.

Cook D (1984) Subsistence and health in the Lower Illinois Valley: osteological evidence. In *Paleopathology at the Origins of Agriculture*, ed. M. Cohen and G. Armelagos, pp. 235–269. Orlando, FL: Academic Press.

Crevel RWR and Saul JAT (1992) Linoleic acid and the immune response. *European Journal of Clinical Nutrition* **46**, 847–855.

Dettwyler KA (1995) A time to wean: the hominid blueprint for the natural age of weaning in modern human populations. In *Breastfeeding. Biocultural Perspectives*, ed. P. Stuart-Macadam and K.A. Dettwyler, pp. 39–73. Chicago: Aldine de Gruyter.

Dinarello CA (1989) Interleukin-1 and biologically related cytokines. *Advances in Immunology* **44**, 153–169.
Donovan SM and Odle J (1994) Growth factors in milk as mediators of infant development. *Annual Reviews in Nutrition* **14**, 147–167.
Duggan MB and Milner RDG (1986) Energy cost of measles infection. *Archives of Disease in Childhood* **61**, 436–439.
Eveleth PB and Tanner JM (1990) *Worldwide Variation in Human Growth*, 2nd edn. Cambridge: Cambridge University Press.
Fearon DT and Locksley RM (1996) The instructive role of innate immunity in the acquired immune response. *Science* **272**, 50–54.
Freij L and Wall S (1977) Exploring child health and its ecology: the Kirkos study in Addis Ababa. *Acta Paediatrica Scandinavica* **267**, *Supplement*, 1–180.
Good MF (1994) Antigenic diversity and MHC genetics in sporozoite immunity. *Immunology Letters* **41**, 95–98.
Goode H, Waldron T and Rogers J (1993) Bone growth in juveniles: a methodological note. *International Journal of Osteoarchaeology* **3**, 321–323.
Goodman A, Lallo J, Armelagos G and Rose J (1984) Health changes at Dickson Mounds, Illinois (A.D. 950–1300). In *Paleopathology at the Origins of Agriculture*, ed. M. Cohen and G. Armelagos, pp. 271–305. Orlando, FL: Academic Press.
Gowan M and Mundy GR (1986) Action of recombinant interleukin 1, interleukin 2, and interferon-gamma on bone resorption in vitro. *Journal of Immunology* **136**, 2478–2482.
Grimble RF (1992) Dietary manipulation of the inflammatory response. *Proceedings of the Nutrition Society* **51**, 285–294.
Hayward A (1986) Ontogeny of the human immune system. In *Human Growth. A Comprehensive Treatise*, ed. F. Falkner and J.M. Tanner, pp. 364–379. New York: Plenum Press.
Hoffman-Goetz L (1988) Lymphokines and monokines in protein-energy malnutrition. In *Nutrition and Immunology*, ed. R.K. Chandra, pp. 9–23. New York: R. Liss.
Hoppa R (1992) Evaluating human skeletal growth: an Anglo-Saxon example. *International Journal of Osteoarchaeology* **2**, 275–288.
Hoppa RD and Gruspier KL (1996) Estimating diaphyseal length from fragmentary subadult skeletal remains: implications for palaeodemographic reconstructions of a Southern Ontario ossuary. *American Journal of Physical Anthropology* **100**, 341–354.
Houle VM, Schroeder EA, Laswell SC and Donovan SM (1996) Small intestinal disaccharidase activities are up-regulated, whereas peptidase activity is decreased by orally administered insulin-like growth factor-I in the neonatal piglet. *FASEB Journal* **10**, A728.
Hummert J and Van Gerven DP (1983) Skeletal growth in a medieval population from Sudanese Nubia. *American Journal of Physical Anthropology* **60**, 471–478.
Inhorn MC and Brown PJ (1998) *The Anthropology of Infectious Disease*. New York: Gordon & Breach Science Publishers.

Ishimi Y, Miyaura C, Jin CH, Akatsu T, Abe E, Nakamura Y, Yamaguchi A, Yoshiki S, Matsuda T, Hirano T, Kishimoto T and Suda T (1990) IL-6 is produced by osteoblasts and induces bone resorption. *Journal of Immunology* **145**, 3297–3303.

Jantz R and Owsley D (1984a) Temporal changes in limb proportionality among skeletal samples of Arikara Indians. *Annals of Human Biology* **11**, 157–163.

Jantz R and Owsley D (1984b) Long bone growth variation among Arikara skeletal populations. *American Journal of Physical Anthropology* **63**, 13–20.

Johnston F (1962) Growth of the long bones of infants and young children at Indian Knoll. *American Journal of Physical Anthropology* **20**, 249–254.

Kirkwood BR (1991a) Diarrhea. In *Disease and Mortality in Sub-Saharan Africa*, ed. R.G. Feachem and D.T. Jamison, pp. 134–157. Oxford: Oxford University Press.

Kirkwood BR (1991b) Acute respiratory infections. In *Disease and Mortality in Sub-Saharan Africa*, ed. R.G. Feachem and D.T. Jamison, pp. 158–172. Oxford: Oxford University Press.

Lallo J (1973) The skeletal biology of three prehistoric American Indian societies from Dickson Mounds. PhD thesis, University of Massachusetts, Amherst.

Lang T, Lafaix C, Fassin D, Arnaut I, Salmon B, Baudon D and Ezekiel J (1986) Acute respiratory infections: a longitudinal study of 151 children in Burkina Faso. *International Journal of Epidemiology* **15**, 553–561.

Lentner C (1981) Breast milk. In *Geigy Scientific Tables*, vol. 1 *Units of Measurement, Body Fluids, Composition of the Body, Nutrition*, ed. C. Lentner, pp. 213–216. Basle: Ciba-Geigy Ltd.

Lindskog U (1987) Child health and household water supply: an intervention study from Malawi. Linkoping University Medical Dissertation no. 259. Linkoping University.

Lovejoy C, Russell K and Harrison M (1990) Long bone growth velocity in the Libben population. *American Journal of Human Biology* **2**, 533–541.

Mays S (1995) The relationship between Harris lines and other aspects of skeletal development in adult and juveniles. *Journal of Archaeological Science* **22**, 511–520.

Mensforth R (1985) Relative tibia long bone growth in the Libben and Bt-5 prehistoric skeletal populations. *American Journal of Physical Anthropology* **68**, 247–262.

Merchant V and Ubelaker D (1977) Skeletal growth of the Protohistoric Arikara. *American Journal of Physical Anthropology* **46**, 61–72.

Miles A and Bulman J (1994) Growth curves of immature bones from a Scottish island population of sixteenth to mid-nineteenth century: limb-bone diaphyses and some bones of the hand and foot. *International Journal of Osteoarchaeology* **4**, 121–136.

Miles A and Bulman J (1995) Growth curves of immature bones from a Scottish island population of sixteenth to mid-nineteenth century: shoulder girdle, ilium, pubis and ischium. *International Journal of Osteoarchaeology* **5**, 15–27.

Nabarro D (1984) Social, economic, health and environmental determinants of nutritional status. *Food and Nutrition Bulletin* **6**, 18–32.

Neumann CG and Harrison GG (1994) Onset and evolution of stunting in infants and children. Examples from the Human Nutrition Collaborative Research Support Program. Kenya and Egypt studies. *European Journal of Clinical Nutrition* **48**, Supplement 1, S90–S102.

Offringa PJ and Boersma ER (1987) Will food supplementation in pregnant women decrease neonatal morbidity? *Human Nutrition: Clinical Nutrition* **41C**, 311–315.

Ogra PL, Fishaut M and Theodore C (1979) Immunology of breast milk: maternal neonatal interactions. In *Human Milk: Its Biological and Social Value*, ed. S. Freier and A.I. Eidelman, pp. 115–124. Amsterdam: Excerpta Medica.

Owsley D and Jantz R (1985) Long bone lengths and gestational age distributions of post-contact period Arikara Indian perinatal infant skeletons. *American Journal of Physical Anthropology* **68**, 321–328.

Pagezy H and Hauspie RC (1989) Growth retardation in weight and catch-up growth following infection by measles and whooping cough among babies of rural Zaire (Lake Tumba). *International Journal of Anthropology* **4**, 87–101.

Prentice AM and Prentice A (1995) Evolutionary and environmental influences on human lactation. *Proceedings of the Nutrition Society* **54**, 391–400.

Prentice A, Prentice AM and Lamb WH (1985) Mastitis in rural Gambian mothers and the protection of the breast by milk antimicrobial factors. *Transactions of the Royal Society of Tropical Medicine and Hygiene* **79**, 90–95.

Price JS, Oyajobi BO and Russell RGG (1994) The cell biology of bone growth. *European Journal of Clinical Nutrition* **48**, *Supplement* 1 S131–S149.

Ribot I and Roberts C (1996) A study of non-specific stress indicators and skeletal growth in two mediaeval subadult populations. *Journal of Archaeological Sciences* **23**, 67–79.

Rowland MGM, Cole TJ and Whitehead RG (1977) A quantitative study into the role of infection in determining nutritional status in Gambian village children. *British Journal of Nutrition* **37**, 441–450.

Rowland MGM, Rowland SGJG and Cole TJ (1988) Impact of infection on the growth of children from 0 to 2 years in an urban West African community. *American Journal of Clinical Nutrition* **47**, 134–138.

Saunders SR (1992) Subadult skeletons and growth related studies. In *Skeletal Biology of Past Peoples: Research Methods*, ed. S.R. Saunders and M.A. Katzenberg, pp. 1–20. New York: Wiley-Liss.

Saunders SR and Hoppa RD (1993) Growth deficit in survivors and non-survivors: biological mortality bias in subadult skeletal samples. *American Journal of Physical Anthropology* **36**, 127–151.

Saunders SR, Hoppa RD and Southern R (1993) Diaphyseal growth in a nineteenth century skeletal sample of subadults from St. Thomas' Church, Bellville, Ontario. *International Journal of Osteoarchaeology* **3**, 265–281.

Schelp FP (1994) Epidemiological factors in malnutrition and infection. In *Nutrition in a Sustainable Environment*, Proceedings of the XVth International Congress of Nutrition, IUNS, Adelaide, ed. M.L. Wahlqvist, A.S. Truswell, R. Smith and P.J. Nestel, pp. 488–490. London: Smith-Gordon.

Sciulli P (1994) Standardization of long bone growth in children. *International*

Journal of Osteoarchaeology **4**, 257–259.

Smith TA, Lehmann D, Coakley C, Spooner V and Alpers MP (1991) Relationships between growth and acute lower-respiratory infections in children aged < 5 y in a highland population of Papua New Guinea. *American Journal of Clinical Nutrition* **53**, 963–970.

Solomons NW (1993) Pathways to the impairment of human nutritional status by gastrointestinal pathogens. *Parasitology* **107**, S19–35.

Steel DM and Whitehead AS (1994) The major acute phase reactants: C-reactive protein, serum amyloid P component and serum amyloid A protein. *Immunology Today* **15**, 81–88.

Stloukal V and Hanáková H (1978) Die Lange der Langsknochen altslawischer Bevolkerungen – Unter besonderer Berucksichtigung von Wachstumsfragen. *Homo* **29**, 53–69.

Stuart-Macadam P and Dettwyler KA (1995) *Breastfeeding. Biocultural Perspectives*. Chicago: Aldine de Gruyter.

Sundick R (1978) Human skeletal growth and age determination. *Homo* **29**, 228–249.

Tomkins AM (1981) Nutritional status and severity of diarrhoea among pre-school children in rural Nigeria. *Lancet* **1**, 860–862.

Tomkins AM (1986) Protein-energy malnutrition and risk of infection. *Proceedings of the Nutrition Society* **45**, 289–304.

Tomkins AM and Watson F (1989) *Malnutrition and Infection. A Review*. London: London School of Hygiene and Tropical Medicine.

Tomkins AM, Garlick PJ, Schofield WN and Waterlow JC (1983) The combined effects of infection and malnutrition on protein metabolism in children. *Clinical Science* **65**, 313–324.

Ulijaszek SJ (1990) Nutritional status and susceptibility to infectious disease. In *Diet and Disease*, ed. G.A. Harrison and J.C. Waterlow, pp. 137–154. Cambridge: Cambridge University Press.

Ulijaszek SJ (1997) Transdisciplinarity in the study of undernutrition–infection interactions. *Collegium Antropologicum* **21**, 3–15.

Ulijaszek SJ and Strickland SS (1993) *Nutritional Anthropology. Prospects and Perspectives*. London: Smith-Gordon.

van Lerberghe W (1989) *Kasongo. Child Mortality and Growth in a Small African Town*. London: Smith-Gordon.

Walton PE and Cronin MJ (1989) Tumour necrosis factor-alpha inhibits growth hormone secretion from cultured anterior pituitary cells. *Endocrinology* **125**, 925–929.

Waterlow JC (1988) Observations on the natural history of stunting. In *Linear Growth Retardation in Less Developed Countries*, ed. J.C. Waterlow, pp. 1–12. New York: Raven Press.

Whitehead RG and Paul AA (1984). Growth charts and the assessment of infant feeding practices in the western world and in developing countries. *Early Human Development* **9**, 187–207.

y'Edynak G (1976) Long bone growth in western Eskimo and Aleut skeletons. *American Journal of Physical Anthropology* **45**, 569–574.

8 *What can be done about the infant category in skeletal samples?*

SHELLEY R. SAUNDERS AND LISA BARRANS

Introduction

The high risk of death amongst the very young has long been recognised by researchers and general observers. Some of the first statistical evidence for high childhood mortality comes from John Graunt's examination of the bills of mortality for the city of London in the 1600s (Lancaster 1990). While the bills were originally used as a means of warning against increases in plague, Graunt took particular note of the high rate of mortality in the earliest years of life, concluding that 36% of all births in London ended in death before the age of 6 years. Shortly after, E. Halley observed that almost half (43%) of all babies born in Breslau around the year 1690 had died before their sixth birthday (Lancaster 1990).

While most of the early work on infant and child mortality comes from studies of European preindustrial cities, high death rates in early life are identified as common for all but the most industrialised populations (Wrigley 1969; Knodel 1983; Lancaster 1990). Even today, many millions of infants die in their first year of life. In industrialised countries, there has been a progressive reduction in infant mortality (as well as fertility) only since the late 19th or early 20th century. This is referred to as the demographic transition. Some countries have achieved infant mortality rates as low as 6 per 1000 live births (United Nations Statistical Office 1994) but rates in most underdeveloped and developing countries are significantly and sometimes many times higher (Hassan 1986; United Nations Statistical Office 1994). The demographic transition is, in fact, considerably restricted; more than half of the world's nations still have very high infant mortality rates and high fertility rates (Hassan 1986).

The period from birth to about 3 years of age is also the most crucial for proper growth (Beaton 1992). Growth velocities, particularly for the skeleton, are at their highest immediately after birth and up to 12 months, when they begin to decline dramatically (Johnston 1978). After 3 years, the pattern of childhood growth is established as advanced or retarded. If there is stunting in the early years, this is converted to a state of being stunted during the childhood growth years (Beaton 1989). The rapid rate of growth means that nutritional needs are highest during the early years and the developing infant is more susceptible to malnutrition as well as to infections as physiological maturation progresses (see also King and Ulijaszek, Chapter 7, this volume).

The crucial importance of the early years of childhood would suggest that researchers would be particularly interested in studying immature individuals from archaeological skeletal samples. But counts of juvenile skeletons from prehistoric and historic cemeteries are commonly very low, introducing a bias into skeletal samples (Jackes 1992; Saunders 1992; Guy *et al.* 1997). Several authors have discussed the relative importance of three different factors in producing the underrepresentation of immature individuals. These are cultural beliefs about infants and children that influence mortuary behaviour, the effect of taphonomic processes causing differential preservation of immature bones, and incomplete archaeological recovery due to biased excavation techniques. There are those who have argued that differential preservation is the most important cause of the underenumeration of immature skeletons (Guy *et al.* 1997), while others have been more comprehensive in their explanations (Saunders 1992; Nawrocki 1995; Hoppa and Gruspier 1996; Paine and Harpending 1998).

This chapter examines the demographic category of 'infants' in archaeological skeletal samples and evaluates the potential for deriving useful information about past societies from examinations of these skeletons. In demographic studies, the frequency of infant mortality is an important component in measuring the overall success of a population. Infant mortality is used as a proxy measure for the social and sanitary conditions of a community and, as indicated above, is a sensitive indicator of the effects of malnutrition (Roth 1992). If it is often the case that infant skeletons do not survive, are not buried in main burial areas, or are hard to find during excavation, is it possible to put forward any conclusions about them when trying to reconstruct past populations? Despite claims to this effect by a few investigators, there are at least some examples of skeletal samples with substantial proportions of immature individuals (Cook and Buikstra 1979; Owsley and Jantz 1985; Farwell and Molleson 1993; Saunders *et al.* 1993b; Sperduti *et al.* 1997; Hutchins 1998). There are also existing bioarchaeological studies that have attempted to find evidence for proximate determinants of infant and child mortality in skeletal samples (Lallo *et al.* 1977; Mensforth *et al.* 1978; Cook and Buikstra 1979). The entire area of growth research in past populations seeks to elucidate the health effects of morbidity amongst different groups. In the following discussion, we examine some of the reasons why researchers investigate variability in infant mortality in current and recent human populations and the implications these studies have for bioarchaeology. We propose that future scientific developments in osteological research may offer a more optimistic outlook to the reconstruction of past populations than is currently held.

Defining infants demographically and in skeletal studies

While various measures of infant mortality have been defined by demographers, the simplest and most frequently used is the ratio of deaths of infants under 1 year of age during a given year to total live births in the same year.[1] But there is variation in the definition of a live birth. In some countries, a newborn infant who does not survive the first 24 hours after birth may be treated as an abortion and not counted as either a live birth or a death. In other countries, infants who die within the first 3 days before registration are registered as 'presented dead', while elsewhere, infants who die before registration may be registered as stillborn but statistically counted as live births. This raises the question of what to do about the counting of deaths of infants that are not live born. There are problems with proper definitions and the ability of researchers to detect and identify foetal deaths. Properly counting infant mortality is, in fact, difficult for modern demographers, particularly when working in countries with poor registration, and is especially difficult for historical demographers working with early handwritten records (Knodel 1988). For skeletal research, where workers do not have access to independent records of mortality, the survival of parturition must somehow be assessed directly from bones by some biological means.

The United Nations Statistical Office (1994) *Demographic Yearbook* defines foetal death as death prior to the complete expulsion or extraction from its mother of a product of conception, irrespective of the duration of the pregnancy. The death is identified by the fact that after such separation the foetus does not breathe or show any other evidence of life such as beating of the heart, pulsation of the umbilical cord, or definite movement of voluntary muscles. Late foetal deaths are those at 28 or more completed weeks of gestation. These are synonymous with the term stillbirth (use of this term is now discouraged in demographic studies). The term perinatal

[1] This definition ignores further refinements available to demographers who can calculate mortality in cohorts by year, quarter or month of birth and calculate weighted infant mortality rates (Pressat 1972). This is done when looking for large temporal variations in infant mortality such as seasonality in births and deaths. See the discussion below on the possibility of detecting seasonality in mortality from skeletal samples.

has been recommended by the Study Group on Perinatal Mortality set up by the World Health Organization (Shapiro *et al.* 1968). The International Conference for the Eighth Revision of the International Classification of Diseases adopted the recommendation that the perinatal period be defined 'as extending from the 28th week of gestation to the seventh day of life' (Shapiro *et al.* 1968). Noting that several countries considered as late foetal deaths any foetal death at 20 weeks or longer gestation, the conference agreed to accept a broader definition of perinatal death, which extends from the 20th week of gestation to the 28th day of life (Shapiro *et al.* 1968).

Is it possible to determine from an infant skeleton whether the individual survived birth? Archaeological information may sometimes help and there are examples of adult females who must have died during parturition where the infant skeleton was discovered in the birth canal at the time of excavation. Otherwise, detailed estimates of age are necessary. Farwell and Molleson (1993) noted that a foetus of less than 28 weeks is unlikely to survive independently, even with modern technology, so that careful age determinations of foetal and neonate-size skeletons may help. There are some useful standards for estimating developmental age from the teeth that provide fairly detailed estimates of late foetal and neonatal individuals (Storey 1986; Farwell and Molleson 1993; Jantz and Owsley 1994). However, researchers must take care to use standards based on samples of known sex and age and not those derived from foetuses whose ages were initially estimated (Deutsch *et al.* 1985). This has also been a problem with age-at-death estimates based on size (Fasekas and Kosa 1978) but more recent work has begun to deal with this problem (Sellier *et al.* 1997).

Yet even if perinatal age can be estimated fairly closely this does not confirm that an infant was live born or dead at birth. Is such confirmation possible with skeletons? One approach may be to look for evidence of a neonatal line or birth line in developing tooth enamel. This line of arrested growth reflects the birth process (Eli *et al.* 1989; Skinner and Dupras 1993). But the infant must survive at least for a short period after birth for the line to form. A live born infant dying shortly after birth would presumably not develop a line. Consequently, unless there are other biochemical methods of detecting the physiological signs of life in bone or dental tissues it would not be possible to identify a certain proportion of live born infants in any skeletal sample.

Is it even important to answer this question? Until recently, many researchers of skeletons did not even use the demographic definition of infant. 'Infant' has been used as a generic term, applied to babies of the first year after birth as well as to young children. The importance of the definition relates to the changing risks of mortality observed within the first year. As is explained below, high proportions of neonatal deaths during the first year of life are not expected in non-industrialised countries or past populations. When all perinatal deaths are counted, these may be proportionately very high and possibly reflect severe problems affecting the health of pregnant women (for a possible example, see Storey 1986). But detecting high proportions of perinatal deaths in cemetery samples is a daunting task, given the sampling problems mentioned above.

Variation in infant mortality

Infant mortality is usually divided into months and even days in order to get a better understanding of the factors underlying infant death (Pharoah and Morris 1979). Under optimal conditions for infants, seen today in some developed countries, deaths in the first month are attributed to causes preceding, or associated with birth such as obstetrical trauma, congenital defects, and other developmental problems. These are referred to as endogenous causes of death. Under poorer conditions, neonates may also die of exogenous causes such as vulnerability to infectious diseases. Because endogenous causes of death are limited mainly to very early infancy, demographers divide infancy into the 'neonatal' period, or, the first four weeks after birth, and the 'postneonatal' period, from one month to the end of the first year. Postneonatal mortality is almost entirely attributed to exogenous or environmental conditions. In the 19th century, William Farr observed that in large towns and cities in England mortality at one or more months of age and beyond increased dramatically under unsanitary conditions (Pharoah and Morris 1979).

An earlier paper by Saunders (1992) noted that there is a broad but mistaken assumption amongst bioarchaeologists that mortality rates will always be highest at birth and slowly decline thereafter so that researchers often expect mainly newborn deaths in any mortality sample. This author pointed out that postneonatal mortality (in total) exceeded neonatal mortality in industrialised countries until the 1930s (Forfar and Arneil 1978) and that this situation still applies in developing countries. Since it is, in fact, high postneonatal mortality that reflects poor social/environmental conditions for infant survival we might extrapolate the situation to historic and prehistoric societies. This was illustrated by Saunders and colleagues (1995) in their study of a skeletal sample from a 19th century church cemetery in Canada. Estimates of the ages of infant skeletons (with a careful attempt to eliminate preterm foetuses) resulted in 74% postneonates and 26% neonates. The proportions proved to be virtually identical with those reported in the parish registers of the church and were also supported by independent reconstructions of sanitary, social and environmental conditions during the period of cemetery use (Herring *et al.* 1991).

In 20th century industrialised countries, where postneonatal mortality is low, total perinatal mortality (the sum of foetal deaths and neonatal deaths) will almost always exceed postneonatal mortality (Forfar and Arneil 1978). But in circumstances where living conditions are poor, even with the poor health of pregnant mothers, postneonatal mortality can still exceed perinatal mortality. Knodel (1988) examined infant and adult mortality from the parish records of 14 German villages spanning the early 18th to the early 20th centuries. When compared with perinatal mortality, the percentage of postneonatal mortality out of total infant mortality in the villages ranged from 36% to 61%. Knodel observed exceptionally high neonatal mortality in several villages whether or not foetal deaths were included and associated this with evidence that babies were not breastfed or were weaned prior to one month in those villages. He also related fluctuations in postneonatal mortality in the villages to differential feeding practices and varying sanitary conditions (see discussion below on determining feeding practices from skeletal studies).

Knodel's work reflects a long and continuing interest on the part of historical demographers in infant mortality. But, except under unusual circumstances (Johansson and Horowitz 1986; Saunders *et al.* 2000), skeletal researchers usually cannot calculate infant mortality rates because it is not possible to determine the total population at risk, i.e. the number of live born infants. This is unfortunate since the infant mortality rate is calculated independent of the age structure of a population, which allows for comparison of communities that are not demographically stationary (expanding or diminishing) (Pressat 1972). It also means that demographic studies of skeletal series are assessing fertility and not mortality, since increased numbers of dead in a cemetery (without the knowledge of the total population at risk to mortality) reflect increased birth rates (see Johansson and Horowitz 1986). Even if fertility might be estimated, we still face the problem of knowing when our samples of immature skeletons are fully representative. Recently, Paine and Harpending (1998) have shown that incomplete representations of infants in skeletal samples will seriously affect estimates of fertility rates.

Is there any hope? Since the variation in infant mortality rates is influenced by culturally based practices that mediate the social/sanitary conditions for a developing infant and the reproductive behaviour of parents, we can ask whether there are some other approaches to examining the infant skeletons that do survive burial and excavation that can tell us something about population history. Investigations might include searches for differential treatment of one sex over the other during infant care, looking for evidence of feeding practices such as the duration of breastfeeding and the introduction of complementary foods, investigating seasonal patterns of weaning and evidence for seasonality of mortality, trying to determine the cause of death from individual skeletons, and evaluating patterns of skeletal growth during infancy.

Determination of sex and infant mortality

Why do we want to know the sex of a deceased infant in a skeletal study? It is quite clear that the number of males exceeds that of females around the time of conception (primary sex ratio, PSR) and at the time of birth (secondary sex ratio, SSR). Despite the potential effect of sex-selective practices in some societies on the secondary sex ratio, strictly defined as the number of male live births per 100 female live births, a number of authors report that the *natural* SSR is 1.06 (Madrigal 1996; Zonta *et al.* 1996). Recent embryological research has demonstrated a clear biological basis for the excess of males in the PSR and the *natural* SSR (Clarke and Mittwoch 1995).

But it is also widely accepted that male foetuses are more susceptible to stress and more likely to die under stressful conditions during pregnancy and after birth. Studies of the sex ratios of foetal deaths have shown that females are more likely to survive late gestation than males, but with overall improvements in medical care and living conditions in developed countries this differential is disappearing (Stinson 1985; Ulizzi and Zonta 1994). Consequently, there is an environmental component to prenatal and neonatal mortality that differentially affects males. But identifying the specific causes of prenatal stresses that might influence male mortality is difficult (Madrigal 1996).

The very existence of excess male mortality at birth under good conditions indicates that natural selection might still be acting on human populations (Zonta *et al.* 1996). In addition, sex ratios of mortality in infancy and childhood can be differentially affected by preferential parental care of one sex over another. In fact, studies of sex differentials in postnatal responses to environmental stress have been inconsistent because male children are usually given preferential treatment in many societies (Stinson 1985). Selective neglect can serve as an effective substitute for family size limitation and, since patrilineal descent is the more common pattern of inheritance in many societies, males are usually preferred. Behavioural observations have shown that the preferred sex receives better food and better quality care.

Presumably, researchers of past populations would be looking for evidence of variation in differential sex ratios of their samples. Differentials may not become evident in infant samples (under 12 months) but perhaps they would if all who died under 5 years of age were examined. In fact, inferring preferential treatment of one sex over another is not so easy, even with documented mortality data because of incompleteness in recording the sex of perinatal deaths (Wall 1981, cited in Knodel 1988). In addition, how does one subtract out the effects of excess male mortality due to stress when the proportions of female mortality are lower in a sample? Where stress is high and mortality levels are high, levels of male mortality are increased (Preston 1976). When good-quality samples exist, skeletal researchers may perhaps infer preferential treatment of males when proportions of female mortality are high but cannot infer preferential treatment of females (historically uncommon anyway) when male mortality levels are high. In his study of sex differentials of child mortality taken from parish registers for a number of 18th and 19th century German villages, Knodel observed that girls experienced a slight mortality advantage close to what might be expected in the absence of any preferential treatment of children of either sex (Knodel 1988).

The major issue for skeletal biologists is their ability to determine the sex of infant skeletons. Sexual dimorphism at birth and during the first year is somewhat magnified because of the influence of higher testosterone levels in males, which peak around the time of prenatal sexual differentiation and then decrease throughout childhood and until puberty (Saunders 1992). While efforts to separate infants' skeletons by sex features of dimorphism have sometimes attained accuracy levels of better than 90%, these are still not good enough to detect minor differences in sex ratios. In addition, one would want to avoid sex determination methods that rely on size differences between the sexes since mortality samples may already be biased due to small males who were at risk of dying (Saunders 1992). In Knodel's study, those German villages experiencing the highest rates of overall infant mortality also showed the highest rates of excess male mortality (average ratio of 1.25), consistent with the observation that greater stress produces greater sex differentials in mortality. Under conditions of high stress but with strong preferential treatment for males the sex ratio of mortality for individuals under 5 years of age will be reduced.

There is now the potential to determine the sex of archaeological skeletons using DNA analysis (Hummel and Herrmann 1994; Faerman *et al.* 1995; Stone *et al.* 1996; Yang *et al.* 1998). But this is far from being a foolproof method readily available to bioarchaeologists. The cost of testing large samples is still prohibitive and problems with extraction and contamination of archaeological specimens have not been overcome (Saunders and Yang 1999). In order to avoid the high likelihood of error (there is a choice of 50% for either sex) several independent tests would need to be done on every specimen. But we might foresee the day when such rigorous testing would be the norm. Yet, we would still face the strong constraint that adequate sampling places on assessing sex ratios in skeletal samples, both adult as well as subadult.

One means of overcoming the problems would be to investigate the relationship between sex and evidence of health or disease, either macroscopically or microscopically. A statistically sufficient sample of infant male and female skeletons whose sex was determined could be compared for indicators of morbidity such as microscopic enamel defects or possibly even for causes of death (see below for a discussion of this topic). Even with an expected mortality bias affecting males, evidence of higher morbidity in the females would be meaningful, supporting the hypothesis of a sex bias in parental care. If the prevalence of morbidity was always higher in males, even in samples where sex discrimination in child care was expected, then we would have to accept the interpretation that mortality bias has too great an effect on reconstructions from skeletal samples.

Detection of feeding practices

There is demographic evidence of a relationship between the duration of nursing and birth spacing in human populations (Gauthier and Henry 1958; Katzenberg *et al.* 1996). It is also widely believed that with the development of agriculture, weaning from breast milk occurred earlier and thereby increased fertility rates. But despite clear demonstrations from recent populations of the complexity of causation in human fertility rates (Campbell and Wood 1994; Katzenberg *et al.* 1996) it still seems worth while for bioarchaeological researchers to investigate infant feeding practices in past populations.

Guy and colleagues (1997) have argued that infants' bones are usually missing from cemeteries and thus it is impossible to put forward conclusions on the effects of weaning in past populations. This presumes that complete samples of infant and juvenile skeletons are absolutely necessary for such an analysis. In fact, a number of researchers have argued that it is possible to detect evidence of a physiological response to weaning if the age of occurrence of specific morbidity markers is homogeneous within skeletal *samples.* A great deal of attention has been given to dental enamel hypoplasias as examples of such a response (for a review, see Katzenberg *et al.* 1996). But this interpretation of the meaning of macroscopic defects of enamel has recently been called into question (Hillson and Bond 1997; Saunders and Keenleyside 1999).

In a study of skeletons and parish records from a 19th century Canadian cemetery (St Thomas' Anglican Church cemetery, Belleville, Ontario), Herring and co-workers (1998) were able to carry out an analysis of the pattern of infant deaths that assisted in the detection of infant feeding practices. Earlier demographic work has shown that cumulative infant mortality rises more steeply than expected in the early months of life if breastfeeding is rare or discontinued soon after birth (Knodel and Kintner 1977). Herring and colleagues showed that calculations of cumulative infant mortality both from a 19th century church cemetery sample and from (corrected) parish registers[2] were consistent with the interpretation that complementary foods were introduced by around 5 months of age (Herring *et al.* 1998). This interpretation has received further support from more recent work on another set of records from 19th century Belleville (L.A. Sawchuk and S.D.A. Burke unpublished results). Nevertheless, this analysis was possible only through the presence of a complete set of church records covering the entire period of cemetery use, so that the total birth rate could be calculated, a situation that is not available to most other skeletal studies.

But there are other more direct ways of approaching this problem. Katzenberg and colleagues (1996) have reviewed the background to the development of chemical methods for reconstructing feeding practices from skeletal samples. Initially, efforts were directed toward investigating changing strontium/calcium (Sr/Ca) ratios in juvenile skeletons to detect the introduction of weaning foods such as cereals that are higher in strontium. However, since the infant digestive system has a lower capacity to discriminate against Sr, bone Sr/Ca ratios must be corrected for this relative discrimination capacity in the diet. Diagenesis or chemical exchange in the burial environment may also homogenise trace element levels in an assemblage of bones, thereby reducing and confusing the ability to detect true biological variability in Sr/Ca ratios that would have meaning concerning infant diets. While there may be ways of overcoming methodological problems with trace element analyses (Sillen 1986) more recently researchers have focused on stable isotopes in bone and teeth to address this question.

[2] Estimates of total birth were derived from baptisms. These estimates require some correction to eliminate later baptisms of individuals who survive infancy.

A pioneering study by Fogel and colleagues (1989) demonstrated that ratios of stable isotopes of nitrogen could be used to detect the consumption and/or loss of breast milk in the diet. This is because nitrogen isotope ratios reflect trophic level differences such that an organism higher on the food web (such as a carnivore) will have tissues that are enriched in the heavier isotope of nitrogen, ^{15}N. Nursing babies are consuming their mother's tissue and so they should be enriched in ^{15}N compared to their mothers. Subsequently, a number of investigations have been directed toward documenting relatively high ^{15}N levels in very young infant skeletons and then a decrease in older infants at what might have been the beginning of weaning (Katzenberg 1991; Tuross and Fogel 1994; White and Schwarcz 1994; Katzenberg and Pfeiffer 1995). A major goal is to detect temporal changes in the length of time of breastfeeding that might have an impact on birth spacing, fertility levels and, therefore, population change. One value of this kind of approach is that it provides point data on individuals. It becomes possible to detect those individuals who might never have been breastfed for health reasons (Katzenberg 1991; Katzenberg and Pfeiffer 1995; Herring *et al.* 1998). On the other hand, because nitrogen isotope ratios are assessed in infant and juvenile skeletons there is again, a potential effect from mortality bias. For example, if we see that isotope ratios are high in most infant skeletons aged under 1 year but drop in infants who lived to more than 1 year, how do we know that this reflects the normal pattern of infant feeding? Individuals who died at a young age might have been breastfed for a longer time simply because they were sick. In addition, the process of the introduction of non-breast milk liquids and foods is a complex behaviour that may take several years. Ideally, researchers wish to track the timing and age-related rate of the weaning process (Schurr 1997). Schurr (1997) has attempted to model the process in a sample of juveniles from a precontact horticultural site in Ohio, controlling for the lag time in collagen formation and therefore nitrogen incorporation by sampling the ends of long bones that should represent the diet consumed at the time of death. The ^{15}N values follow the expected trends of an initial increase after birth, a peak close to 1 year of age and a gradual decline thereafter.

More recently, Wright and Schwarcz (1998) have examined both carbon and oxygen stable isotope ratios from the carbon dioxide component of the apatite in tooth enamel. The archaeological sample comes from Middle Preclassic to Late Postclassic occupations at Kaminaljuyú, Guatemala. These authors used ^{13}C values to identify the introduction of solid foods to infants and children and ^{18}O levels to monitor the decline in breastfeeding. Stable carbon isotope analysis is commonly used to measure maize consumption in New World peoples because it uses the C4 photosynthetic pathway that provides for an enrichment of the heavier stable isotope ^{13}C. Recent experimental work has suggested that the carbon used to construct bone collagen is derived with preference from dietary proteins, while the carbon in bone mineral represents the bulk average of all dietary components (Ambrose and Norr 1993; Tieszen and Fagre 1993). Oxygen in body water is enriched in the heavy isotope ^{18}O, relative to water consumed and since human breast milk is formed from the body water pool, it is heavier in ^{18}O than the water imbibed by the mother.

Wright and Schwarcz measured isotopic ratios in fully formed tooth crowns from adults. They found that the teeth which developed at older ages are more enriched in ^{13}C and more depleted in ^{18}O than teeth which developed earlier. The ^{13}C values changed from first molars to premolars, indicating a change after 2 years of age, when premolars begin to mineralise, while ^{18}O values changed significantly only between third and first molars, suggesting that breastfeeding continued during the period of premolar formation. This study is important in that it attempts to track physiological changes during the growth period of individuals who survived to adulthood.

A more detailed approach in any attempt to reconstruct individual physiological histories (true life history analysis in skeletal samples) would be to carry out microdissections of earlier- and later-forming enamel or dentine within individual teeth to compare isotopic ratios over days, weeks or months instead of years. The precision of timing of tooth formation is made possible by registering enamel formation from the birth event, which is recorded by the neonatal line (FitzGerald 1995). Theoretically, there is potential for this kind of microanalysis in bone as well, if it were possible to accurately determine 'bone age' (chronological timing of bone formation) in a given domain of bone (Frost 1987; Lazenby 1992).

Potentially, a microanalysis of feeding history could then be compared to other information on the morbidity history of an individual or individuals. Knowing the detailed age of formation of a tooth crown, having a thorough history of changing isotopic ratios within daily, weekly or monthly increments of enamel, and a comparison of this information to the dated occurrence of microscopic enamel defects should provide a wealth of information about the relationship between culturally modified feeding practices and levels of health. This kind of research requires analysis of adult skeletons – individuals who had survived at least into their twenties, thirties or beyond – to avoid the problems of 'mortality bias' when focusing on those who only survived a few month or years (Wood *et al.* 1992; Saunders and Hoppa 1993).

Seasonality and mortality

It has long been recognised that mortality risks are higher at some times of the year than at others and investigations of seasonal fluctuations in mortality have been of interest to historical demographers. Seasonal analyses provide partial insight into the kinds of disease causing infant mortality, but may also detect periods during the year when communities experienced particular hardships.

Knodel's (1988) study of parish records from 18th and 19th century German villages observed a pronounced summer mortality peak during the last half of the 19th century. This was true for both neonatal and postneonatal mortality, although more marked for the latter. High summer mortality has been found for a number of historical populations in both temperate and subtropical zones and has been linked to the risks of gastrointestinal infection during periods when the opportunity for contamination of water and solid foods is high and breast milk is supplemented or replaced (Boatler 1983; Cheney 1984; Sawchuk *et al.* 1985; Sawchuk 1993).

Another common pattern of seasonal increases in infant mortality has been found to occur in the winter and early spring. This has been thought to reflect a difference in the importance of respiratory diseases in northern climates versus gastrointestinal diseases in southern ones (Wrigley and Schofield 1981). However, winter and early spring infant mortality may also be correlated with fluctuating periods of famine and general hardship (Moffat 1992).

Gastrointestinal infections and respiratory disease are two of the most common sources of mortality in infant and child populations (Lancaster 1990). Gastroenteritis is known to be the more important of the two, having been identified as the major cause of infant mortality in developing and developed countries up to recent times (Lancaster 1990). However, respiratory disease has been found to be more common in some African populations where breastfeeding may continue for some months after birth but living conditions are poor and rates of infection are high (Saunders and Hoppa 1993). Certainly, there is more complexity in the predisposing risk factors associated with infant mortality than may initially meet the eye or than what we may be able to infer from seasonal variation.

With the benefit of nominative records such as parish registers, which provide information on baptisms as well as burials, historical demographers can also check that seasonal fluctuations in mortality are not influenced by seasonal fluctuations in births (Knodel 1988). A high proportion of deaths in a certain month could be an artefact of a preponderance of births in a particular month. This problem would be virtually impossible to overcome with skeletal samples unless it were possible to determine calendric dates of births. In addition, seasons vary in intensity, time and length of occurrence from place to place and may also show variation from year to year, so that calendric information would be absolutely necessary to overcome the problem of variability at individual sites.

The literature on season of death determination from human burial sites is relatively sparse. Human burials usually take place over long periods of time (centuries or generations) so that only in some circumstances could a cemetery or group of burials be associated with archaeological, artefactual, or environmental evidence of a seasonally occupied site (Spence 1988). Most likely, researchers would want to have direct information on the season of death from the interment location of a burial or from the actual remains of the individual.

There have been suggestions that seasonality of death can be inferred from the orientation of burials, the presence of certain animal bones, artefact types, pollen remains, stomach contents, and even insect pupal cases from historic burials almost 200 years old (Gilbert and Bass 1967; Osgood 1970; Gruber 1971; Binford 1978) but we are not aware of any such studies that have attempted to relate these seasonal patterns to infant mortality.

One direct study of seasonality of mortality from human remains is that of White (1993), who examined sections of hair from Sudanese Nubian mummies for their isotope ratios. She compared ^{13}C ratios in segments of hair taken next to the scalp with those from further along the shaft to look for shifts in food consumption. Hair segments close to the scalp would indicate diet just prior to death, while segments further along the shaft would reflect earlier shifts in food composition. The results indicated evidence for seasonal crop scheduling that involves the cultivation of C3 plants (wheat, barley, and most fruits and vegetables) in the winter and cultivation of the hardier C4 plants (sorghum and millet) in the more difficult summer period. A greater proportion of individuals showed equilibration to a C4 diet, suggesting that most people died during the mid to late summer (including one child aged 1 year at death), a pattern that is also found in contemporary Sudanese populations. This study is significant in its apparent ability to detect a relationship between body chemistry and mortality risk. But it necessarily relies on isotopic comparisons of tissue formed at different periods of time, not a specific pinpointing of the calendrical time of death.

Timing the calendrical point of death in skeletal samples does not appear possible at this stage. There are no other clear methods for determining seasonality from bone or dental tissue. Layering of dental cementum in mammalian teeth has been used in archaeological seasonality studies (Monks 1981) but not without problems. It has also been hypothesised that seasonal variation in growth might produce differences in the density of cells, cemental composition and mineralisation of teeth but this layering has been very difficult to visualise and evaluate in human teeth (Hillson 1996). Even if some unequivocal indicators of seasonality could be found in hard tissues we would still be waiting for the momentous discovery of how to identify calendrical start points of development in past peoples. But if the ultimate goal of the investigation of seasonality in infant mortality samples is to investigate the circumstances leading to death, then perhaps attention should be directed toward scientific advancements in that direction.

Morbidity and cause of death

As many investigators are well aware, the ultimate cause of a person's death may not be as scientifically meaningful as the circumstances that led up to death. For example, eventual expiration from pneumonia does not identify influenza or some other disease that preceded and promoted the demise. Bioarchaeological research on past populations wishes to identify the factors that contributed to mortality but it has been recognised that most morbid conditions will not leave their mark in bone (and that they will be extremely hard to recognise even in mummified tissues).

The chief hazards to the foetus and the neonatal infant are genetic anomalies, other developmental anomalies, birth trauma and the state of health of the mother. Very poor living conditions can adversely affect pregnant mothers and dramatically increase the rate of perinatal deaths. By the end of the first 4 weeks, there remain two chief problems to the infant – adequate nutrition and freedom from infection, especially of the bowel (Lancaster 1990). The leading causes of infant and child mortality in developing countries today are acute respiratory infections and diarrhoeal diseases. Gastrointestinal infection has been referred to as the 'Great Pestilence' because of its pervasiveness as a scourge upon infants (Lancaster 1990). Viewed from another perspective, in all modern populations, the incidence of gastrointestinal infection is highest in the first year of life (Harries 1989) and it remains an important cause of infant mortality for developing and developed countries to this day (Slutsker *et al.* 1996; al-Nahedh 1997; Gessner and Wickizer 1997; Jaffar *et al.* 1997; Kempe *et al.* 1997; Read *et al.* 1997).

It would enhance our efforts at the reconstruction of past populations if it were possible to identify the disease agents of acute respiratory and gastrointestinal infections in archaeological skeletons. Unfortunately the sources of these diseases are diverse, including viruses, bacteria and parasites. On the other hand, modern epidemiological information points to some identifiable differences in infection patterns of gastroenteritis from different pathogenic sources that might be extrapolated to geographical regions and environmental conditions. This might narrow the search path at least for the 'Great Pestilence' rather than instituting any broadly based palaeopathological investigations. For example, rotaviruses are currently the most common viral sources of diarrhoeal disease in most temperate zone counties and their frequency of infection shows winter peaks (Harries 1989; Northrup and Flanigan 1994). Common bacterial sources of gastroenteritis seen in the first year of life in developing countries are enterotoxigenic and invasive *Escherichia coli*, *Salmonella*, *Shigella*, *Campylobacter* and *Vibrio cholerae* (Northrup and Flanigan 1994).

Modes of transmission and clinical symptoms vary with different pathogens. For example, *Shigella* is currently commonly recognised as the major source of gastrointestinal disease in children aged 6 months to 10 years (Northrup and Flanigan 1994). *Shigella* is relatively resistant to food acids and requires only a small inoculum to cause disease. Consequently, it is easily transmitted. Of the protozoans that cause gastroenteritis, *Cryptosporidium*, a coccidian intracellular parasite, is the source of a substantial proportion of diarrhoeal enteritides in infants in developing countries today (Northrup and Flanigan 1994).

The recent development of a relatively precise research field of 'archaeomolecular' biology (ancient DNA research) in which the DNA of deceased humans or the DNA of their disease pathogens is recovered, amplified and sequenced, offers some hope for 'cause of death' research in bioarchaeology (Spigelman and Lemma 1993; Salo *et al.* 1994; Baron *et al.* 1996; Taylor *et al.* 1996). Very recently, there have been reports of the amplification of DNA fragments of enteric bacteria from a mastodon, human coprolite samples and a bog body (Fricker *et al.* 1997; Reinhard 1998; Rhodes *et al.* 1998). However, none of these DNA samples came from hard tissues. A difficulty for skeletal research is that most gastrointestinal diseases do not spread to the circulatory system where they will be found in bone. But the studies of mummies, bog bodies and other kinds of recovered organic samples may offer promise for this area of research. Dental tissues offer some hope in that pulp cavities are well protected and might retain pathogenic DNA under certain circumstances.

While it cannot be assumed that gastrointestinal diseases will always be the major source of infant mortality in past populations, their great prevalence in recent human populations suggests that they should be a source of investigation in bioarchaeology. One can envisage interesting investigative omparisons of sedentary vs. hunting populations in the past where there would be expectations of higher gastrointestinal disease in the sedentary groups. In addition, a recent focus on certain inflammatory and metabolic diseases in infant archaeological samples, reflective of nutritional problems, shows that it is possible to learn more about death-related diseases from this age category (Schultz 1995, 1996, 1998; Ortner 1998; Ortner and Mays 1998).

Infant mortality and the study of growth patterns in the past

Growth-related studies of archaeological skeletal samples assume that the differences among populations reflect the health status and level of well being of those populations (Johnston 1969; Saunders *et al.* 1993b; Larsen 1997). While the general pattern of juvenile growth in past populations appears to conform to that of modern populations, a large number of studies over the past several decades have reported evidence of growth retardation in a diverse array of archaeological samples, suggesting stress was widespread both prehistorically and historically (Lovejoy *et al.* 1990; Hoppa 1992; Saunders 1992; Saunders *et al.* 1993b; Miles and Bulman 1994; Larsen 1997). Only one reported historical sample appears to approach the same magnitude of skeletal growth as the 20th century sample of radiographs from children in the Denver Growth Study, the modern sample that is most often used for comparisons (Maresh 1970). The long bones of infants and children from the 19th century Anglican church of St Thomas in Belleville, Ontario are only slighter shorter for age than those of the modern series (Saunders *et al.* 1993b).

But there are a number of problems that still plague efforts to investigate juvenile growth in past human populations, including methodological barriers and diverse sources of biological variability. The measurements of modern Denver children reported by Maresh (1970) represent a single, limited sample of individuals who were possibly hypernourished, especially during infancy. The measurements come from radiographs, producing a magnification effect that should be accounted for (Miles and Bulman 1994). Chronological age in skeletal samples must be estimated from dental development but reference standards for dental age estimation present their own limitations (Smith 1991; Saunders *et al.* 1993a; Liversidge 1994; Miles and Bulman 1994) so that consistent use of the same reference standards is not found in the literature. It has been demonstrated quite clearly that different populations experience different rates of dental development (Jantz and Owsley 1994; Tompkins 1996), confounding the reliance on a 20th century set of North American tooth formation standards that are presumed to be devoid of influence from environmental stress (Moorrees *et al.* 1963a,b).

A focus on the telescoped period of very rapid, almost linear growth during infancy might help researchers to deal with some of these problems. It is more likely that adverse factors will produce retardation in the first 1 or 3 years because growth rates are highest in infancy (Martorell 1995). Modern studies of living infants and children in low income countries have found that linear skeletal growth faltering is usually detectable at about 6 months of age, reaching a nadir by the second year, after which 'catch-up' growth may occur if conditions are not too severe (Martorell *et al.* 1994).

Until recently, very few osteoarchaeology researchers have concentrated on infants' skeletal growth in archaeological samples despite the fact that infants will constitute the largest proportion of well-preserved subadult samples. Some workers, however, have reported on archaeological samples of American aboriginal populations that are shorter than 20th century American Whites in the first months and years (Mensforth *et al.* 1978; Mensforth 1985; Lovejoy *et al.* 1990). Jantz and Owsley (1994) argued, on the basis of their work with precontact and postcontact Arikara samples, that population variation in dental development can explain these differences. Yet there are reports of growth retardation appearing around 3 or 4 months of age in White European samples aged using the American White dental standards (Farwell and Molleson 1993; Miles and Bulman 1994). It remains to be shown whether there are more localised population differences in dental development (between American Whites and northwest European Whites) and/or secular trends in dental development that might still explain the differences. If it can be shown clearly that the long bones of one sample are shorter than those of individuals of comparable age in another sample, this would allow for inferences about related causes of mortality (acute vs. chronic). Consequently, Jantz and Owsley (1994) controlled for variation in dental development rates within the Arikara sample and observed differences between pre- and postcontact groups that they attribute to changes in nutritional stress. Saunders and colleagues (1993b) observed similar bone lengths in modern American Whites and 19th century Canadians and thus inferred that mortality in the historical sample was attributable to acute conditions. In the future, a more comprehensive approach to growth-related research in past populations would involve the concurrent investigation in samples of infant bone mass (Huss-Ashmore *et al.* 1982), pathological data (Ribot and Roberts 1996), other skeletal elements besides the long bones (Miles and Bulman 1995) and all of those aspects that have been discussed in the previous sections of this chapter. For example, those workers who have claimed to identify growth retardation among the infant remains of archaeological skeletal samples cite changing infant feeding patterns as the causal explanation for their observations.

Summary and conclusions

This chapter has shown that by evaluating the demographic, biological and medical literature on current and recent human populations the bioarchaeologist can identify determining factors affecting mortality and morbidity in specific age categories. In this case, we have examined infants, and the crucial role that infant mortality plays in signifying the quality of life in current and past societies.

While it is clear that underrepresentation of infants is a common problem for archaeological skeletal samples, there are probably a number of approaches that can be taken with partial samples to reconstruct health conditions. Researchers need to be aware of the demographers' definitions of various age classes and the attendant problems with identifying those categories in population comparisons. While it is rarely possible to derive accurate infant mortality rates (IMR) from skeletal samples, there are a number of methods that have been devised to assess the representativeness and quality of samples (Paine 1989; Jackes 1992; Hoppa 1996). This chapter highlights some of the exciting developments now taking place in bioarchaeological research and offers suggestions for new ways to approach interpopulation comparisons of infant skeletons in samples.

A reliable determination of sex of infant skeletons would expand the scope of retrievable information, particularly for the relationship between sex and stress levels as well as the possibility of detecting preferential care of one sex over the other. Currently, an active body of research on dietary reconstruction suggests that it might be possible to specify the detailed feeding history of individuals, particularly during the crucial period of infancy when dramatic dietary changes take place. This work could help to resolve debate over the relationship between breastfeeding, weaning and skeletal evidence of physiological stress. Recent developments in ancient DNA research offer the expectation of specifying the infectious diseases causing death in past populations. We suggest that the focus of some of this research should be on the identification of infectious diarrhoeal disease occurring during infancy, since infant mortality is such a useful marker of overall population well being. Finally, a comprehensive approach to reconstructing infant growth in past populations places the focus on the most susceptible portion of the population and would help to confirm inferences about population well being made from the other areas of research mentioned above.

Although it is still considerably difficult to 'decode' the hard tissues of archaeological samples, probably two of the most promising and exciting areas of research are 'cause of death' studies and dietary reconstruction. However, a comprehensive approach to analysing infant skeletal samples can work to piece together a picture of the influence of the two most important factors affecting infant growth and health, diet and disease. We would argue that research into the skeletal biology of infant morbidity, mortality and growth has a future that, though challenging, offers opportunities for examining the lives of past peoples in some detail.

References

al-Nahedh N (1997) Infant mortality in the rural Riyadh region of Saudi Arabia. *Royal Society for Health* **117**, 106–109.

Ambrose SH and Norr L (1993) Experimental evidence for the relationship of the carbon isotope ratios of whole diet and dietary protein to those of bone collagen and carbonate. In *Prehistoric Human Bone: Archaeology at the Molecular Level*, ed. J.B. Lambert and G. Grupe, pp. 1–37. Berlin: Springer-Verlag.

Baron H, Hummel S and Herrmann B (1996) *Mycobacterium tuberculosis* complex DNA in ancient human bones. *Journal of Archaeological Science* **23**, 667–671.

Beaton GH (1989) Small but healthy? Are we asking the right question? *Human Organization* **48**, 30–39.

Beaton GH (1992) The Raymond Pearl memorial lecture, 1990: Nutritional research in human biology: changing perspectives and interpretations. *American Journal of Human Biology* **4**, 159–178.

Binford LR (1978) *Nunamiut Ethnoarchaeology*. New York: Academic Press.

Boatler JF (1983) Patterns of infant mortality in the Polish community of Chappell Hill, Texas, 1895–1944. *Human Biology* **55**, 9–18.

Campbell KL and Wood JW (1994) *Human Reproductive Ecology: Interactions of Environment, Fertility, and Behavior*. New York: Academy of Sciences.

Cheney RA (1984) Seasonal aspects of infant and childhood mortality: Philadelphia, 1865–1920. *Journal of Interdisciplinary History* **14**, 561–585.

Clarke CA and Mittwoch U (1995) Changes in the male to female ratio at different stages of life. *British Journal of Obstetrics and Gynecology* **102**, 677–679.

Cook DC and Buikstra JE (1979) Health and differential survival in prehistoric populations: prenatal dental defects. *American Journal of Physical Anthropology* **51**, 649–664.

Deutsch D, Tam O and Stack MV (1985) Postnatal changes in size, morphology and weight of developing postnatal deciduous anterior teeth. *Growth* **49**, 202–217.

Eli I, Sarnat H and Talmi E (1989) Effect of the birth process on the neonatal line in primary tooth enamel. *Pediatric Dentistry* **11**, 220–223.

Faerman M, Filon D, Kahila G, Greenblatt CL, Smith P and Oppenheim A (1995) Sex identification of archaeological human remains based on amplification of the X and Y amelogenin alleles. *Gene* **167**, 327–332.

Farwell DE and Molleson TI (eds.) (1993) *Excavations at Poundbury* 1966–80, vol. 2, *The Cemeteries*. Dorset Natural History and Archaeological Society, Monograph Series, no. 11.

Fasekas IG and Kosa F (1978) *Forensic Foetal Osteology*. Budapest: Akademia Kiado.

FitzGerald CM (1995) Tooth crown formation and the variation of enamel microstructural growth markers in modern humans. PhD thesis, University of Cambridge.

Fogel M, Tuross N and Owsley DW (1989) Nitrogen isotope tracers of human lactation in modern and archaeological populations. Carnegie Institution, Annual Report of the Director; Geophysical Laboratory.

Forfar JO and Arneil GC (1978) *Textbook of Paediatrics*. New York: Churchill Livingstone.

Fricker EJ, Spigelman M and Fricker CR (1997) The detection of *Escherichia coli* DNA in the ancient remains of Lindow Man using the polymerase chain reaction. *Letters in Applied Microbiology* **24**, 351–354.

Frost H (1987) Secondary osteon populations: an algorithm for determining mean bone tissue age. *Yearbook of Physical Anthropology* **30**, 221–238.

Gauthier E and Henry L (1958) *La Population de Crulai, Paroisse Normande*. Paris: L'Institut National d'Études Demographiques.

Gessner BD and Wickizer TM (1997) The contribution of infectious diseases to infant mortality in Alaska. *Pediatric Infectious Diseases Journal* **16**, 773–779.

Gilbert BM and Bass WM (1967) Seasonal dating of burials from the presence of fly pupae. *American Antiquity* **32**, 534–535.

Gruber JW (1971) Patterning in death in a late prehistoric village in Pennsylvania. *American Antiquity* **36**, 64–76.

Guy H, Masset C and Baud C (1997) Infant taphonomy. *International Journal of Osteoarchaeology* **7**, 221–229.

Harries AD (1989) Gastro-intestinal infections. *Practitioner* **233**, 75–78.

Hassan AH (1986) The effectiveness of a home visiting program to reduce infant mortality using a 'risk approach'. MSc thesis, McMaster University.

Herring DA, Saunders SR and Boyce G (1991) Bones and burial registers: infant mortality in a 19th century cemetery from Upper Canada. *Council for Northeast Historical Archaeology Journal* **20**, 54–70.

Herring DA, Saunders SR and Katzenberg MA (1988) Investigating the weaning process in past populations. *American Journal of Physical Anthropology* **105**, 425–439.

Hillson S (1996) *Dental Anthropology*. Cambridge University Press.

Hillson S and Bond S (1997) Relationship of enamel hypoplasia to the pattern of tooth crown growth: a discussion. *American Journal of Physical Anthropology* **104**, 89–104.

Hoppa RD (1992) Evaluating human skeletal growth: an Anglo-Saxon example. *International Journal of Osteoarchaeology* **2**, 275–288.

Hoppa RD (1996) Representativeness and bias in cemetery samples: implications for palaeodemographic reconstructions of past populations. PhD dissertation, Department of Anthropology, McMaster University.

Hoppa RD and Gruspier KL (1996) Estimating diaphyseal length from fragmentary subadult skeletal remains: implications for palaeodemographic reconstructions of a southern Ontario ossuary. *American Journal of Physical Anthropology* **100**, 341–354.

Hummel S and Herrmann B (1994) Y-chromosome DNA from ancient bones. In *Ancient DNA*, ed. B. Herrmann and S. Hummel, pp. 205–217. New York: Springer-Verlag.

Huss-Ashmore R, Goodman AH and Armelagos GJ (1982) Nutritional inference from paleopathology. In *Advances in Archaeological Method and Theory*, ed. M.B. Schiffer, pp. 395–474. New York: Academic Press.

Hutchins LA (1998) Standards of infant long bone diaphyseal growth from a late nineteenth century and early twentieth century almshouse cemetery. MA thesis, University of Wisconsin-Milwaukee.

Jackes M (1992) Paleodemography: problems and techniques. In *The Skeletal Biology of Past Peoples: Research Methods*, ed. S.R. Saunders and M.A. Katzenberg, pp. 189–224. New York: Wiley-Liss.

Jaffar S, Leach A, Greenwood AM, Jepson A, Muller O, Ota MO, Bojang K, Obaro S and Greenwood BM (1997) Changes in the pattern of infant and childhood mortality in upper river division, the Gambia, from 1989 to 1993. *Tropical Medicine and International Health* **2**, 28–37.

Jantz RL and Owsley DW (1994) Growth and dental development in Arikara children. In *Skeletal Biology in the Great Plains: Migration, Warfare, Health, and Subsistence*, ed. D.W. Owsley and R.L. Jantz, pp. 247–258, Washington, DC: Smithsonian Institution Press.

Johansson SR and Horowitz S (1986) Estimating mortality in skeletal populations: influence of the growth rate on the interpretation of levels and trends during the transition to agriculture. *American Journal of Physical Anthropology* **71**, 233–250.

Johnston FE (1969) Approaches to the study of development variability in human skeletal populations. *American Journal of Physical Anthropology* **31**, 335–341.

Johnston FE (1978) Somatic growth of the infant and preschool child. In *Human Growth, 2 Postnatal Growth*, ed. F. Falkner and J.M. Tanner, pp. 91–116. London: Plenum Press.

Katzenberg MA (1991) Stable isotope analysis of remains from the Harvie family. In *The Links that Bind: The Harvie Family Nineteenth Century Burying Ground*, ed. S.R. Saunders and R. Lazenby, pp. 65–69. Occasional Papers in Northeastern Archaeology, no. 5. Dundas, Ontario: Copetown Press.

Katzenberg MA and Pfeiffer S (1995) Nitrogen isotope evidence for weaning age in a nineteenth century Canadian skeletal sample. In *Bodies of Evidence: Reconstructing History through Skeletal Analysis*, ed. A.L. Grauer, pp. 139–160. New York: John Wiley and Sons.

Katzenberg MA, Herring DA and Saunders SR (1996) Weaning and infant mortality: evaluating the skeletal evidence. *Yearbook of Physical Anthropology* **39**, 177–199.

Kempe A, Wise PH, Wampler NS, Cole FS, Wallace H, Dickinson C, Rinehart H, Lezotte DC and Beaty B (1997) Risk status at discharge and cause of death for postneonatal infant deaths: a total population study. *Pediatrics* **99**, 338–344.

Knodel JE (1983) Seasonal variation in infant mortality: an approach with applications. *Annales de Demographie Historique* 208–230.

Knodel JE (1988) *Demographic Behavior in the Past: A Study of Fourteen German Village Populations in the Eighteenth and Nineteenth Centuries.* Cambridge: Cambridge University Press.

Knodel J and Kintner H (1977) The impact of breast feeding patterns on the biometric analysis of infant mortality. *Demography* **14**, 391–409.

Lallo JW, Armelagos GJ and Mensforth RP (1977) The role of diet, disease, and physiology in the origin of porotic hyperostosis. *Human Biology* **49**, 471–483.

Lancaster HO (1990) *Expectations of Life: A Study in the Demography Statistics, and History of World Mortality.* New York: Springer-Verlag.

Larsen CS (1997) *Bioarchaeology: Interpreting Behaviour from the Human Skeleton.* Cambridge: Cambridge University Press.

Lazenby RA (1992) A geometric and microradiographic study of functional adaptation in the human skeleton. PhD Dissertation, McMaster University.

Liversidge HM (1994) Accuracy of age estimation from developing teeth of a population of known age (0–5.4 years). *International Journal of Osteoarchaeology* **4**, 37–46.

Lovejoy CO, Russell KF and Harrison ML (1990) Long bone growth velocity in the Libben population. *American Journal of Human Biology* **2**, 533–542.

Madrigal L (1996) Sex ratio in Escazú, Costa Rica, 1851–1901. *Human Biology* **68**, 427–436.

Maresh MM (1970) Measurements from roentgenograms. In *Human Growth and Development*, ed. R.W. McCammon, pp. 157–90. Springfield: Charles C. Thomas.

Martorell R (1995) Promoting healthy growth: rationale and benefits. In *Child Growth and Nutrition in Developing Countries*, ed. D. Pelletier and H. Alderman, pp. 15–31. Ithaca, NY: Cornell University Press.

Martorell R, Khan LK and Schroeder DG (1994) Reversibility of stunting: epidemiological findings in children from developing countries. *European Journal of Clinical Nutrition* **48**, Suppl. 1, S45–S57.

Mensforth RP (1985) Relative tibia long bone growth in the Libben and Bt-5 prehistoric skeletal populations. *American Journal of Physical Anthropology* **68**, 247–262.

Mensforth RP, Lovejoy CO, Lallo JW and Armelagos GJ (1978) The role of constitutional factors, diet and infectious disease in the etiology of porotic hyperostosis and periosteal reactions in prehistoric infants and children.

Medical Anthropology **2**, 1–59.
Miles AW and Bulman JS (1994) Growth curves of immature bones from a Scottish Island population of sixteenth to mid-nineteenth century: limb bone diaphyses and some bones of the hands and feet. *International Journal of Osteoarchaeology* **4**, 121–136.
Miles AW and Bulman JS (1995) Growth curves of immature bones from a Scottish Island population of sixteenth to mid-nineteenth century: shoulder girdle, ilium, pubis and ischium. *International Journal of Osteoarchaeology* **5**, 15–28.
Moffat T (1992) Infant mortality in an Aboriginal community: an historical and biocultural analysis. MA thesis, McMaster University.
Monks GG (1981) Seasonality studies. In *Advances in Archaeological Method and Theory*, ed. M.B. Schiffer, pp. 177–240. New York: Academic Press.
Moorrees CFA, Fanning EA and Hunt EE (1963a) Age variation of formation stages for ten permanent teeth. *Journal of Dental Research* **42**, 1490–1502.
Moorrees CFA, Fanning EA and Hunt EE (1963b) Formation and resorption of three deciduous teeth in children. *American Journal of Physical Anthropology* **21**, 205–213.
Nawrocki SP (1995) Taphonomic processes in historic cemeteries. In *Bodies of Evidence: Reconstructing History through Skeletal Analysis*, ed. A.L. Grauer, pp. 49–68. New York: Wiley-Liss.
Northrup RS and Flanigan TP (1994) Gastroenteritis. *Pediatrics Review* **15**, 461–472.
Ortner DJ (1998) *Workshop X: Human Skeletal Disease with an Emphasis on Disease Caused by Malnutrition.* Paleopathology Association Meetings, Salt Lake City, Utah.
Ortner DJ and Mays S (1998) Dry-bone manifestations of rickets in infancy and early childhood. *International Journal of Osteoarchaeology* **8**, 45–55.
Osgood C (1970) *Ingalik Material Culture.* Yale University Publications in Anthropology no. 22.
Owsley DW and Jantz RL (1985) Long bone lengths and gestational age distributions of postcontact period Arikara Indian perinatal infant skeletons. *American Journal of Physical Anthropology* **68**, 321–328.
Paine RR (1989) Model life table fitting by maximum likelihood estimation: a procedure to reconstruct paleodemographic characteristics from skeletal age distributions. *American Journal of Physical Anthropology* **79**, 51–61.
Paine RR and Harpending HC (1998) Effect of sample bias on paleodemographic fertility estimates. *American Journal of Physical Anthropology* **105**, 231–240.
Pharoah PD and Morris JN (1979) Postneonatal mortality. *Epidemiology Review* **1**, 170–183.
Pressat R (1972) *Demographic Analysis: Methods, Results, Applications.* Chicago: Aldine.
Preston SH (1976) *Mortality Patterns in National Populations: With Special Reference to Recorded Causes of Death.* New York: Academic Press.
Read JS, Troendle JF, Klebanoff MA (1997) Infectious disease mortality among infants in the United States, 1983 through 1987. *American Journal of Public Health* **87**, 192–198.

Reinhard K (1998) *Survey of Prehistoric Parasites from South America.* Paleopathology Association Meetings, Salt Lake City, Utah.

Rhodes AN, Urbance JW, Youga H, Corlew-Newman H, Reddy CA, Klug MJ, Tiedje JM and Fisher DC (1998) Identification of bacterial isolates obtained from intestinal contents associated with 12,000–year-old mastodon. *Applied Environmental Microbiology* **64**, 651–658.

Ribot I and Roberts C (1996) A study of non-specific stress indicators and skeletal growth in two mediaeval subadult populations. *Journal of Archaeological Science* **23**, 67–79.

Roth EA (1992) Applications of demographic models to paleodemography. In *Skeletal Biology of Past Peoples: Research Methods*, ed. S.R. Saunders and M.A. Katzenberg, pp. 175–188. New York: Wiley-Liss.

Salo WL, Aufderheide AC, Buikstra JE and Holcomb TA (1994) Identification of *Mycobacterium tuberculosis* DNA in a Pre-Columbian Peruvian mummy. *Proceedings of the National Academy of Sciences, USA* **91**, 2091–2094.

Saunders SR (1992) Subadult skeletons and growth related studies. In *Skeletal Biology of Past Peoples: Research Methods*, ed. S.R. Saunders and M.A. Katzenberg, pp. 1–20. New York: Wiley-Liss.

Saunders SR and Hoppa RD (1993) Growth deficit in survivors and non-survivors: biological mortality bias in subadult skeletal samples. *Yearbook of Physical Anthropology* **36**, 127–152.

Saunders SR and Keenleyside A (1999) Enamel hypoplasia in a Canadian historic sample. *American Journal of Human Biology*, in press.

Saunders SR and Yang D (1999) Sex determination: XX or XY from the human skeleton. In *Chilled to the Bone: Case Studies in Forensic Anthropology*, ed. S. Fairgrieve. Springfield: Charles C. Thomas, in press.

Saunders SR, De Vito C, Herring DA, Southern R and Hoppa RD (1993a) Accuracy tests of tooth formation age estimations for human skeletal remains. *American Journal of Physical Anthropology* **92**, 173–188.

Saunders SR, Hoppa RD and Southern R (1993b) Diaphyseal growth in a nineteenth century skeletal sample of subadults from St. Thomas' Church, Belleville, Ontario. *International Journal of Osteoarchaeology* **3**, 265–281.

Saunders SR, Herring DA and Boyce G (1995) Can skeletal samples accurately represent the living populations they come from? The St. Thomas' Cemetery Site, Belleville. In *Bodies of Evidence: Reconstructing History Through Skeletal Analysis*, ed. A.L. Grauer, pp. 69–99. New York: Wiley-Liss.

Saunders SR, Herring DA, Sawchuk L, Boyce G, Hoppa RD and Klepp S (2000) The St. Thomas' Anglican Church Cemetery Project. In *History of Health and Nutrition in the Western Hemisphere*, ed. R. Steckel and J. Rose. Cambridge: Cambridge University Press, in press.

Sawchuk LA (1993) Societal and ecological determinants of urban health: a case study of prereproductive mortality in 19th century Gibraltar. *Social Sciences and Medicine* **76**, 875–892.

Sawchuk LA, Herring DA and Waks L (1985) Evidence of a Jewish advantage: a study of infant mortality in Gibraltar, 1840–1959. *American Anthropologist* **87**, 616–625.

Schultz M (1995) The role of meningeal diseases in the mortality of infants and children in prehistoric and historic populations. [Abstract]. *American Journal of Physical Anthropology Supplement* **20**, 192.

Schultz M (ed.) (1996) *Advances in Paleopathology and Osteoarchaeology* vol. I, Göttingen: Verlag Erich Goltze.

Schultz M (1998) Differences in infant health patterns in prehistory: a contribution to etiology and epidemiology of infant diseases. Paper presented at the XIIth European Meeting of the Paleopathology Association, Prague-Pilsen, Czech Republic.

Schurr MR (1997) Stable nitrogen isotopes as evidence for the age of weaning at the Angel Site: a comparison of isotopic and demographic measures of weaning age. *Journal of Archaeological Science* **24**, 919–927.

Sellier P, Tillier A-M and Bruzek J (1997) The estimation of the age at death of perinatal and postnatal skeletons: methodological reassessment and reliability. *American Journal of Physical Anthropology Supplement* **24**, 208.

Shapiro S, Schlesinger ER and Nesbitt Jr, REL (1968) *Infant, Perinatal, Maternal, and Childhood Mortality in the United States.* Cambridge, MA: Harvard University Press.

Sillen A (1986) Biogenic and diagenetic Sr/Ca in Plio-Pleistocene fossils of the Omo Shungura Formation. *Paleobiology* **12**, 311–323.

Skinner M and Dupras T (1993) Variation in birth timing and location of the neonatal line in human enamel. *Journal of Forensic Sciences* **38**, 1383–1390.

Slutsker L, Bloland P, Steketee RW, Wirima JJ, Heymann DL and Breman JG (1996) Infant and second-year mortality in rural Malawi: causes and descriptive epidemiology. *American Journal of Tropical Medicine and Hygiene* **55**, 1 supplement, 77–81.

Smith BH (1991) Standards of human tooth formation and dental age assessment. In *Advances in Dental Anthropology*, ed. M.A. Kelley and C.S. Larsen, pp. 143–168. New York: Wiley-Liss.

Spence MW (1988) The human skeletal material of the Elliott Site. *Kewa* **88**(4), 10–20.

Sperduti A, Bondioli L, Prowse TL, Salomone F, Yang D, Hoppa RD, Saunders SR and Macchiarelli R (1997) Reconstructing life conditions of the juvenile population of Portus Romae in 2nd–3rd cent. AD. Paper presented at Canadian Association for Physical Anthropology Meetings, London, Ontario.

Spigelman M and Lemma E (1993) The use of the polymerase chain reaction (PCR) to detect *Mycobacterium tuberculosis* in ancient skeletons. *International Journal of Osteoarchaeology* **3**, 137–143.

Stinson S (1985) Sex differences in environmental sensitivity during growth and development. *Yearbook of Physical Anthropology* **28**, 123–148.

Stone AC, Milner GR, Paabo S and Stoneking M (1996) Sex determination of ancient human skeletons using DNA. *American Journal of Physical Anthropology* **99**, 231–238.

Storey R (1986) Perinatal mortality at pre-Columbian Teotihuacan. *American Journal of Physical Anthropology* **69**, 541–548.

Taylor GM, Crossey M, Saldanha J and Waldron T (1996) DNA from *Mycobac-*

terium tuberculosis identified in mediaeval human skeletal remains using polymerase chain reaction. *Journal of Archaeological Science* **23**, 789–798.

Tieszen LL and Fagre T (1993) Effect of diet quality and composition on the isotopic composition of respiratory CO_2, bone collagen, bioapatite and soft tissues. In *Prehistoric Human Bone: Archaeology at the Molecular Level*, ed. J.B. Lambert and G. Grupe, pp. 121–155. Berlin: Springer-Verlag.

Tompkins RL (1996) Human population variability in relative dental development. *American Journal of Physical Anthropology* **99**, 79–102.

Tuross N and Fogel ML (1994) Stable isotope analysis and subsistence patterns at the Sully site. In *Skeletal Biology in the Great Plains: Migration, Warfare, Health and Subsistence*, ed. D.W. Owsley and R.L. Jantz, pp. 283–289. Washington, DC: Smithsonian Institution Press.

Ulizzi L and Zonta LA (1994) Sex ratio and selection by early mortality in humans: fifty-year analysis in different ethnic groups. *Human Biology* **66**, 1037–1048.

United Nations Statistical Office (1994) *Demographic Yearbook*. New York: United Nations.

White CD (1993) Isotopic determination of seasonality in diet and death from Nubian mummy hair. *Journal of Archaeological Science* **20**, 657–666.

White CD and Schwarcz HP (1994) Temporal trends in stable isotopes for Nubian mummy tissues. *American Journal of Physical Anthropology* **93**, 165–187.

Wood JW, Milner GR, Harpending HC and Weiss KM (1992) The osteological paradox: problems in inferring prehistoric health from skeletal samples. *Current Anthropology* **33**, 343–358.

Wright LE and Schwarcz HP (1998) Stable carbon and oxygen isotopes in human tooth enamel: identifying breastfeeding and weaning in prehistory. *American Journal of Physical Anthropology* **106**, 1–18.

Wrigley EA (1969) *Population and History*. New York: McGraw-Hill.

Wrigley EA and Schofield RS (1981) *The Population History of the English, 1541–1871*. Cambridge, MA: Harvard University Press.

Yang DY, Eng B, Waye JS, Dudar JC and Saunders SR (1998) Technical note: improved DNA extraction from ancient bones using silica-based spin columns. *American Journal of Physical Anthropology* **105**, 539–543.

Zonta LA, Astolfi P and Ulizzi L (1996) Early selection and sex composition in Italy: a study at the regional level. *Human Biology* **68**, 415–426.

9 *Sources of variation in estimated ages at formation of linear enamel hypoplasias*

ALAN H. GOODMAN AND RHAN-JU SONG

Introduction

Precise understanding of the pattern and timing of dental development is key to a number of basic and applied research areas. If one assumes that 'dental age' corresponds closely to chronological age, dental age is often used as a proxy for chronological age in contemporaneous and bioarchaeological studies of health and nutrition. In order to better understand the evolution of events associated with human tooth formation, one needs to know the degree of variation in tooth formation within, and among, human populations and primate species. The importance of understanding tooth crown development timing is also fundamental to studies of linear enamel hypoplasias (LEHs). The purpose of this chapter is to explore some of the issues involved in estimating an individual's age at formation of LEHs.

Enamel hypoplasias are a class of developmental defect of enamel that is morphologically distinguished by a macroscopically observable area of decreased enamel thickness (Sarnat and Schour 1941; Goodman *et al.* 1980; Fig. 9.1). These developmental defects are direct consequences of a temporary disruption to the ameloblasts during their primary function of secreting enamel matrix (Shawashy and Yaeger 1986). Because of their permanence and direct aetiological link to disrupted ameloblastic function, enamel hypoplasias have been intensively studied as retrospective evidence of ameloblastic disruption. Moreover, because secretory phase ameloblasts are sensitive to a wide variety of 'stressors' or 'insults' such as undernutrition, trauma, infection and temperature extremes, a resulting enamel hypoplasia may ultimately be related to these underlying conditions and processes (Kreshover 1960; Cutress and Suckling 1982).

A wealth of studies use enamel hypoplasias as a mirror onto overall levels of physiological perturbation and environmental causes of that

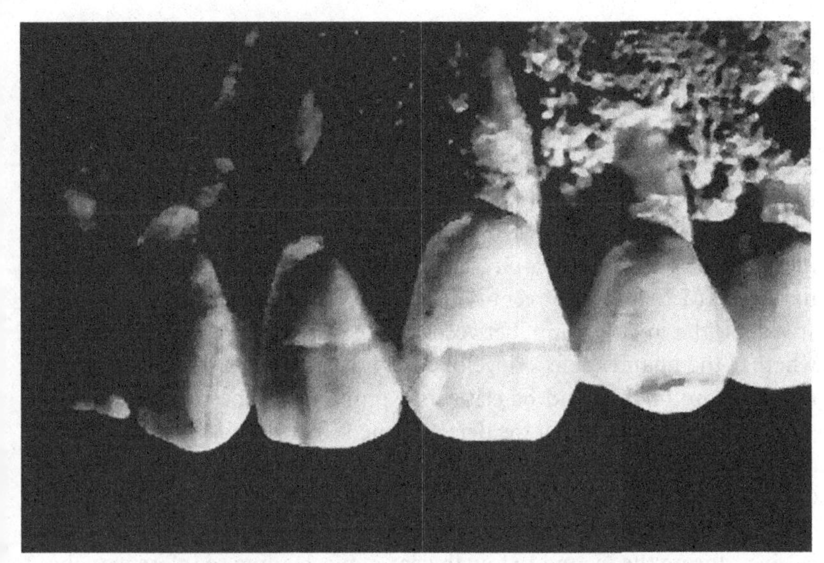

Fig. 9.1. Photographic example of chronological enamel hypoplasias on anterior maxillary permanent teeth. Note that the position of the hypoplasias varies on the different tooth crowns, presumably due to variation in the time of initial crown development and the rate of subsequent development. The lateral incisor and canine LEH is further toward the occlusal surface than is the defect on the premolar, probably because these teeth began development before the premolar.

perturbation during tooth crown development (Goodman and Rose 1990, 1991). In clinical research, for example, enamel hypoplasias have been compared between individuals with prenatal brain damage and a control group in order to determine whether the brain damage might be related to some yet undiscovered prenatal growth disruption (Judes *et al.* 1985). Similar clinical research has been undertaken in comparing low and normal birth weight individuals (Seow and Perham 1990), nutritionally supplemented and non-supplemented individuals (Goodman *et al.* 1991), and individuals with and without a history of infectious disease (Hall 1989). Finally, particularly intense research has taken place in bioarchaeology, in which the frequency of enamel defects has been used to discern levels of physiological disruption in different past groups. For example, Cohen and Armelagos's (1984) volume *Paleopathology at the Origins of Agriculture* includes over 15 reports on enamel hypoplasia in early agricultural groups. The number of studies of enamel developmental defects, the majority of which are applications such as these, is now around 1000.

Most enamel hypoplasias run linearly around part, or all, of tooth crowns, roughly parallel to the cemento-enamel junction, and perpendicular to the long axis of teeth (Fig. 9.1). This linear form reflects the chronological development of tooth crowns, also seen in histological section by the location of striae of Retzius and surface perikymata (Hillson and Bond 1997). Because of this characteristic of most LEHs, they were once referred to as chronological enamel hypoplasia (Sarnat and Schour 1941).

Because there are so few methods to retrospectively assess the age of individuals at the time of a physiological event, understanding an individual's age at LEH development is critical. For example, in the above-noted study of the relationship between brain damage and LEH, knowing whether the enamel defect developed before birth, and if so exactly how much before birth, could be critical to pinpointing the trimester, month, week or day during which the defect formed. In bioarchaeological studies, as well as other population studies, it is common to reconstruct the distribution of LEH by age at development. Many studies of the last two decades have shown peak periods of LEH in permanent incisors and canines during the second to fourth year of life, leading to inferences about the degree of weaning or postweaning stress (Corruccini *et al.* 1985; Goodman *et al.* 1987). However, imprecision in estimating the age at formation muddles our ability to test this assertion.

In this chapter we outline and explore factors that may affect the accuracy of estimates of age at LEH formation. We first present a brief historical review of methods for estimation of age at formation of defects and the construction, from these data, of patterns of LEH formation in populations. We then categorise possible sources of variation in estimation of age at formation of an LEH. Finally, we focus on three factors: tooth size variation, the length of time in which appositional or cuspal enamel is formed (and, thus, not amenable to macroscopic observation), and the choice of developmental standard. These factors were chosen because their effects can be estimated and they may be the three most important sources of variation. The consequences of these factors are illustrated with applications to LEH data from the Classic Period Maya from Altun Ha, Belize (see Song 1997) and a historical sample, the Hamann–Todd osteological collection from greater Cleveland, Ohio (see Goodman 1988).

A brief history of methods for estimating age at LEH formation

The timing of LEH has been central to the larger history of research into dental development. Smith (1991), in her exceptional review of the development of tooth chronologies, notes that hypoplastic banding of teeth probably 'inspired' many early pictorial charts of tooth formation. This observation is clear from a perusal of early works on tooth development. For example, Logan and Kronfeld (1933) and Kronfeld (1935) reproduced a picture of an individual with an enamel hypoplasia assumed to have developed in the first year. Their point in so doing was to show that the upper lateral incisors commence development after the other anterior teeth: the upper central incisors are the only anterior teeth without hypoplastic involvement on the incisal surface. Just as the degree to which anatomical information on development provided valid information on development timing, it was assumed that the location of enamel hypoplasias did the same. The location of enamel hypoplasias was especially important for noting relative development of teeth at the time of a common event. The relative position of the 'birth line' in histological section is an excellent example of this. In a sense, this work has provided inspiration for recent histological methods of timing dental events (Bromage and Dean 1985; Macho and Wood 1995).

While dental texts noting the timing of tooth formation were common by the 18th century, the first extensive table of tooth formation events is generally credited to Legros and Magitot (1880, 1881), soon followed by Black (1883) and Berten (1895). All of these authors produced tables and graphs that illustrated the development of the different teeth by zones (the more occlusal/incisal zones developing earlier) to the later developing cervical zones (Smith 1991). In general, methods are not well articulated and sample sizes are not made explicit. One is left to assume that the charts were based on small sample sizes, with a great deal of extrapolation between ages.

Although these early works correctly identified general patterns of development, variation among these standards in the estimated chronological age of events concerned Logan and Kronfeld (1933). For example, Black (1883) estimated that second molars began calcification around the sixth year, whereas Legros and Magitot (1880, 1881) estimated that calcification began around the third year. In an effort to add greater precision and certainty to the age at tooth formation, Logan and Kronfeld (1933) initially studied 25 individuals. Of these, 19 died before 2 years of age and only three were older than 4.5 years at death. Later on, an additional five individuals were added. Of greatest concern is the low number of individuals over the age of 2 years, and especially older than 4.5 years. In addition, like all studies based on postmortem examination, results may be biased by being based on individuals who died, possibly from conditions that may also affect the rate of tooth development.[1] Thus, while Logan and Kronfeld's research is an improvement over the prior generation of research, it remains limited. Through its influence on the standard of Massler and co-workers (1941), it is the basis for nearly all research into the timing of LEHs.

In two publications in 1941, among the two most widely cited in dental research, dental timing and enamel hypoplastic chronologies were again intimately related. In the paper on developmental timing, Massler *et al.* (1941) extrapolated from Logan and Kronfeld (1933). Concurrently written, the classic enamel hypoplasia paper of Sarnat and Schour (1941) was the first to explicitly present a group pattern of enamel defects based on translating the location of defects on tooth crowns to ages at formation. Sarnat and Schour (1941) estimated the age at formation of enamel hypoplasias for 60 individuals with hypoplasias from the Chicago area. They recorded the age at development of defects to monthly periods, extrapolating from the Massler *et al.* (1941) standard. According to these authors, most defects occurred before the first year of life and were of a few months' duration.

Despite the authors' obvious enthusiasm for using enamel hypoplasias as kymographic records of stress, little follow-up research was done over the next quarter century. However, on the basis of the fact that their work is widely cited in dental texts (Goodman 1988), this lack of replicability seems to be more due to a general acceptance of the universality of their results than lack of enthusiasm for their work.

An additional factor that may have contributed to the non-replicability of the Sarnat and Schour (1941) study is lack of details regarding their method of translating from LEH location on tooth crowns to age at formation. In 1966, Swärdstedt completed a Swedish dental thesis on the pattern of enamel hypoplastic defects found at Westerhus, a Mediaeval population from Jämtland/Mid-Sweden. In addition to being one of the first applications of enamel hypoplasias to bioarchaeology, this study clarified the method of translating enamel hypoplasia locations to age at formation. Swärdstedt (1966) explicitly diagrammed the location of enamel

[1] Long-term illness and undernutrition may slow dental eruption and also to some degree dental formation rates. In the practice of estimating ages at LEH formation, research is done both on samples who died, possibly due to long-term illness and undernutrition, and living individuals. Thus the possibility of error is introduced; however, it is difficult to estimate its magnitude. In one regard, the older, postmortem-based standards may be more appropriate for archaeological populations, and radiographic standards may be more appropriate for living populations. This idea, however, is further complicated in LEH studies because these are often based on survivors of early stress.

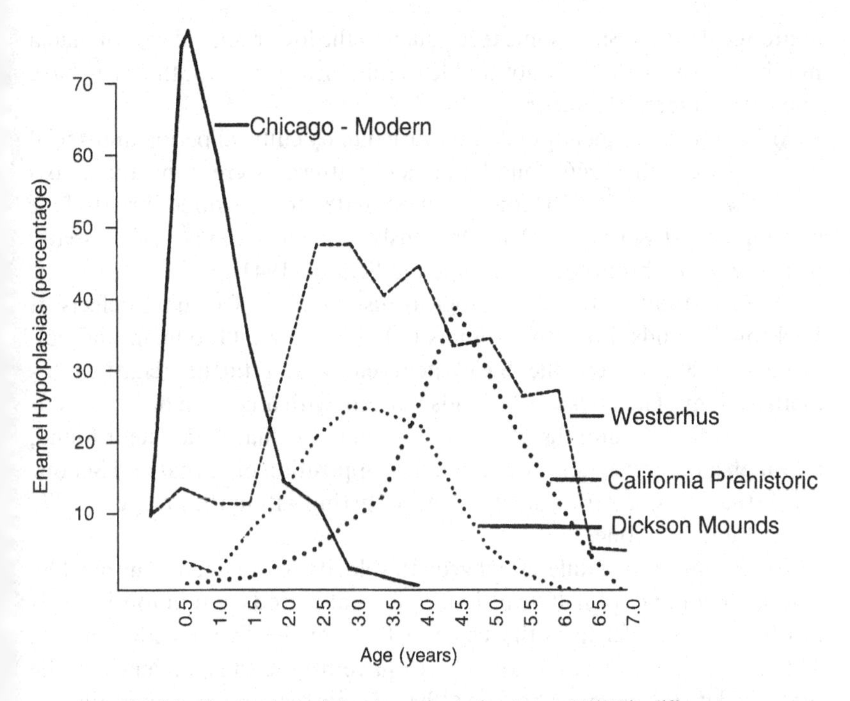

Fig. 9.2. Comparison of the chronological pattern of LEH in different populations. The Chicago-Modern sample is from Sarnat and Schour (1941). The peak in LEH is within the first year. Westerhus is from Swärdstedt (1966), California Prehistoric is from Schultz and McHenry (1975) and Dickson Mounds (Lewiston, Illinois, USA) is from Goodman *et al.* (1980). These four populations are among the first studied and represent some of the variation that is seen with different methods and different standards of use.

defects against an age at formation (Fig. 9.2). Like Sarnat and Schour (1941), he used the Massler *et al.* (1941) ages at beginning and end of crown formation as 'anchors'. He then divided tooth crowns into half-year developmental zones. Thus, a tooth such as a lower central incisor, which develops between birth and 4 years, was divided into eight zones. Because severe occlusal attrition is often found on archaeological teeth, Swärdstedt measured the locations of LEHs from the cemento–enamel junction (CEJ).

Half-year zones varied in width across teeth as a function of the number of zones per tooth and crown height, with crown heights based on previously published data. Interestingly, zones within teeth are not of similar width. Although this point is not discussed by Swärdstedt (1966), and its rationale is hard to reconstruct from the initial work of Massler *et al.* (1941), it appears that Swärdstedt may have divided crowns into anatomical thirds. Since some teeth, such as the lower central incisor, had a number of zones that did not divide evenly into thirds, tooth thirds have unequal numbers of zones.[2]

Although the frequency of defects varied in by cultural period and social class, Swärdstedt (1966) found that age patterns were constant across groups with a peak in LEH formation estimated to be around three to four developmental years (Fig. 9.3). Obviously, this chronology varies greatly from the prior chronology of Sarnat and Schour (1941).

In their study of the chronological distribution of enamel defects at Dickson Mounds, Lewiston, Illinois (AD 950–1300), Goodman and colleagues (1980) redrew the Swärdstedt chart and further clarified his methodology. The Dickson Mounds pattern, with a peak in defect formation at around 30 months of age, was similar to what Swärdstedt found, although a bit earlier, and countered the supposition of Sarnat and Schour (1941) that their pattern – most defects occurring within the first year of life – was a universal one.

The assumption of differential growth velocity within teeth is implied by a variable number of half-year developmental periods within tooth thirds in what we will refer to as the 'chart method' or the 'Swärdstedt method'. This assumption, however, appears to have little correspondence to the primary data (Goodman and Rose 1990). To correct for these limitations, a series of regression equations were formulated to translate age at development of an LEH from its location, starting and ending age at crown calcification and crown height (Table 9.1; Goodman and Rose 1990). The basic forms of this linear (constant velocity) regression equation is:

$$\text{age at formation} = -[(1/\text{velocity}) \times \text{distance of LEH from CEJ}] + \text{age at crown completion}$$

In this equation velocity is in mm/year and is thus related to tooth crown height and the total time of tooth development. All ages are in years and should be expressed as developmental ages. The 'velocity term' is negative and inverse (years/mm).

Although this 'rationalises' the methodology and makes clear the assumptions upon which it is based, the assumptions themselves are not challenged. The Massler chart method has been used in ageing LEHs in

[2] Swärdstedt's allocation of half-year developmental zones to tooth crown thirds is more complicated still. For some teeth with a number of half-year developmental periods evenly divisible by 3, Swärdstedt still assigned a variable number of half-year periods to different thirds. For the upper central incisor and canine, the incisal third 'received', at the expense of the middle third, an extra half-year developmental period. Conversely, the lower canine's cervical third received an extra half-year period at the expense of its incisal third.

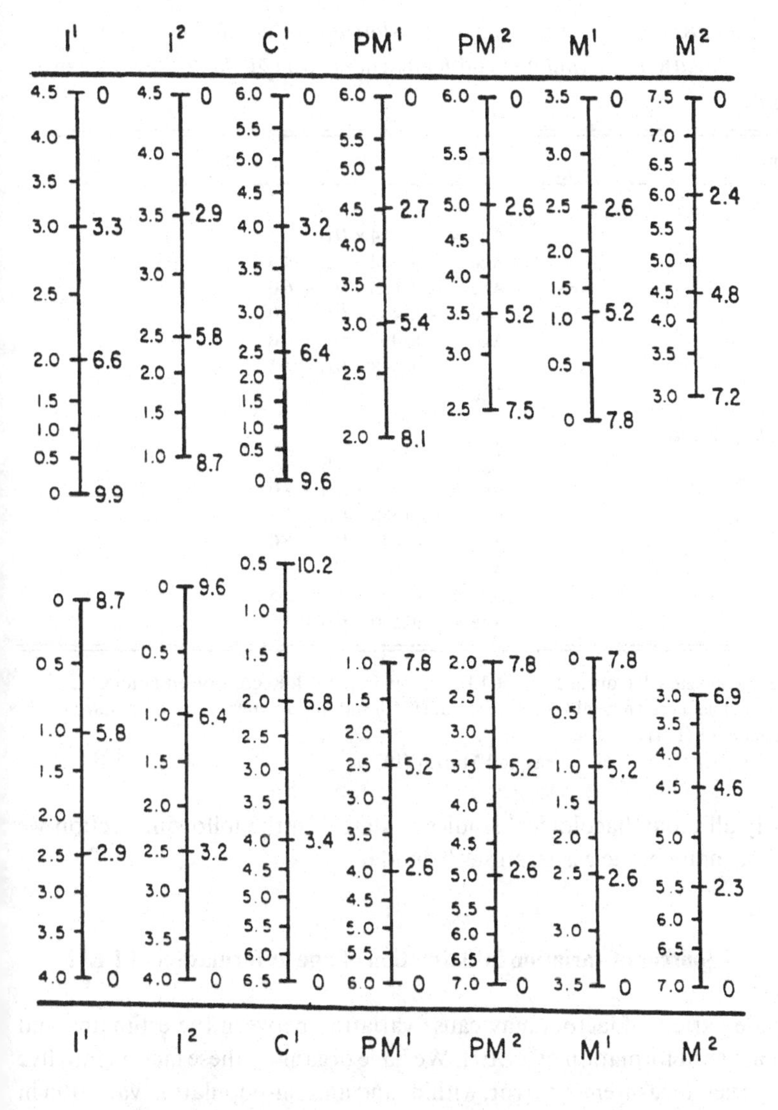

Fig. 9.3. Swärdstedt chart for estimating the age at development of enamel hypoplasias. The numbers to the right of each line are distances from the cemento-enamel junction to the LEH (in mm). The numbers to the left of each line are the corresponding ages at formation of an LEH in years. The crown heights are from Swärdstedt (1966) and the beginning and ending of crown development are adopted from Massler *et al.* (1941; also see Fig. 9.2). Super- and subscripts denote upper and lower jaws, respectively. (From Goodman *et al.* 1980, modified from Swärdstedt 1966.)

Table 9.1. *Regression equations to translate age at development of an LEH from its location, starting and ending age at crown calcification and crown height*

Tooth	Formulae[a]
Maxillary teeth	
I1	Age = − (0.454 × *Ht*) + 4.5
I2	Age = − (0.402 × *Ht*) + 4.5
C	Age = − (0.625 × *Ht*) + 6.0
P1	Age = − (0.494 × *Ht*) + 6.0
P2	Age = − (0.467 × *Ht*) + 6.0
M1	Age = − (0.448 × *Ht*) + 3.5
M2	Age = − (0.625 × *Ht*) + 7.5
Mandibular teeth	
I1	Age = − (0.460 × *Ht*) + 4.0
I2	Age = − (0.417 × *Ht*) + 4.0
C	Age = − (0.588 × *Ht*) + 6.5
P1	Age = − (0.641 × *Ht*) + 6.0
P2	Age = − (0.641 × *Ht*) + 7.0
M1	Age = − (0.449 × *Ht*) + 3.5
M2	Age = − (0.580 × *Ht*) + 7.0

[a]Age, age in years; Ht, distance of the LEH in mm from CEJ. Regression equations are based on mean crown heights of Swärdstedt (1966) and the crown formation standard of Massler *et al.* (1941).
Based in part on work of Murray and Murray (1989).

nearly all bioarchaeological studies to date.[3] In the following section we outline many of the assumptions behind it.

Sources of variation in estimation of age at formation of LEH

A wide variety of factors may cause variation between the estimated and actual age at formation of a LEH. We have organised these factors into five categories: measurement error, within- and among-population variation in developmental timing, within- and among-population variation in crown heights, interpretation of tooth histology and development, and choice of developmental standard (Table 9.2).

[3] As we discuss below, a few authors have tried various corrections to the general method. For example, Hodges and Wilkinson (1990) have demonstrated the importance of population variation in crown heights and Song (1997) and Wright (1994, 1997) have both corrected for hidden cuspal enamel (discussed on pp. 221–224) and have 'shortened' the time of development of one or more canines.

Table 9.2. *Possible sources of variation and types of error in estimating the age at formation of linear enamel hypoplasias. Sources of error and variation focused upon in this chapter are shown in bold print*

Measurement error
Instrumentation and operator errors
Obliteration of CEJ anchor point
Measurement from center versus borders of LEH
Variation in developmental timing
Population variation in developmental timing
Sex differences in developmental timing
Within-population individual timing variation
Crown height variation
Population variation in crown height
Sex differences in crown height
Within population individual size variation
Correction related to enamel histology and developmental pattern
Changes in velocity of enamel surface extension
Buried cuspal enamel
Choice and interpretation of developmental standard

Measurement error refers to any variation in measured location of an enamel defect compared to the actual location (with location typically expressed as distance between the CEJ and the defect). While of obvious importance, especially when a defect is to be matched to a specific life event, this error is likely to be relatively minor compared to others noted below. In our experience, a typical error in measurement is of the order of 0.1–0.2 mm, which equals about a month of developmental time in permanent teeth and about a week for faster-developing deciduous teeth. Furthermore, at least part of this error is random and, therefore, may have little effect on population parameters.

Within- and among-population variation in tooth crown calcification is also likely and underexplored (Tompkins 1996). This variation may be characterised as overall advancement or delay, or variation in pattern of relative dental development (some teeth advanced or delayed relative to other teeth). Overall advancement in dental eruption has been proposed for females versus males (Demirjian 1986; Moorrees *et al.* 1963) better nourished individuals (Garn *et al.* 1965) and in African populations (Evelyth and Tanner 1976). The most easily interpreted results are of females vs. males. The extent of these differences may be around 2–3% (Gleiser and Hunt 1955; Garn *et al.* 1958; Demirjian and Levesque 1980;

Smith 1991). The cause of the difference may in part be due to the larger average size of teeth in males. If so, then the difference is not a sex difference per se (Moss and Moss-Salentijn 1977; also see Blakey and Armelagos 1985). While these population differences may be kept in mind when considering the accuracy of estimated ages at formation of enamel defects, at this time they are too small and too poorly understood to be systematically evaluated. More precise data are required before we can adequately address the importance of population variation in developmental timing.

In the following section, we focus on three sources of variation in timing of LEH development. One of these factors, tooth crown size variation, has previously been studied by Hodges and Wilkinson (1990). We review their results, extend their analysis, and then focus on two additional sources of variation: buried cuspal enamel and choice of developmental standard.

Buried cuspal enamel

Enamel matrix formation begins occlusally at the dentine–enamel border. As ameloblasts retreat toward the eventual surface of the tooth, cervically located ameloblasts begin forming the enamel matrix. The synchronised and combined action of these secretory ameloblasts forms a developing front of enamel, shaped like a loop in cross-section (Fig. 9.4). The striae of Retzius in this incipient enamel form partial circles (or loops) which begin and end in the inner enamel. Because these striae never reach the surface of the tooth, any disruption in enamel formation at the time of their formation may only be detected histologically, and not by surface observation.

The question for LEH analyses concerns the length of time in which domed enamel is formed, or 'How much time passes between initial crown formation and the first striae that reach the outer (sleeve) enamel?' At the present time, the few histological studies that address this have produced variable estimates. Early research by Bromage and Dean (1985) ascribed a period of 6 months for appositional enamel in permanent incisors. However, in reviewing the findings since then, primarily in the work of Dean and Beynon (1991), Bullion (1987), Bromage and Dean (1985), and FitzGerald (1995), this half-year period appears to be an underestimation (Table 9.3).

According to Bullion (1987), on average about 150 striae characterise incisors, over 180 striae are in permanent canine enamel, and molars average 120–150 striae (also see Hillson and Bond 1997). From these totals, the first 20–35 striae (approximately 15–20%) are 'hidden' in incisors and canines, while significantly more, the first 50–80 striae (approximately

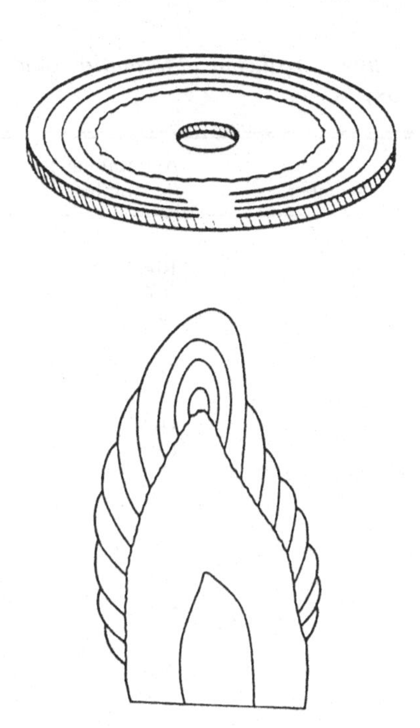

Fig. 9.4. Illustration of cuspal enamel (reproduced from Aiello and Dean 1990, with permission).

30–50%), are not detectable on molar surfaces (Hillson 1996; Hillson and Bond 1997). The results of histological analyses indicate that the majority of appositional times, the time of formation of dome enamel, fall between the range 8–12 months (Table 9.3).

We are aware of only two LEH studies that have corrected for hidden period of appositional enamel (Wright 1994, 1997; Song 1997). Both studied ancient Maya samples. In her work, Wright estimated 1 year of non-visible 'buried' enamel for all tooth types. Song incorporated times from the survey table (Table 9.3; also see FitzGerald 1995: Table 12.5) to arrive at estimations ranging between 7 months (lower I1) to 10 and 12 months for all other teeth (Song 1997: Table 6.2.2).

What is the effect of correcting for this initial period of hidden development? To illustrate this correction, dental data from the Hamann–Todd human osteological collection were selected for study (Goodman 1988). The Hamann–Todd collection is made up of ethnically diverse individuals who died in the Cleveland area (northern Ohio) around the end of the 19th century and first quarter of the 20th century. Because no-one claimed them

Table 9.3. *A survey of estimated appositional enamel formation times from the literature (see FitzGerald 1995)*

Tooth source	Sample size	Appositional period (in months)
Maxillary I1		
Boyde (1963)	1	10.4
Bullion (1987) – mod.	3	9.0
Bullion (1987) – arch.	1	8.7
FitzGerald (1995)	2	11.4
Maxillary I2		
Bullion (1987) – mod.	4	11.6
Bullion (1987) – arch.	1	10.4
FitzGerald (1995)	2	12.8
Maxillary C		
Bullion (1987) – mod.	4	8.6
Dean and Beynon (1991)	1	12.3
FitzGerald (1995)	2	17.3
Mandibular I1		
Bromage and Dean (1985)	10	6.0
Bullion (1987) – mod.	6	7.5
Bullion (1987) – arch.	1	8.1
FitzGerald (1995)	3	6.3
Mandibular I2		
FitzGerald (1995)	*3*	*10.3*
Mandibular C		
Bullion (1987) – mod.	2	9.2
Bullion (1987) – arch.	1	9.5
FitzGerald (1995)	3	16.1

mod., modern; arch., archaeological.

and they underwent postmortem examination, individuals are generally considered to have been from the lower socioeconomic strata of society. Our data include the locations of moderate to severe hypoplasias on unworn mandibular canines and maxillary central incisors, the most frequently hypoplastic teeth. A total of 79 hypoplasias were found on the canines and 60 on the incisor.

On the basis of the regression equations, we plotted the estimated age at formation of an LEH on the *y*-axis vs. its location from the CEJ on the *x*-axis (Fig. 9.5). Formation ends at the CEJ at 4.5 years, on the basis of the development standard of Massler *et al.* (1941), and starts at the cusp,

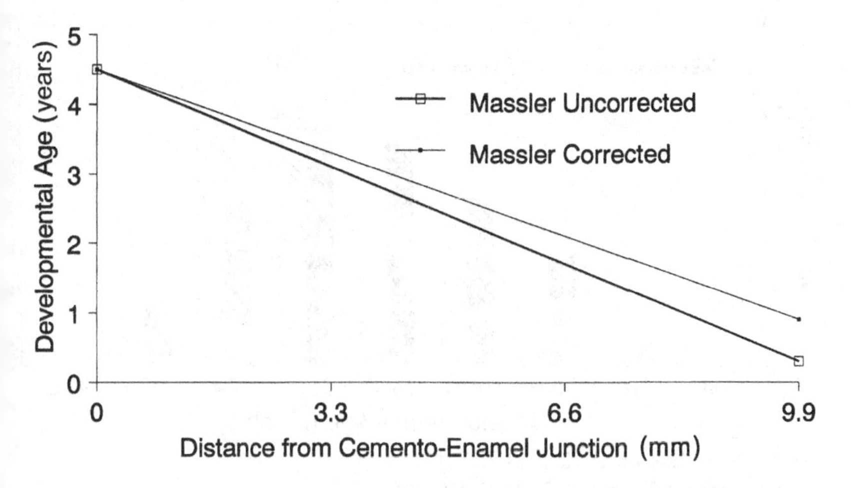

Fig. 9.5. The effect of a 6 month correction for buried cuspal enamel on the estimated age at development of an LEH on the upper central incisor.

9.9 mm from the CEJ, on the basis of the crown heights used initially by Swärdstedt (1966).

The effect of correcting for buried cuspal enamel is to change the length of time that the external crown forms and, thus, to change the 'velocity' term in the equation. Defects toward the cusp will be aged relatively older, while those toward the CEJ will be similarly aged. For example, using the conservative estimate of 6 months of hidden enamel, as originally proposed by Bromage and Dean (1985), yields a regression line that is 6 months advanced from the uncorrected line at the incisal tip. As the CEJ anchor point has not changed, the distance (in years) between the corrected and uncorrected regression lines decrease as they approach the CEJ (Fig. 9.5). In other words, the correction affects LEH ageing by a maximum of 6 months incisally, and then has a progressive decrease in consequence.

Applying the 6 month correction to the Hamann–Todd sample pushes back the mean age of an LEH by less than 3 months. For the upper central incisor, this correction pushes back the mean age at development from 2.51 to 2.73 years, and, for the lower canine, this correction pushes back the mean age at development from 4.10 to 4.30 years. As anticipated, the effect of the correction is asymmetrical, greater for earlier developing defects and least for later developing defects. Aged by the uncorrected method, the earliest incisor defect is aged to 0.64 years, compared to 1.07 years with the correction. Conversely, the latest defect for the incisor is aged at 3.91 and 3.97 years, respectively, by the two methods.

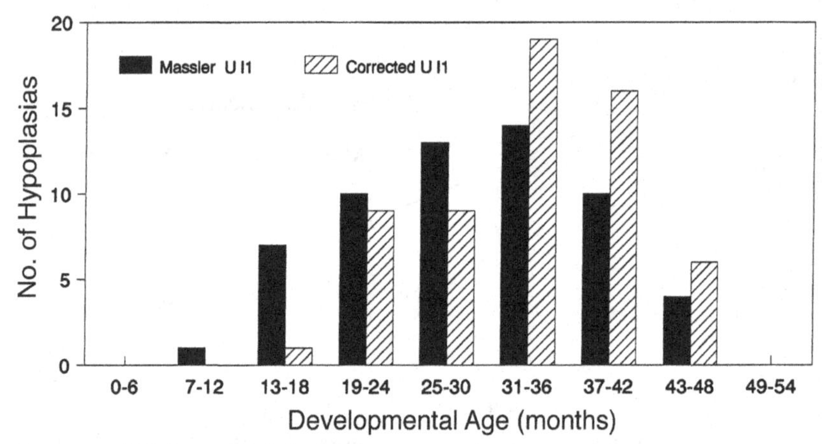

Corrected Mean = 2.73 (±0.74) yrs
Uncorrected Mean + 2.51 (±0.72) yrs

Fig. 9.6. The chronology of LEH by half-year developmental periods for the Hamann–Todd population with and without a 6 month correction for buried cuspal enamel.

The effect that the buried cuspal enamel correction has on the half-year distribution of enamel defects suggests a slight shift to older defects (Fig. 9.6). With the uncorrected method, eight defects are aged at 18 months or earlier, whereas only one defect is so aged by the corrected method. However, the peak half-year period is the same, 31–36 months of developmental age (Fig. 9.6).

A second test of correcting for hidden cuspal enamel involves re-examining the estimated mean age at development of enamel defects found in the Maya of Altun Ha, Belize (Song 1997; Table 9.4). Table 9.4 presents the results with and without correcting for hidden cuspal enamel. Corrections are for either 0.8 or 1.0 years, a rounded approximation of the averages from Table 9.3. The effect of the correction is to increase the mean age at formation of an LEH from a minimum of 0.24 years to 0.37 years (approximately 3 months to 4.5 months).

The change in estimated age at LEH formation based on hidden cuspal enamel are directional and biologically justified. We recommend that the correction should be implemented in future studies. Remaining questions concern the size of the correction and whether it should be used both with standards based on postmortem examination and those based on radiographic appearance of calcification.

Table 9.4. *The effect of correcting for hidden cuspal enamel on the estimated age (in years) of formation of LEH for the Ancient Maya of Altun Ha, utilising mean crown height standards of Swärdstedt (1966) and developmental timing of Massler et al. (1941)*

Tooth	Uncorrected: Equation A (no cuspal, Swärd, Massler)[a]	Equation A (Mean LEH age)	Cuspal enamel time (yrs) (Song)	Ext. dev. time (yrs) (Massler w/o cuspal)	Equation C (Cuspal, Swärd, Massler)	Equation C (Mean LEH age)	Diff. (yrs)
Upper							
I1	− 0.455x + 4.5	2.83	0.8	3.7	− 0.374x + 4.5	3.13	0.3
I2	− 0.402x + 4.5	3.19	1.0	2.5	− 0.287x + 4.5	3.56	0.37
C	− 0.625x + 6.0	3.78	1.0	5.0	− 0.521x + 6.0	4.15	0.37
P3	− 0.494x + 6.0	4.56	0.8	3.2	−0.395x + 6.0	4.85	0.29
P4	− 0.467x + 6.0	4.51	0.8	2.7	− 0.360x + 6.0	4.86	0.35
M1	− 0.449x + 3.5	2.44	0.8	2.7	− 0.371x + 3.5	2.68	0.24
M2	− 0.625x + 7.5	5.94	0.8	3.7	− 0.496x + 7.5	6.21	0.27
Lower							
I1	− 0.460x + 4.0	2.32	0.6	3.4	− 0.391x + 4.0	2.57	0.25
I2	− 0.417x + 4.0	2.47	0.8	3.2	− 0.333x + 4.0	2.77	0.3
C	− 0.588x + 6.5	4.31	1.0	5.0	− 0.490x + 6.5	4.67	0.36
P3	− 0.641x + 6.0	3.79	0.8	4.2	− 0.538x + 6.0	4.15	0.36
P4	− 0.641x + 7.0	5.35	0.8	4.2	− 0.538x + 7.0	5.62	0.27
M1	− 0.449x + 3.5	2.37	0.8	2.7	− 0.346x + 3.5	2.63	0.26
M2	− 0.580x + 7.0	5.28	0.8	3.2	− 0.464x + 7.0	5.63	0.35

cuspal, correction for cuspal enamel; w/o cuspal; without this correction; *x*, crown height in mm; Swärd, Swärdstedt (1966) standard; Massler, Massler *et al.* (1941) standard; Song, according to Song (1997); Diff., difference in mean age of formation (equation C − equation A).

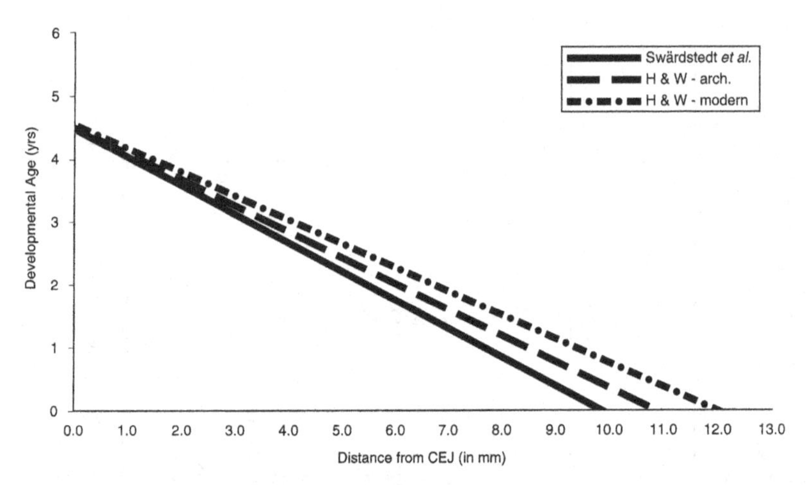

Fig. 9.7. The effect of correcting for larger upper central incisors on the estimated age at formation of an LEH. Three regression lines are shown. The darkest (bottom) line is based on the Swärdstedt tooth size, the intermediate line on tooth size based on the archaeological sample of Hodges and Wilkinson (1990) and the upper line on the modern sample tooth size of Hodges and Wilkinson (1990).

Crown height variation

The regression equations (Table 9.1) are also based on crown heights that were originally used in the Swärdstedt study. These are referenced to a study by Krogh-Poulsen (1950). Unfortunately, these crown heights have consistently been found to be small, and thus their use in the regression equation may have consequence, too, for the estimated age at formation of LEH. In the following, two types of variation are discussed: variation within populations and variation among populations.

Hodges and Wilkinson (1990) measured crown sizes for two samples: individuals with hypoplasias from prehistoric and historic skeletal collections from the eastern USA, and recently extracted teeth from donors living near urban areas in the northeastern and midwestern USA. They found that the maxillary central incisor increased from the 9.9 mm estimate in the original Swärdstedt method, to 10.86 mm in the archaeological sample and 11.99 in the contemporaneous sample. The change in mandibular canine is slightly less dramatic: from 10.2 mm in the original method to 10.92 mm and 11.78 mm in the archaeological and contemporaneous samples, respectively.

Assuming that larger teeth do not take longer to develop than shorter teeth of the same type produces a regression equation with a slower velocity but the same starting point. Showed graphically (Fig. 9.7) with the

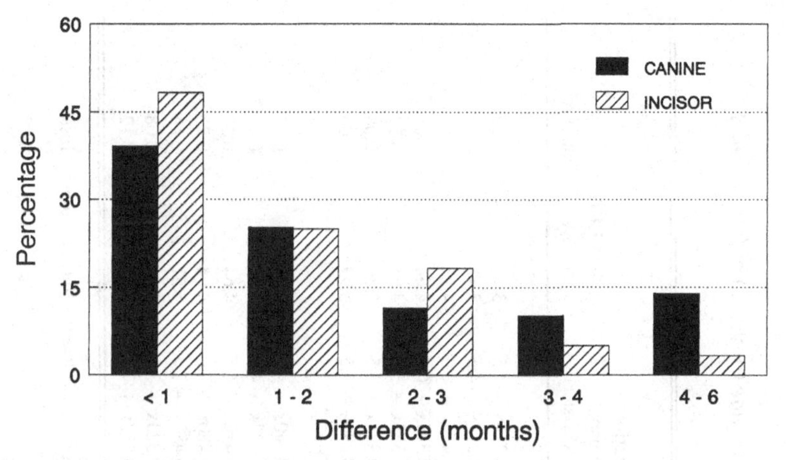

Fig. 9.8. The variation introduced in estimated ages of individual LEHs when the individual tooth size is substituted for the population's mean tooth size.

Hodges and Wilkinson incisor crown heights, it is clear that the majority of the variation occurs with earlier developing defects. The consequence of this correction is precisely similar in form to the hidden cuspal enamel correction.

Hodges and Wilkinson (1990) found that using the population's own crown heights may significantly change the estimated age at formation of an LEH. Interestingly, their half-year peak age at formation of an enamel defect does not change for the archaeological sample, and changes by just a half-year for the larger-toothed, contemporaneous sample.

Hodges and Wilkinson (1990) also corrected for individual crown heights and found that this has little consequence on population parameters over the population correction. We concur that correcting for individual crown heights has little consequence for estimated age at formation of an LEH. With the Hamann–Todd sample, we examined the extent of variation between the individual and population crown height estimations (Fig. 9.8). The mean difference was 1.35 months for the upper central incisor and 1.99 months for the lower canine. As we expected, the consequence on the population distribution of LEH is negligible.

Similar to the crown heights found by Hodges and Wilkinson (1990), the Altun Ha crown heights are consistently larger than those originally used by Swärdstedt (Table 9.5). Only one crown height was actually smaller, the maxillary first molar, and many crowns were over 1.0 mm larger, including

Table 9.5. *The effect of correcting for mean population crown height on the estimated age (in years) of formation of LEH for the Ancient Maya of Altun Ha*

Tooth	Uncorrected equation A (no cuspal, Swärd, Massler)	Equation A (mean LEH age)	Swärd (Mean height)	AH (Mean height)	Crown Height diff.	Equation B (no cuspal, AH Massler)	(Mean LEH age)	Diff. (yrs)
Upper								
I1	– 0.455x + 4.5	2.83	9.9	11.77	1.87	– 0.314x + 4.5	3.1	0.27
I2	– 0.402x + 4.5	3.19	8.7	10.44	1.74	– 0.239x + 4.5	3.41	0.22
C	– 0.625x + 6.0	3.78	9.6	11.51	1.91	– 0.434x + 6.0	4.15	0.37
P3	– 0.494x + 6.0	4.56	8.1	8.56	0.46	– 0.374x + 6.0	4.64	0.08
P4	– 0.467x + 6.0	4.51	7.5	8.11	0.61	– 0.333x + 3.5	4.63	0.12
M1	– 0.449x + 3.5	2.44	7.8	7.27	– 0.53	– 0.371x + 3.5	2.36	– 0.08
M2	– 0.625x + 7.5	5.94	7.2	7.46	0.26	– 0.496x + 7.5	5.99	0.05
Lower								
I1	– 0.460x + 4.0	2.32	8.7	9.99	1.29	– 0.340x + 4.0	2.54	0.22
I2	– 0.417x + 4.0	2.47	9.6	9.91	0.31	– 0.323x + 4.0	2.51	0.04
C	– 0.588x + 6.5	4.31	10.2	12.03	1.83	– 0.416x + 6.5	4.64	0.33
P3	– 0.641x + 6.0	3.79	7.8	8.75	0.95	– 0.480x + 6.0	4.04	0.25
P4	– 0.641x + 7.0	5.35	7.8	8.62	0.82	– 0.487x + 7.0	5.51	0.16
M1	– 0.449x + 3.5	2.37	7.8	8.35	0.55	– 0.323x + 3.5	2.45	0.08
M2	– 0.580x + 7.0	5.28	6.9	7.39	0.49	– 0.433x + 7.0	5.4	0.12

cuspal, correction for cuspal enamel; w/o cuspal; without this correction; x, crown height in mm; Swärd, Swärdstedt (1966) standard; Massler, Massler *et al.* (1941) standard; AH, Altun Ha; Diff., difference in mean age of formation (equation B – equation A).

Table 9.6. *The effect of correcting for mean population crown height length and hidden cuspal enamel on the estimated age (in years) of formation of LEH for the Ancient Maya of Altun Ha*

Tooth	Uncorrected equation A (no cuspal, Swärd, Massler)	Uncorrected equation A (Mean LEH age)	Corrected equation A (cuspal, AH, Massler)	Corrected (Mean LEH age)	Diff. (yrs)
Upper					
I1	– 0.455x + 4.5	2.83	– 0.314x + 4.5	3.35	0.52
I2	– 0.402x + 4.5	3.19	– 0.239x + 4.5	3.72	0.53
C	– 0.625x + 6.0	3.78	– 0.434x + 6.0	4.49	0.68
P3	– 0.494x + 6.0	4.56	– 0.374x + 6.0	4.91	0.35
P4	– 0.467x + 6.0	4.51	– 0.333x + 6.0	4.94	0.43
M1	– 0.449x + 3.5	2.44	– 0.371x + 3.5	2.62	0.18
M2	– 0.625x + 7.5	5.94	– 0.496x + 7.5	6.26	0.32
Lower					
I1	– 0.460x + 4.0	2.32	– 0.340x + 4.0	2.76	0.44
I2	– 0.417x + 4.0	2.47	– 0.323x + 4.0	2.81	0.34
C	– 0.588x + 6.5	4.31	– 0.416x + 6.5	4.95	0.64
P3	– 0.641x + 6.0	3.79	– 0.480x + 6.0	4.35	0.56
P4	– 0.641x + 7.0	5.35	– 0.487x + 7.0	5.75	0.4
M1	– 0.449x + 3.5	2.37	– 0.323x + 3.5	2.69	0.32
M2	– 0.580x + 7.0	5.28	– 0.433x + 7.0	5.72	0.44

cuspal, correction for cuspal enamel; w/o cuspal; without this correction; x, crown height in mm; Swärd, Swärdstedt (1966) standard; Massler, Massler *et al.* (1941) standard; AH, Altun Ha; Diff., difference in mean age of formation (corrected – uncorrected).

five out of six of the anterior teeth. A longer crown has the effect of increasing the estimated age at formation of an LEH. The effect of using the overall greater population crown heights is to change the average age at formation of an LEH from – 0.08 years (upper first molar) to + 0.37 years (upper canine). Most corrections are less than 2 months, but interestingly the largest corrections are for canines.

What, then, is the consequence of simultaneously correcting for both hidden cuspal enamel and population crown heights? Using the Altun Ha data, one is able to see that the consequences are nearly additive (Table 9.6). The estimated mean age at formation of an LEH increases from 0.18 years to 0.68 years. The average adjustment tends to be between 4 and 6 months and both these corrections are most important in the ageing of early defects.

Choice of developmental standard

Mainly due to historical precedent, nearly all estimated ages at formation of LEHs are based on the tooth development diagram of Massler and co-workers (1941). As noted above, this standard is based on small sample sizes and individuals who died during tooth development. More recent developmental standards are based on a radiographic study of the degree of dental development, often with repeat radiographs of the same individuals. With the exception of the small study by Trodden (1982) that is based on native Canadian children, all of the modern growth studies are based on populations of European descent.

The key questions are, first, how much effect a change to a more recent developmental standard might have on the chronology of enamel hypoplasias and, secondly, which standard, if any, is most justified? As with skeletal growth research, can one standard be chosen as a universal standard for all studies?

Since the 1950s, over 10 studies have published data on the radiographic appearance of permanent crown development. All of them are generally lacking data on initial crown formation, especially for the earlier developing anterior teeth. These studies do, however, provide estimates of mean ages at termination of crown formation. To illustrate the importance of the choice of developmental standard in estimating the age at formation of LEH, we focus on variation among developmental standards in age at completion of crown calcification.

On the basis of a sampling of the most frequently referenced standards, sexes combined and averaged, the range of variation for upper central incisor completion extends from 3.3 to 5.5 years (Table 9.7). The Massler *et al.* (1941) incisor mean, at 4.5 years, is toward the centre of the distribution. The earliest-forming estimate is from the Finnish study by Haavikko (1974), at 3.3 years, and the latest two are from Moorrees, Fanning and Hunt (MFH) (Moorrees *et al.* 1963) for Boston area children and Trodden (1982) for her small sample of Canadian Indian children, at 5.1 and 5.5 years, respectively.

The lower canine standards are even more variable. While there is a general lag in canine development compared to the incisor, the standards are inconsistent in relative degree of delay. In other words, the relative development of the canine vs. the incisor shifts from standard to standard. The MFH canine develops relatively early compared to the incisor. At the other extreme, the Massler *et al.* (1941) standard includes a relatively early-developing incisor and the last-developing canine.

Applying different standards for age at crown completion to the

Table 9.7. *Comparison of ages (in years) at ending of crown formation of the upper central incisor and lower canine in different developmental standards*

Standards (Ref.)	Upper central incisor	Mandibular canine
Haavikko (1974)	3.3	4.3
Anderson *et al.* (1976)	3.7	4.5
Daito *et al.* (1990)	3.8	4.4
Nolla (1960)	4.5	5.8
Gustafson and Koch (1974)	4.5	6.0
Massler *et al.* (1941)	4.5	6.5
Moorrees *et al.* (1963)	5.1	3.8
Trodden (1982)	5.5	5.2

Table 9.8. *Central tendencies of estimated ages (in years) at formation of LEH on maxillary central incisors from Hamann–Todd using different developmental standards*

Standard (Ref.)	Mean	Median	Maximum	Minimum
Haavikko (1974)	2.07	2.06	2.93	0.90
Anderson *et al.* (1976)	2.26	2.25	3.24	0.95
Fass (1969)	2.51	2.50	3.63	1.02
Nielsen and Ravn (1976)	2.57	2.56	3.72	1.03
Massler *et al.* (1941)	2.73	2.72	3.97	1.07
Moorrees *et al.* (1963)	3.09	3.08	4.53	1.18

Hamann–Todd data yields some highly variable ages at formation of LEHs (Table 9.8). The Haavikko (1974) standard yields a mean age LEH formation of 2.07 years for maxillary central incisor, whereas the Massler *et al.* (1941) standard is nearly 8 months later at 2.73 years, and the MFH standard is a full year later. Contrary to the hidden cuspal enamel and crown length corrections, the consequence of this standard change is minimal for earlier-developing defects and increases linearly with later developing defects.

The variation introduced by changing standards is even greater for the canine, and of further concern, because of the extreme position of the widely used MFH standard and the Massler (1941) standard (Table 9.9). For the Hamann–Todd data, the MFH standard yields a mean age at formation of LEH on lower canines of 2.70 years. From this earlier standard, the mean ages increase quickly to around 4.0 years for the Gustafson and Koch (1974) 'summary standard' and 4.3 years for the Massler *et al.* (1941) standard.

Table 9.9. *Central tendencies of estimated ages (in years) at formation of LEH on mandibular canines using different developmental standards*

Standard (Ref.)	Mean	Median	Maximum	Minimum
Moorrees *et al.* (1963)	2.70	2.81	3.55	1.28
Haavikko (1974)	2.95	3.07	3.93	1.34
Daito *et al.* (1990)	3.05	3.18	4.09	1.39
Anderson *et al.* (1976)	3.07	3.20	4.11	1.34
Nolla (1960)	3.88	4.06	5.32	1.47
Gustafson and Koch (1974)	4.00	4.19	5.51	1.49
Massler *et al.* (1941)	4.30	4.51	6.05	1.54

Figure 9.9 is a plot of the age distribution of LEHs on the Hamann–Todd canines by half-year developmental periods based on the Massler *et al.* (1941) and MFH estimated ages at crown completion. There is almost no overlap between the distributions: the MFH standard peaks between about 25 and 42 months, whereas the Massler standard yields a peak a full 2 years later.

Implications and recommendations

In this chapter we have endeavoured to determine the consequence of different sources of variation on estimated ages at formation of LEH. It should be clear that the sources of variation focused upon are just three of many possible sources of variation. A major goal of anthropology is to understand this variation.

Whereas more work is required to clarify the precise size of the correction, we feel that some corrections are now justified. These include using population crown heights, correcting for hidden cuspal enamel and moving forward the age at completion of canine crown formation from the extreme ages of the Massler *et al.* (1941) chronology.

Crown height

At present, there appears to be little consequence of correcting for individual variation in crown heights. However, correcting for population variation in crown heights does have a measurable consequence. Where possible, we recommend that LEH regression equations should be based on mean population crown heights. The correction we have experimented

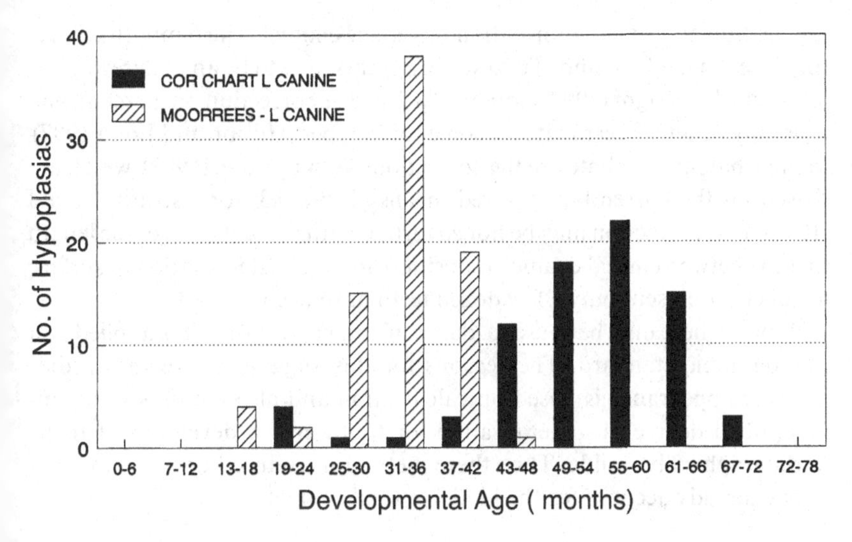

Fig. 9.9. Comparison of the chronology of LEH on lower canines with two different standards: Massler *et al.* (1941) (solid) and Moorrees *et al.* (1963) (hatched).

with in this chapter is to slow the rate of enamel formation. This is a simple correction that has consequences for earlier-developing defects. Whether it is biologically justified is the more important question, and unfortunately is not yet answered with absolute certainty. It may prove true that larger teeth simply grow for a longer period of time, rather than have greater growth velocity, as the current correction employs. If this alternative hypothesis proves true, then a different type of correction will need to be employed.

Another correction, and one that we would recommend for studies in which it is difficult to estimate population crown lengths, is to use a standard with larger teeth. The clearest issue concerned with the factor of crown size is that the existing standard is based on small teeth (Fig. 9.2; Table 9.1). Most studies of which we are aware find anterior teeth to be on the order of 5–15% longer. How much to 'lengthen' the tooth crown size standard is a question we leave for further study.

Hidden cuspal enamel

Correcting for hidden (dome or cuspal) enamel is biologically justified and significantly increases the ages of early hypoplasias. Here, also, further research is needed to better gauge the length of time between initial enamel apposition and the first formation of surface enamel. The 6 month correction used in the Hamann–Todd study appears now to be an underestimate. Conversely, Wright (1997) employed a full year correction on all teeth, and this may be an overestimate for some tooth types (Hillson and Bond 1997). Somewhat intermediate are the corrections used by Song (1997), which are based on the currently measured means (Table 9.3; see also FitzGerald 1995). This correction may be honed further with more studies of the length of time between initial enamel apposition and the first formation of surface enamel. At present only a few dozen teeth have been studied.

Finally there may be cause to 'dampen' this correction if it is applied to a radiographic standard. The reasons for this suggestion are that radiographic appearance is based on calcification and also requires some unquantified degree of mineralisation for tooth crown development to be radiographically visible. Thus these method-introduced delays may, in a sense, already account for the hidden cuspal enamel.

Developmental standards for the incisor and canine

The Massler *et al.* (1941) upper central incisor standard is generally close to, but perhaps somewhat delayed in comparison with, other developmental standards. Conversely, the Massler *et al.* (1941) age at completion of the lower canine is extremely late in comparison with other standards. It may, therefore, be reasonable to adjust the Massler *et al.* (1941) canine forward. The question is 'how much to adjust?'

Wright (1994, 1997) and Song (1997) have recently adjusted forward from 6.5 to 4.5 years the age at completion of the crown of the lower canine. This adjustment better correlates the temporal pattern of LEHs found on the canine with those found on incisor teeth. This adjustment, however, puts the canine completion before that found in most standards. We, therefore, recommend an intermediate adjustment forward of 1 year to 5.5 years at completion of the lower canine crown. It is probably also advisable to similarly adjust forward the upper canine's age at crown formation by 1 year, from 6.0 to 5.0 years. This recommendation, however, should be viewed as only temporary.

LEH, peak stress and weaning

A number of researchers have suggested that the peak period of LEH formation on permanent teeth might reflect weaning stress (e.g. Corruccini *et al.* 1985). Goodman and co-workers (1987) found that the peak in LEH formation in a contemporary sample of Mexican children was only slightly delayed from the mean age at documented termination of breast-feeding. Most studies of ancient groups find a peak age of hypoplasia well into the fourth or fifth year and there is disagreement as to whether this peak might correspond to age at termination of breastfeeding. Although this age is rather late, recent work such as that of Wright and Schwarcz (1998) suggests that breastfeeding might continue for this long.

Furthermore, the corrections we suggest above further push back the mean age at formation of LEHs, and thus increases the 'gap' between ages at completion of weaning and the peak age at LEH formation. We therefore caution against interpreting LEH patterns as reflecting the age at completion of weaning without additional sources upon which to suggest this inference.

Deciduous teeth

Although we have not addressed deciduous tooth LEHs, the study has some clear implications for the primary dentition. Most studies report much smaller variations in the time of deciduous crown formation. The standards problem may therefore be less serious. However, less is known of crown height variation; correcting for population crown height is recommended too for deciduous teeth, although sample sizes will obviously be restricted due to the requirement for unworn tooth crowns.

On the basis of the angle of accentuated striae of Retzius, including neonatal lines, it is certain that a greater percentage of deciduous tooth enamel development occurs as dome vs. sleeve enamel. Thus, correcting for dome/cuspal enamel is also likely to have great consequence for deciduous tooth LEH chronologies. In particular, it is not yet known with certainty at what approximate age the first striae reach the external surface of teeth. Finally, the location of the neonatal line provides an important potential landmark (Skinner 1992).

Implications beyond studies of LEH timing

The variation shown by enamel development chronologies is an issue that plagues studies beyond enamel hypoplasias. Depending on what standard is used, hominids can be made to resemble pongids, or to be seen as having a very different developmental pattern. In a creative study of the pattern of enamel defects, Owsley and Jantz (1983) found systematic variation between the Arikara (native North American) developmental pattern and that of the Moorrees *et al.* (1963) standard. However, it is not clear at this point whether the variation they found is due to different population patterns of development or the choice of developmental standard. Whereas these authors found many differences in pattern of development compared to the standard, a different standard would have yielded an entirely different set of deviations.

A similar problem plagues hominid/pongid comparisons. Dean and Wood (1981: 116) have produced a chart that compares a generalised pattern of calcification of humans and pongids. Their human canine crown completes calcification at around the same time as the incisors, which is in agreement with some, but not all, of the standards we have reviewed (see Table 9.7). This interpretation sets off the lengthened age of canine crown formation in pongids vs. hominids. It is noteworthy that the Massler *et al.* (1941) chronology resembles the pongid pattern better than it does the human pattern. The point is that the degree of variation among human standards is so great that it may obscure phylogenetic comparison.

Conclusions

We have pointed out significant sources of variation in the estimated age at formation of enamel defects. Clearly, more research is needed into these and other sources of variation. However, we do not wish to suggest that there is no value in recording the position of enamel defects on tooth crowns and trying to estimate their age at development. Our intention is exactly the opposite. Precise location of LEHs and precise location of accentuated striae of Retzius are required to tap the potential of LEH analyses, and these data may also be key to understanding the relative pattern of development of different teeth.

References

Aiello I and Dean C (1990) *An Introduction to Human Evolutionary Anatomy*. New York: Academic Press.

Anderson DL, Thompson GW and Popovich F (1976) Age of attainment of mineralization stages of the permanent dentition. *Journal Forensic Sciences* **21**, 191–200.

Berten J (1895) Hypoplasie des Schmelzes. *Deutsche Monatschrifte für Zahnheil* **13**,

425, Sept; 483, October.
Black GV (1883) Lines of contemporaneous calcification of the teeth. *Illinois State Dental Society Transactions*, frontispiece.
Blakey ML and Armelagos GJ (1985) Deciduous enamel defects in prehistoric Americans from Dickson Mounds: prenatal and postnatal stress. *American Journal of Physical Anthropology* **66**, 371–380.
Boyde A (1963) The structure and development of mammalian enamel. PhD thesis, University of London.
Bromage TG and Dean MC (1985) Re-evaluation of the age at death of Plio-Pleistocene fossil hominids. *Nature* **317**, 525–528.
Bullion SK (1987) The biological application of teeth in archaeology. PhD thesis, University of Lancaster.
Cohen M and Armelagos GJ (eds.) (1984) *Paleopathology at the Origins of Agriculture*. New York: Academic Press.
Corruccini RS, Handler JS and Jacobi KP (1985) Chronological distribution of enamel hypoplasias and weaning in a Caribbean slave population. *Human Biology* **57**, 699–711.
Cutress TW and Suckling GW (1982) The assessment of non-carious defects of enamel. *International Dental Journal* **32**, 117–122.
Daito M, Kawahara S, Tanaka T, Nishihara G and Hieda T (1990) Calcification of the permanent anterior teeth observed in panoramic radiographs. *Journal of Osaka Dental University* **24**, 63–85.
Dean MC and Beynon AD (1991) Histological reconstruction of crown formation times and initial root formation times in a modern human child. *American Journal of Physical Anthropology* **86**, 215–228.
Dean MC and Wood B (1981) Developing pongid dentition and its use for aging individual crania in comparative cross-sectional growth studies. *Folia Primatologia* **36**, 111–127.
Demirjian A (1986) Dentition. In *Human Growth, A Comprehensive Treatise*, ed. F. Falkner and J.M. Tanner, 2nd edn, pp. 269–298. New York: Plenum Press.
Demirjian A and Levesque G-Y (1980) Sexual differences in dental development and prediction of emergence. *Journal of Dental Research* **59**, 1110–1122.
Eveleth PB and Tanner JM (1976) *Worldwide Variation in Human Growth*. Cambridge: Cambridge University Press.
Fass EN (1969) A chronology of growth of the human dentition. *Journal of Dentistry for Children* **36**, 391–401.
FitzGerald CM (1995) Tooth crown formation and the variation of enamel microstructural growth markers in modern humans. PhD thesis, University of Cambridge.
Garn SM, Lewis AB and Kerewsky S (1965) Genetic, nutritional, and maturational correlates of dental development. *Journal of Dental Research* **44**, 228–242.
Garn SM, Lewis AB, Koski K and Polacheck DL (1958) The sex differences in tooth calcification. *Journal of Dental Research* **37**, 561–567.
Gleiser I and Hunt EE (1955) The permanent mandibular first molar: its calcification, eruption and decay. *American Journal of Physical Anthropology* **13**, 253–283.

Goodman AH (1988) The chronology of enamel hypoplasias in an industrial population: a reappraisal of Sarnat and Shour (1941, 1942). *Human Biology* **60**, 781–791.

Goodman AH and Rose JC (1990) The assessment of systematic physiological perturbations from developmental defects of enamel and histological structures. *Yearbook of Physical Anthropology* **33**, 59–110.

Goodman AH and Rose LC (1991) Dental enamel hypoplasias as indicators of nutritional status. In *Advances in Dental Anthropology*, ed. C. Larsen and M. Kelley, pp. 279–293. New York: Alan R. Liss.

Goodman AH, Armelagos GJ and Rose JC (1980) Enamel hypoplasias as indicators of stress in three prehistoric populations from Illinois. *Human Biology* **52**, 515–528.

Goodman AH, Martinez C and Chavez A (1991) Nutritional supplementation and the development of linear enamel hypoplasias in children from Tezonteopan, Mexico. *American Journal of Clinical Nutrition* **53**, 773–781.

Goodman AH, Allen LH, Hernandez GP, Amador A, Arriola LV, Chavez A and Pelto GH (1987) Prevalence and age at development of enamal hypoplasias in Mexican children. *American Journal of Physical Anthropology* **72**, 7–19.

Gustafson G and Koch G (1974) Age estimation up to 16 years of age based on dental development. *Odontologisk Revy* **25**, 297–306.

Haavikko K (1974) Tooth formation age estimated on a few selected teeth. *Proceeding of the Finnish Dental Society* **70**, 15–19.

Hall R (1989) The prevalence of developmental defects of tooth enamel (DDE) in a pediatric hospital department of dentistry population. *Advances in Dental Research* **3**, 114–119.

Hillson S. (1996) *Dental Anthropology*. Cambridge: Cambridge University Press.

Hillson S and Bond S (1997) Relationship of enamel hypoplasia to the pattern of tooth crown growth: a discussion. *American Journal of Physical Anthropology* **104**, 89–103.

Hodges D and Wilkinson D (1990) Effect of tooth size on the ageing and chronological distribution of enamel hypoplastic defects. *American Journal of Human Biology* **2**, 553–560.

Judes H, Eli I, Jaffe M, Attias D and Jagerman K (1985) The histological examination of primary enamel as a possible diagnostic tool in developmental disturbances. *Journal of Pedodontics* **10**, 68–75.

Kreshover SJ (1960) Metabolic disturbance in tooth formation. *Annals of the New York Academy of Sciences* **85**, 161–167.

Krogh-Poulsen W (1950) *Taendernes Morfologi.* Kobenhavn: Ejnar Munksgaard.

Kronfeld R (1935) Postnatal development and calcification of the anterior permanent dentition. *Journal of the American Dental Association* **22**, 1521–1536.

Legros C and Magitot E (1880) *The Origin and Formation of the Dental Follicle.* Translated from the French by M.S. Dean. Chicago: Jansen, McClurg and Co.

Legros C and Magitot E (1881) Recherches sur l'évolution du follicle dentaire chez les mammifères. Paris: Librairie Germer Ballière.

Logan WHG and Kronfeld R (1933) Development of the human jaws and surrounding structures from birth to the age of fifteen years. *Journal of the*

American Dental Association **20**, 379–427.

Macho GA and Wood B (1995) The role of time and timing in hominid dental evolution. *Evolutionary Anthropology* **4**, 17–31.

Massler M, Schour I and Poncher HG (1941) Development pattern of the child as reflected in the calcification patterns of the teeth. *American Journal of Diseases of Children* **62**, 33–67.

Moorrees CF, Fanning EA and Hunt E (1963) Age variation of formation stages for ten permanent teeth. *Journal of Dental Research* **42**, 1490–1502.

Moss ML and Moss-Salentijn L (1977) Analysis of developmental processes possibly related to human dental sexual dimorphism in permanent and deciduous canines. *American Journal of Physical Anthropology* **46**, 407–414.

Murray SR, Murray KA (1989) Computer softward for hypoplasia analysis. *American Journal of Physical Anthropology* **78**, 277–278.

Nielsen HG and Ravn JJ (1976) A radiographic study of mineralization of permanent teeth in a group of children aged 3–7 years. *Scandinavian Journal of Dental Research* **84**, 109–118.

Nolla CM (1960) The development of the permanent teeth. *Journal of Dentistry for Children* **27**, 254–266.

Owsley D and Jantz R (1983) Formation of the permanent dentition in Arikara Indians: timing differences that affect dental age assessments. *American Journal of Physical Anthropology* **61**, 467–471.

Peirce CN (1984) Calcification and decalcification of the teeth. *Dental Cosmos* **26**, 449–455.

Sarnat BG and Schour I (1941) Enamel hypoplasias (chronic enamel aplasia) in relationship to systemic diseases: a chronological, morphological and etiological classification. *Journal of the American Dental Association* **28**, 1989–2000.

Schawashy M and Yaeger J (1986) Enamel. In *Orban's Oral History and Embryology*, ed. S.N. Behaskar, pp. 45–100. St Louis, MO: C.V. Mosby.

Schultz PD and McHenry H (1975) Age distribution of enamel hypoplasias in prehistoric California Indians. *Journal of Dental Research* **54**, 913.

Seow WK and Perham S (1990) Enamel hypoplasia in prematurely-born children: a scanning electron microscopic study. *Journal of Pedodontics* **14**, 235–239.

Skinner M (1992) Gestation length and location of the neonatal line in human enamel. *Journal of Paleopathology Monographic Publication* no. **2**, pp. 41–50.

Smith BH (1991) Standards of human tooth formation and dental age assessment. In *Advances in Dental Anthropology* ed. M. Kelley and C.S. Larsen, pp. 143–168. New York: Wiley-Liss, Inc.

Song RJ (1997) Developmental defects of enamel in the Maya of Altun Ha, Belize: implications for Ancient Maya childhood health. MA thesis, Trent University.

Swärdstedt T (1966) *Odontological Aspects of a Medieval Population from the Province of Jamtland/Mid-Sweden.* Stockholm: Tiden Barnangen, AB.

Tompkins R (1996) Human population variability in relative dental development. *American Journal of Physical Anthropology* **99**, 79–102.

Trodden B (1982) A radiographic study of the calcification and eruption of the permanent teeth in Inuit and Indian children. *National Museum of Canada Archaeological Survey of Canada Paper* no. 112.

Wright LE (1994) The sacrifice of the earth? Diet, health, and inequality in the Pasion Maya Lowlands. PhD dissertation, University of Chicago.

Wright LE (1997) Intertooth pattern of hypoplasia expression: implications for childhood health in the Classic Maya collapse. *American Journal of Physical Anthropology* **102**, 223–247.

Wright LE and Schwarcz HP (1998) Stable carbon and oxygen isotopes in human tooth enamel: identifying breastfeeding and weaning in prehistory. *American Journal of Physical Anthropology* **106**, 1–18.

10 *Reconstructing patterns of growth disruption from enamel microstructure*

SCOTT W. SIMPSON

Introduction

The earliest years of life present the greatest risk of sickness and death. An understanding of the causes of age-specific morbidity and mortality in the youngest cohort can produce a very useful model of the patterns of adaptation in a population. Palaeoepidemiological studies rely on the hard tissues as the primary medium of analysis. In the study of past disease episodes, we must rely on individuals that lived with, or recovered from, some pathological state, thus retaining a record of this homeostatic imbalance. Accurate data of the timing, duration and frequency of metabolic disruptions are necessary to provide insight into their cause. Growth disruptions that occurred during the first years of life may leave a characteristic signature in the tissues developing at that time. Teeth are useful because they develop over a very characteristic schedule, are formed incrementally, do not remodel, and have unique metabolic demands during formation.

In this chapter, the temporal patterning and relationship between two types of structural defect in the enamel crown (pathological striae of Retzius and surface defects) are explored in a New World lineage. These data can be used to develop hypotheses of the frequency and types of disease that affected the study group. The population lived in the southeastern USA (Florida) between AD 1 and AD 1704. During this period, marked changes occurred within Native American population. In the prehistoric period (AD 1–AD 1565), the indigenes made a transition from gathering–hunting economic base to one increasingly focused on maize agriculture occurring around AD 600 (Larsen 1982). In the middle of the 16th century, this area was the focus of European (primarily Spanish) exploration and missionisation. Thus ultimately led to the displacement, extinction, or incorporation of the native groups. Within this area and during this time, two significant changes occurred: the adoption of intensive agriculture and Spanish missionisation. The implications of these changes are great, with both causing a change in the dietary base, population density, pathogen exposure and psychological stress. A number of the biological aspects of the health changes in these groups have been documented elsewhere (especially caries, stature, enamel hypoplasia, long bone cross-sectional geometry, etc.; see Larsen 1982, 1990, 1993, 1999).

An extensive literature on enamel hypoplasia in archaeological collections exists, including these La Florida samples (Hutchinson and Larsen 1990; Simpson *et al.* 1990; Larsen and Hutchinson 1992). Hypoplasia are transverse surface defects manifesting as either pits or depressed bands in the enamel surface (FDI 1982; Suckling 1989). Internally, the enamel microstructure is not necessarily disrupted and the rods retain a normal microstructural appearance despite having a thinning in the enamel surface. The breadth of the defects ranges from narrow to very broad (spanning dozens of perikymata). The resolution of serial events can be low and it is difficult to determine whether a single broad defect represents a single episode or multiple stress events. The distribution of the surface defects across tooth types is non-random (Goodman and Armelagos 1985). Molars show a very low frequency of defects whereas canines are the most susceptible tooth in the dentition to demonstrating defects, even though these two teeth types develop coincidentally. Within a tooth not all portions are equally likely to show a surface defect, with the cervical and middle thirds of the tooth most likely to be afflicted (Goodman and Armelagos 1985; Goodman and Rose 1990). Although there may be a greater intrinsic resistance to disruption in the cuspal enamel, a more likely explanation is the geometry of the developing enamel (Rose *et al.* 1985; Simpson 1999). The chronology of the formation of surface defects is necessary for palaeoepidemiological studies; the choice of crown formation schedule is subject to debate (Goodman and Rose 1990; Lanphear 1990; Sciulli 1992; Skinner and Goodman 1992; Simpson and Kunos 1998; Simpson 1999). Traditionally, the tooth of analysis (usually the mandibular canine or maxillary central incisor) has been divided into a series of equal breadth transverse segments (Rose 1977; Rose *et al.* 1978, 1985), each representing a common interval of time (Swärdstedt 1966; Schultz and McHenry 1975; Condon 1981; Goodman 1989; Wright 1990; Condon and Rose 1992) commonly using the dental developmental schedule created by Massler and co-workers (1941) (Goodman and Rose 1990; although see Skinner and Goodman 1992). The choice of dental developmental schedule has marked implications for the resulting schedule of defect chronology (Rose *et al.* 1985; Goodman and Rose 1990; Berti and Mahaney 1992; Skinner and Goodman 1992; Simpson and Kunos 1998). Finally, the causes of these defects are many (see Rose *et al.* 1985; Goodman and Rose 1990), with febrile diseases and dietary irregularities or insufficiencies often implicated.

A more recent approach to the analysis of growth disruptions is to analyse pathological striae of Retzius (or Wilson bands) in longitudinally sectioned teeth (Wilson and Shroff 1970; Weber and Eisenmann 1971; Weber *et al.* 1974; Rose 1977, 1979; Rose *et al.* 1978; Whittaker and Richards 1978; Condon 1981; Wright 1990; Marks 1992). Normal Retzius striae appear as regularly spaced dark or brown lines radiating from the dentine-enamel junction (DEJ) to the surface of the tooth. These are not evidence of pathological growth but reflect a circaseptian (6–9 days) period of normal growth (Dean 1987). Each line is a composite growth front formed by ameloblasts of different ages responding systemically to some unknown common cause. Deep within the enamel, the striae are primarily optical artefacts, although they can sometimes be highlighted with mild acid etching, which preferentially erodes enamel along the Retzius line. In the most superficial enamel, the striae of Retzius appear as offsets or bending of the enamel rods (Wilson and Shroff 1970; Osborn 1973; Risnes 1986, 1990; Suckling *et al.* 1989). Thus, a normal Retzius line can have optical, mineralogical and structural characteristics. On occasion, especially dark Retzius stria can be identified. Higher magnification (using either light (LM) or scanning electron microscopy (SEM)) reveals a disruption in the structure of the enamel prisms, with a zone of disorganised crystallites crossing the rods. These bands are generally narrow in breadth (less than 15 μm) and discrete. These pathological striae (PS) can be followed throughout their course from the DEJ to the external surface of the tooth. Although their cause is at present uncertain, it is commonly thought that surface defects and PS are common responses to similar causes (Condon 1981; Goodman and Rose 1990).

Materials and methods

A total of 143 individuals, represented by their mandibular canines, were included in this analysis. A number of additional specimens were excluded from this study because of the extreme degree of attrition or the cervical enamel was damaged. These teeth, collected from 14 archaeological sites or horizons, spanning AD 1 to AD 1704 from northern Florida, USA (Larsen 1993), were divided into four groups (Table 10.1). These groups are early prehistoric, late prehistoric, early contact and mission. The prehistoric groups were defined on the basis of the economic base, the earlier sample relying on hunting and gathering and the later group being primarily

Table 10.1. *Sample composition and distribution of pathological striae*

Site	No. of inds.	No. of w/PS	No. of PS	% afflict	PS/tooth	PS/population
Prehistoric – early						
Mayport Mound	7	5	9	71	1.8	1.3
McKeithen Mound	3	2	4	67	2.0	1.3
Melton Mound A-5	4	2	4	50	2.0	1.0
Wacahoota Mound	1	1	1	100	1.0	1.0
Early – total/mean	15	10	18	*67*	*1.8*	*1.2*
Prehistoric – late						
Holy Spirit	14	2	7	14	3.5	0.5
Browne Mound	6	4	12	67	3.0	2.0
Goodman Mound	1	0	0	0	0.0	0.0
Lake Jackson	4	3	4	75	1.3	1.0
Late – total/mean	25	9	23	*36*	*2.6*	*0.9*
Prehistoric – total/*mean*	40	19	41	*48*	2.2	*1.0*
Early contact						
SM Ossuary	26	14	24	54	1.7	0.9
Mission						
SMDY-SC	30	24	43	80	1.8	1.4
Patale	3	3	10	100	3.3	3.3
Fig Springs	27	23	70	85	3.0	2.6
Santa Maria	8	8	20	100	2.5	2.5
San Luis	9	6	20	67	3.3	2.2
Mission – total/*mean*	77	64	163	*83*	2.5	*2.1*
Sample totals/*means*	143	97	228	*68*	*2.4*	*1.6*

No. of inds, number of individuals in each population. No. w/PS, number of individuals in sample with pathological striae (PS). No. of PS = number of PS observed in each sample. % afflict, percentage of individuals in a sample demonstrating at least one PS. PS/tooth, number of PS per tooth in those teeth demonstrating PS. PS/population, number of PS in a sample divided by the total number of teeth in that sample. This last measurement sedentary agriculturalists with an emphasis on maize. The two historical groups bear clear evidence of European contact. The early contact group (ossuary at Santa Maria), although found in association with a mission, probably samples the reinterred remains of individuals from a traditional, non-mission setting (Simmons *et al.* 1989). The mission sample, representing an indoctrinated native American population living at the mission site, was recovered from a series of mission cemeteries (Santa Maria dos Yamassee, Patale, San Martin de Timucua (Fig Springs), Patale, Santa Maria and San Luis de Talimali). In other writings (Larsen and Hutchinson 1992; Simpson 1999), the mission sample was originally partitioned into early (pre-1670) and late periods (post-1670) but are combined into a single group here.

All mandibular canines were measured using either needlepoint calipers (maximum crown mesio-distal breadth, maximum crown labio-lingual (LL) breadth, cervical LL breadth) or a calibrated eyepiece micrometer on a stereomicroscope (maximum crown height on the labial surface and position of all surface defects). The positions of the cuspal and cervical margins of each defect was measured from the labial enamel cervix. Unworn or lightly worn specimens were then embedded in Araldite resin and oven-cured for 24 hours. Each specimen was longitudinally sectioned along the labio-lingual plane with a Buhler Isomet low speed saw following procedures similar to that of Marks and co-workers (1996). Each face was cleaned with a dilute solution of HCl for 5 seconds and then rinsed. This has the benefit of removing most of the saw striae, cleaning the surface and highlighting the enamel microstructure. One of the blocks was then prepared for removing a thin (approximately 200–250 μm) section by gluing it (with either cyanoacrylate or epoxy) to a petrographic slide. The Buhler saw includes a fitting for holding petrographic slides and the section was removed from this slide-mounted block. The new face was also washed with dilute hydrochloric acid. Each section was photographed, traced and the major measurements taken with either an eyepiece micrometer (cervical breadth, position of the PS from the labial cervix) or a digitising tablet and software (dentine area, enamel area, local enamel thickness, angle between a stria of Retzius and the DEJ, length of the labial DEJ). Each specimen was examined twice (separated by a period of 6 months) and scored for presence and location of PS. The PS present a graded appearance from mild prismatic bending to extreme prismatic disruption (Swanson 1931; Rose 1977; Rose *et al.* 1978; Marks 1992), which may lend itself to differences in scoring by different observers. Although differences in the 'morphological threshold' of PS should exist between independent researchers this should have little effect on the temporal patterning of PS, although it may be reflected in the populational frequency of affliction.

The enamel crown forms by two processes: (1) serial differentiation of ameloblasts from cusp to cervix, which produces the height of the crown and with differentiating odontoblasts together forms the DEJ; and (2) matrix production by ameloblasts, which produce the thickness of the enamel crown. The Retzius lines are regularly formed structures whose form and distribution are a product of these two processes, and can be used to build a chronology of development for the enamel crown. Although the mechanisms producing Retzius lines and circadian prismatic varicosities are unknown, their regularity can be used to identify regional differences in crown formation rate and calculate crown formation duration. Due to the narrow and discrete morphology of the PS, its position along the DEJ can be very carefully measured. Not only does this provide a great resolution of the position of the event that can be used to determine the age of occurrence, it also permits the registration of the PS and surface defects along a common datum, the DEJ. It is necessary to construct a schedule of dental development on the sample of teeth because populational differences in crown formation period have been recorded (e.g. Harris and McKee 1990). It is uncertain whether these differences between the many studies are a consequence of actual biological differences or factors introduced by the sample composition, method of data analysis, or researcher idiosyncrasy (Smith 1991; Simpson and Kunos 1998).

Data description

Site-specific summary metric information and distribution of PS in the mandibular canines can be found in Tables 10.1 and 10.2. The summary crown dimensions are reasonably consistent across the groups. All of the samples had a mean cervical breadth in the range 7.21–7.27 mm except the Santa Maria ossuary with a mean value of 7.10 mm. Other evidence indicates that a majority of the ossuary individuals were females (Simmons 1989), thus explaining their smaller canine crowns. The overall similarity in crown breadth is very indirect support for the biological continuity of the samples. More importantly, it demonstrates that the teeth are structurally and geometrically the same, allowing direct comparison. To account for any individual differences in crown size, the position of all pathological striae were normalised by dividing their position along the DEJ by cervical breadth, resulting in a relative position of the PS. Differences in crown height between the groups may indicate differences in diet (both in terms of

Table 10.2. *Summary descriptive metrics of the dental sample*

	Florida prehistoric			Early contact:	Florida mission:
Mandibular canine	Early (AD 1–600)	Late (AD 600–1450)	Total (AD 1–1450)	Santa Maria Ossuary	Total (AD 1600–1704)
DEJ height Total sample	9.78 ± 2.08 (15)	9.03 ± 1.39 (25)	9.31 ± 1.70 (40)	8.74 ± 1.47 (26)	9.39 ± 1.63 (77)
DEJ height Teeth without PS	10.04 ± 3.16 (5)	8.91 ± 1.36 (16)	9.18 ± 1.90 (21)	8.35 ± 1.67 (12)	8.69 ± 1.91 (13)
DEJ height Teeth with PS	9.65 ± 1.50 (10)	9.23 ± 1.50 (9)	9.46 ± 1.47 (19)	9.06 ± 1.24 (14)	9.57 ± 1.55 (64)
Cervical breadth Total	7.23 ± 0.70 (13)	7.21 ± 0.52 (21)	7.22 ± 0.58 (34)	7.10 ± 0.55 (25)	7.27 ± 0.63 (65)
Cervical breadth Teeth without PS	7.53 ± 0.81 (5)	7.25 ± 0.38 (14)	7.32 ± 0.52 (19)	7.06 ± 0.40 (11)	7.32 ± 0.66 (12)
Cervical breadth Teeth with PS	7.09 ± 0.64 (7)	7.14 ± 0.75 (7)	7.12 ± 0.67 (14)	7.13 ± 0.66 (14)	7.27 ± 0.64 (54)
PS location Absolute	6.32 ± 1.91 (18)	5.63 ± 1.44 (23)	5.93 ± 1.67 (41)	5.75 ± 1.68 (24)	5.87 ± 2.03 (163)
PS location Relative	0.96 ± 0.25 (13)	0.70 ± 0.12 (13)	0.83 ± 0.23 (26)	0.80 ± 0.26 (24)	0.77 ± 0.26 (133)

DEJ height is presented in three ways: (1) summary for all teeth within a subsample; (2) DEJ height of those teeth not showing pathological striae (PS); and (3) DEJ height of those teeth demonstrating at least one PS. DEJ height is measured in the longitudinally sectioned teeth from the cervical margin to the most cuspal extent of the DEJ. Cervical breadth is also presented in three ways: (1) total subsample breadth; (2) breadth of those teeth without a PS; and (3) summary breadth of those teeth with at least one PS. Cervical breadth is measured on the longitudinally sectioned teeth between the most cervical extensions of the unbroken labial and lingual surfaces. Not all specimens had a complete cervical breadth. PS location – absolute is the mean location of all of the PS for that subsample in millimetres as measured from the midlabial cervical margin. PS location – relative is the relative position of each PS, which is calculated by dividing its absolute position by the cervical breadth of that tooth. Numbers in parentheses are sample sizes.

dietary 'grittiness' but also in terms of dietary sugars producing greater rates of postcanine tooth loss and accelerated anterior tooth wear) but also provide indirect evidence about the ages of death of the samples. Although additional studies are necessary it is perhaps safe to say that individuals with shorter crowns had a greater occlusal duration.

In the prehistoric samples, the earlier preagricultural group differs from the later agricultural sample by having a greater frequency of individuals with PS (early, 67%; late, 36%). Of all the individuals with PS, the late prehistoric group has a greater number of PS per tooth than the earlier group (early, 1.8/tooth; late, 2.2/tooth), indicating that, although fewer individuals were severely ill in the later group, when they did get sick they tended to be predisposed to subsequent bouts of sickness. The degree of sample affliction differs slightly for the two groups (early, 1.2 PS/tooth; late, 0.9 PS/tooth), suggesting similar populational values, although it is clear that they achieved similar rates in different ways. For the entire prehistoric sample combined, about half of all individuals had at least one PS; those with PS had on average 2.2 on each of their mandibular canines, producing a population value of 1.0 PS per tooth for all teeth combined.

The early contact sample from the Santa Maria ossuary will be treated separately from the prehistoric and mission groups. The rate of affliction is slightly higher than the summed prehistoric value (prehistoric, 48%; early contact, 54%). The number of PS per afflicted tooth is the same as the early prehistoric sample, lower than the later prehistoric sample, and lower than the combined prehistoric collection and mission samples (early contact, 1.7/tooth; early, 1.8/tooth; late, 2.6/tooth; prehistoric combined, 2.2/tooth; mission, 2.5/tooth). Like the previous prehistoric groups, the population rate of affliction is low, with less than one PS per tooth. The canine teeth in this sample were smaller than all of the other groups, probably a result of the high frequency of females in the sample (Simmons *et al.* 1989; Larsen 1993). Metrically, the individuals with at least one PS have a taller crown height (younger mean age at death), although the cervical breadth is insignificantly different between the two subsamples (Table 10.2; Fig. 10.1).

The mission sample is clearly different from the earlier samples in all measures of frequency of PS, especially populational frequency of affliction (83%), number of PS per afflicted tooth (2.5/tooth), and number of PS per all canines combined (2.1/all teeth) (Fig. 10.2). Not only were more individuals sick in this group, the frequency of illness was high as well. The quality of health declined substantially in the native American population following the arrival of the Spanish and their disruption of indigenous practices. The populational profiles of PS affliction indicate that the Santa Maria materials, although recovered in an historic context within the perimeter of a mission, probably reflect a population living under late prehistoric circumstances.

The absolute and relative mean positions of the PS along the DEJ are different between the samples, the early preagricultural prehistoric collection having the earliest (greater distance from the cervix) mean peak

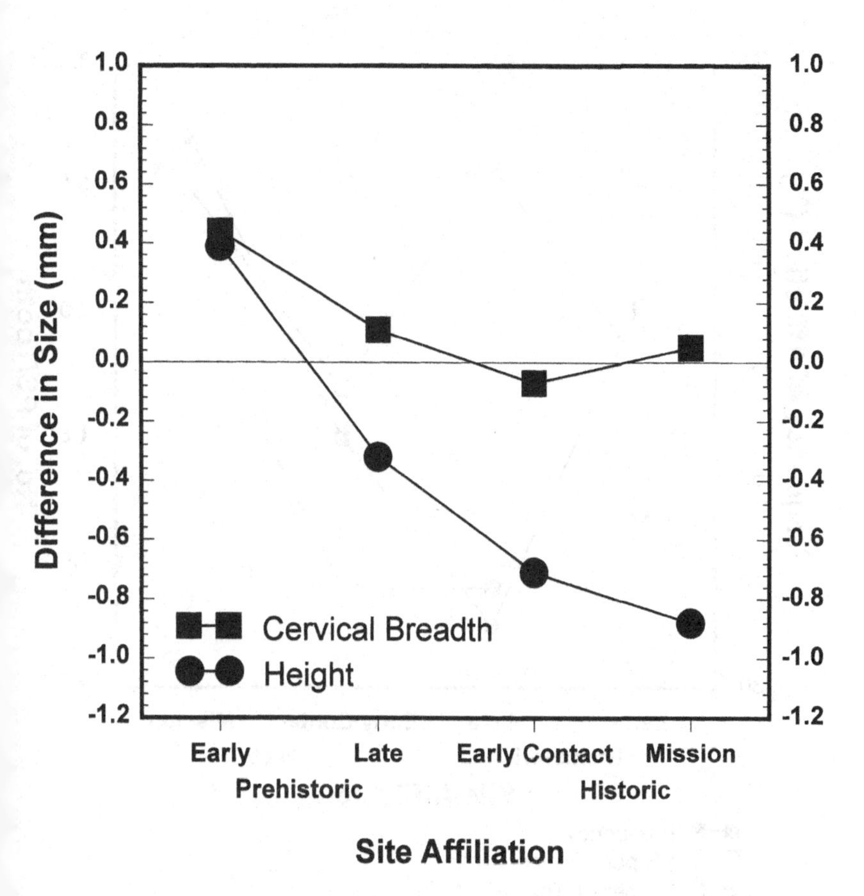

Fig. 10.1. Differences in crown dimension. Differences in crown dimensions are measured in millimetres. Difference is calculated as non-afflicted group dimension minus afflicted group dimension. Except for the early prehistoric group, both normal and pathological subsamples show a common cervical breadth and a greater crown height in the non-pathological subsample.

affliction (Fig. 10.3). This suggests that the patterns of sickness in the preagricultural group differs from that of the later agricultural groups. No similar difference in the distribution of surface defects has been recorded (Hutchinson and Larsen 1990). All of the subsamples, especially the early prehistoric, show a common relationship between the presence of PS and cervical breadth, individuals with at least one PS having smaller cervical breadths than those without a PS (Fig. 10.1, Table 10.2). This could indicate that females (with their smaller canine crowns) are more susceptible to enamel disruption than males or that individuals that are less able to resist extrinsic stressors are also more likely to not achieve their growth potential

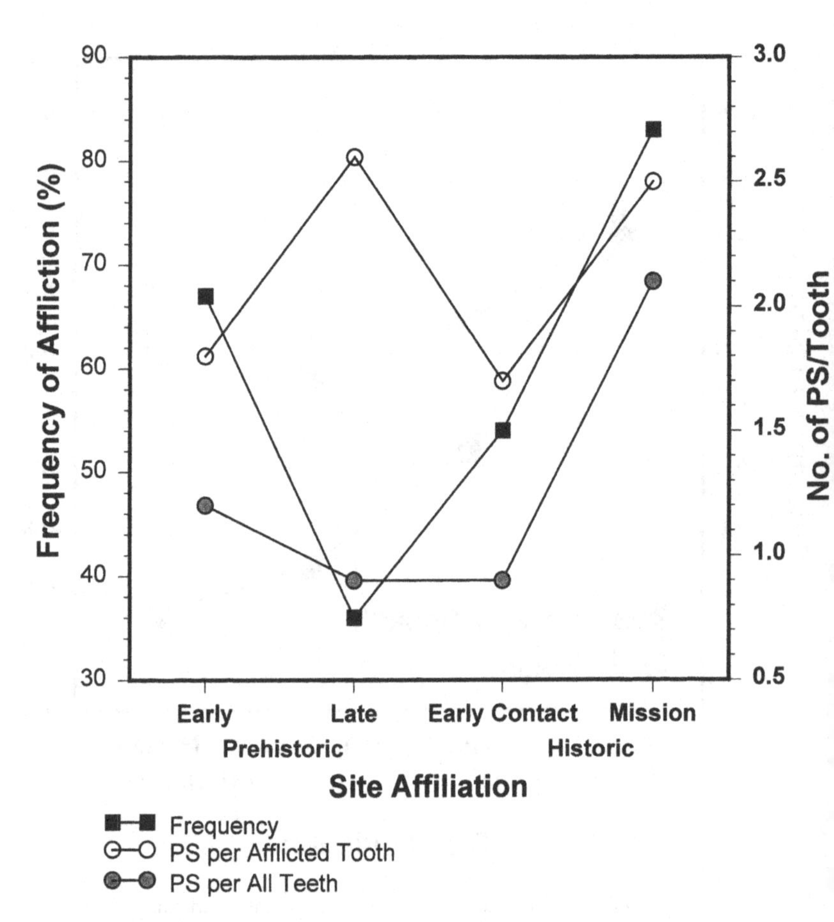

Fig. 10.2. Pathological striae in La Florida. A combination plot including frequency of individuals afflicted (solid squares), average number of PS per afflicted tooth (open circles), and average number of PS for all teeth within a sample (lightly shaded circles). Clearly, the mission collection samples a cohort with a greater degree of morbidity than in earlier groups.

(Simpson *et al.* 1990). Except for the early prehistoric group, there is a relationship between crown height and presence/absence of PS, individuals having a shorter crown with no evidence of PS, suggesting a link between adult age at death and early morbidity. In the early prehistoric sample, individuals with PS have a shorter crown height than those without PS (Table 10.2; Fig. 10.1). This indicates that those individuals who suffered a severe episode of growth disruption lived on average to a greater age (as indicated by degree of attrition) than those without. Although a sampling error cannot be ruled out owing to small sample sizes, this may be evidence

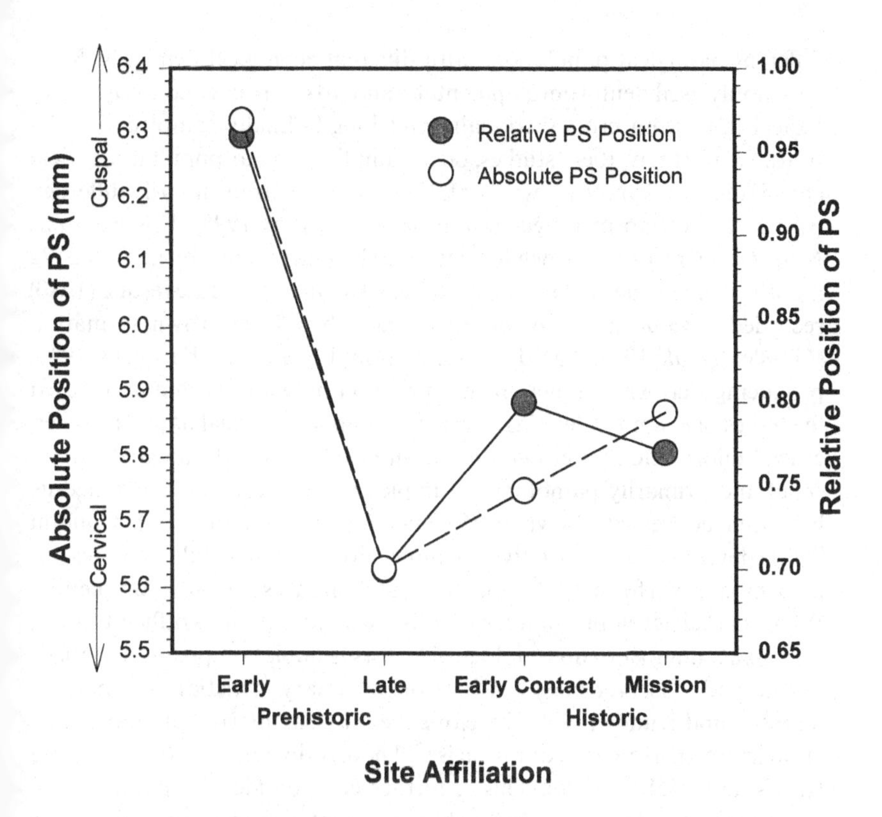

Fig. 10.3. Pathological striae (PS) position. This figure includes the absolute and relative (PS position/cervical breadth) mean positions of PS for each of the study groups. All measures from the labial cervical margin. Note that the early preagricultural prehistoric population has an earlier age of mean affliction, perhaps indicating a different pattern of weaning in gatherer–hunter populations and agriculturalists.

supporting the hypothesis by Meindl and Swedlund (1977), who suggested that those that survive a severe stress event may have a greater subsequent survivorship.

Chronology of development of the mandibular canine

The chronology of canine crown formation is crucial to an understanding of the timing of developmental defects. Fortunately, tooth crowns form in a regular and stereotypic fashion. We can exploit this information to develop an internally consistent schedule of the formation of the enamel crown. This overcomes the limitations of using growth standards developed on different biological populations with different ecological demands. Most commonly used dental developmental standards were created using radiological data from cross-sectionally and longitudinally sampled populations. Surprisingly, these studies, often sampling similar populations, yield chronologies of canine crown formation that are extraordinarily different, with little overlap in ranges (Simpson and Kunos 1998). For example, Nolla (1960) recorded crown formation as being complete by about 68–72 months (female and male means) whereas Demirjian and Levesque (1980) recorded a value of 35–40 months. The schedule of crown formation (Massler *et al.* 1941, based on data from Logan and Kronfeld 1933) producing a crown completion age of 78 months, which is commonly used in studies of enamel defects, is developed from histological data. However, crucial information on crown formation periods was interpolated from a small and primarily pathological sample. The weakness in this schedule has been addressed elsewhere (Skinner and Goodman 1992). Condon (1981) developed an alternative histologically based schedule of mandibular canine crown formation, with an age of 54 months at crown completion. Which dental developmental schedule is most accurate? It is difficult to say, although a number of non-biological factors (sample composition, statistical method, observer bias) can bias the summary statistics (Smith 1991; Simpson and Kunos 1998). Therefore, the range of interpopulational variation in crown formation duration is still poorly documented (although see Harris and McKee 1990). This is further compounded by populational differences in crown size. Do teeth of different size form over the same period? Do ameloblasts differentiate and is enamel matrix deposited at the same rate in all human groups? Again, this is not well known, although even minor differences in either of these factors can result in marked differences in crown formation period.

An internally consistent chronology of the developing canine was developed here using unworn or slightly worn canine crowns from the archaeological sample. The longitudinally sectioned teeth were scored for the number and distribution of striae of Retzius. Knowing that Retzius stria appear at reasonably even intervals (FitzGerald 1998), the distribution of Retzius lines can be arrayed across the duration of crown formation time. Since the density of Retzius lines varies with position along the crown, accommodation of the non-linear schedule of crown height formation must be made (Figs. 10.4 and 10.5) (Gohdo 1982; Shellis 1984; cf. Massler *et al.* 1941; Goodman *et al.* 1987, 1989; Goodman 1988; Goodman and Rose 1990). This reflects a slowing of crown elongation as the cervix is reached, thus producing a greater local density of perikymata and striae of Retzius. Schedules based on a linear relationship between absolute crown position

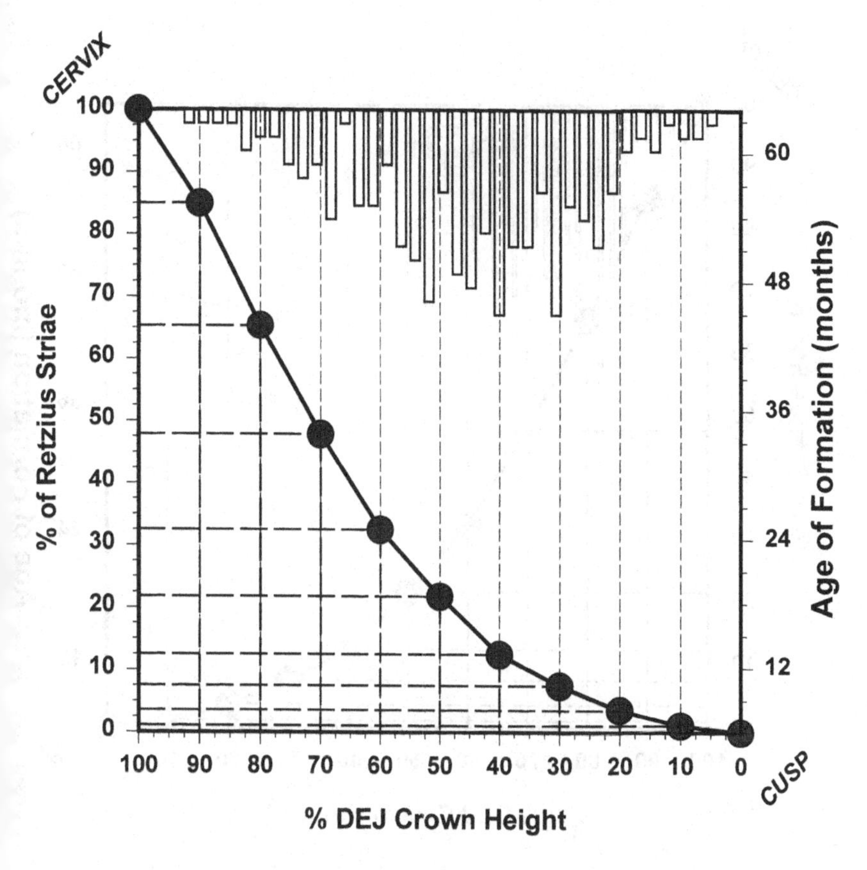

Fig. 10.4. Age distribution of pathological striae (PS). This figure can be used to translate absolute position of the PS along the dentine–enamel junction (DEJ) (distance in millimetres from the cervical margin) into age of occurrence. The upper and lower horizontal axes represent distance in millimetres from the cervix. In this way, the frequency distribution of PS absolute position can be represented. The rate of crown height elongation is non-linear. The filled circles and bold line describe this non-linear relationship between absolute crown height and relative distribution of Retzius lines throughout the entire length of the DEJ. Along the right vertical axis is the age of crown formation. In this way, the absolute timing of the PS can be determined. Approximately 50% of the PS are located in the cuspal half of the crown which forms before the age of 18 months.

and time will be inaccurate. Thus, position along the DEJ can be measured in terms of Retzius lines, which provides a more direct link between position and age of formation. The cuspal portion of the crown contains a substantial amount of enamel formed in the first 15–18 months of life that has no counterpart on the external surface of the tooth. This is represented by Retzius lines that do not intersect the external surface and are therefore

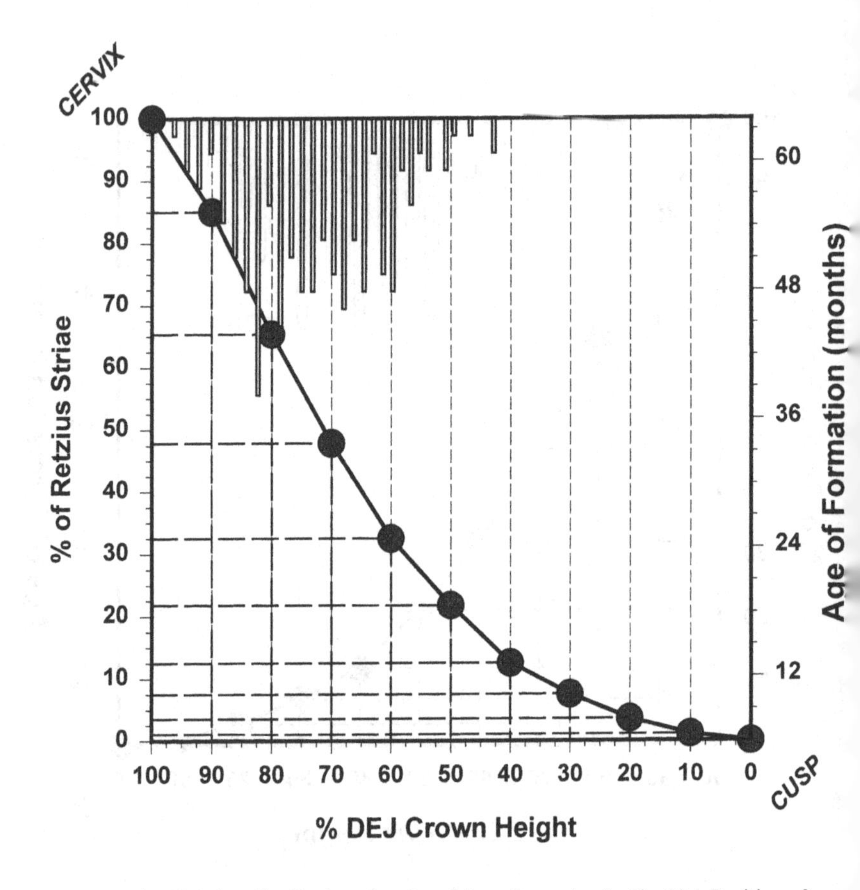

Fig. 10.5 Age distribution of surface defects. Legend as in Fig. 10.4. Position of the surface defects is measured relative to the DEJ using the formula calculating DEJ position from surface position described in text. Note that very few of the surface defects are formed prior to 18 months, with a mean age of occurrence between approximately 36 and 42 months. Surface defects and pathological striae have a very different distribution throughout the tooth and in time.

'hidden' in a complete tooth. The time it takes to form this cuspal enamel has been estimated (Bromage and Dean 1985). Using the histological approach, the Retzius lines can be counted directly and the duration calculated. The period of canine crown formation was compared with recent radiographic (Simpson and Kunos 1998) and histological (D.J. Reid, personal communication) estimates, yielding an age of formation beginning at approximately 3–6 months and terminating at about 66 months for a duration of 60–63 months. This is shorter than the 78 month period used by some researchers (Goodman and Rose 1990; Goodman *et al.* 1992).

By using a small sample of longitudinally sectioned teeth to document the number and distribution of Retzius lines, the developmental chronology of a tooth can be determined. This obviates the need to rely on other dental developmental growth schedules and instead uses a growth standard developed on, and useful for, the teeth of analysis. In addition, populations with different size tooth crowns can be compared. Clearly, this is not possible in studies of living individuals, but in retrospective studies, where histological preparation of the teeth is possible, this approach should be adopted. This method creates a more precise chronology of dental development, especially documentation of the duration of cuspal enamel formation producing the 'hidden' Retzius striae.

Overall, this histological approach takes into account that different portions of the crown form at the different absolute rates, calculates directly the duration of hidden increments, and is independent of crown size.

Comparison of the distribution of surface defects and pathological striae

Each Retzius line is the manifestation of a growth front and these lines arc from the DEJ to the external surface. To make the positions of the surface defects and the PS directly comparable, their position relative to the DEJ must be made. Following the course of a number of striae of Retzius on a series of five teeth (and tested on 20 additional teeth), a function for translating the position of the surface defects to a measure along the DEJ was made. The following function usefully describes this relationship:

$$y_{DEJ} = 0.062 + (x_{EXT} \times 0.902) - (x_{EXT}^2 \times 0.022)$$

where y_{DEJ} equals absolute position along the DEJ from the cervical margin and x_{EXT} equals distance from the labial cervical margin along the external surface of the tooth. All measurements of the surface defects are taken at the cuspal margin of the defect. The function is useful for this sample of canine teeth only and must be recalculated for each different collection. Careful registration of the surface defects and pathological striae makes their temporal distributions compatible and directly comparable.

Combination plots (Figs. 10.4 and 10.5) including relative and absolute distribution of striae of Retzius, age of crown formation, and distribution of PS and surface defects for the entire collection were prepared. These plots clearly show that the age distribution of the PS and surface defects differ markedly. The PS have a flatter distribution spanning a greater portion of the total period of crown formation, whereas surface defects are clustered in the last 70% of the period of crown development. In terms of timing, PS are found primarily in enamel formed before 24 months of age whereas surface defects are found in enamel forming between 24 and 65 months of age and peaking at about 45 months. A major difference in these distributions has to do with the nature of enamel microstructural geometry. Enamel formed in the earliest 12–18 months of life has no surface expression ('hidden increments') and cuspal hypoplasias may be too shallow to be readily identifiable. This explains why surface defects are infrequent in the cuspal half of the tooth, although it does not explain why PS are less prevalent in the cervical third of the crown. PS can appear throughout the tooth although they are often difficult to distinguish from normal stria in the cuspal and cervical reaches of the tooth (Goodman and Rose 1990) and absent in the cuspal areas due to attrition (Rose *et al.* 1985). Although surface defects and PS are often found together, the presence of one does not necessarily indicate the presence of the other (Rose 1977; Condon 1981; Wright 1990). When the two do coincide, the PS are often found at the cuspal edge of the defect, although the PS can be found throughout the defect.

The two enamel irregularities have very different microstructural characteristics. Although surface defects are a transverse thinning of the enamel thickness that may be quite broad, crossing dozens of perikymata, the contributing enamel prisms are not necessarily irregular in form. Pathological striae, by contrast, are irregularities in enamel prism structure. Their breadth is rather narrow (usually less than 15 μm) and their boundaries are discrete. Their thin breadth and discrete boundaries indicates that the duration of the disruption is rather short, with a rapid onset and remission. If enamel forms at approximately 3 μm per day (Schour and Poncher 1937), the defects represent growth disruptions lasting substantially less than a week.

Overall, surface defects and PS have different age distributions and microstructural detail. This suggests that the epidemiology and physiological disruptions producing the defects are different. The proposition that they represent graded responses to similar stimuli (Rose *et al.* 1985) can no longer be supported (Wright 1990).

What produces PS? At the microstructural level the rods are disrupted, as shown by interruption of rod continuity and a 'blending' of adjacent rods (Fig. 10.6). Backscatter electron (BSE) microscopic images do not reveal any differences in density, although structural differences in the organisation of crystallites are apparent. The crystallites in abnormal enamel have indistinguishably different degrees of mineralisation although

Fig. 10.6. Scanning electron micrograph of pathological striae. Note the disruption of the organisation of enamel prisms running from the upper right to the lower left in the image. Solid bar represents 5 μm.

their organisation is obviously irregular. This irregularity is perhaps due to a variety of different causes, including changes in the shape of the Tomes processes or a disruption of the protein scaffold on which the crystallites are organised but not to compositional changes of the contributing mineralised crystallites. Crystals form within the extracellular matrix soon after it is deposited and the axes of the rod and interrod crystallites are determined by the matrix as it is produced by the rod and interrod growth sites along the Tomes process (Nanci *et al.* 1996; Robinson *et al.* 1996). Altering the shape of the Tomes process will change the geometry of matrix deposition, perhaps producing less well-organised enamel. During enamel matrix deposition, amelogenin and non-amelogenin proteins provide the structural framework within which mineralisation occurs. The hydroxyapatite crystals form on a temporary matrix of hydrophobic amelogenins and acidic non-amelogenin proteins that acts as a crystallite-spacing and mineral-diffusing element or template (Doi *et al.* 1984; Deutsch 1989; Robinson *et al.* 1989, 1996; Nanci *et al.* 1996). Normal enamel structure requires the enamel protein structure to produce oriented crystal growth (Simmer and Fincham 1995). Disrupting this template, either as a consequence of modifying its original deposition or by altering the shape of the matrix proteins, produces irregular enamel. Water is crucial during matrix formation. Any factor that challenges normal fluid and electrolyte homeostasis at the gross and cellular levels could disrupt this process. Acute diarrhoea and vomiting can produce extreme dehydration, to the extent that intracellular fluids move into the intercellular spaces causing cellular dysfunction. Weanling diarrhoea producing severe dehydration is a major source of morbidity and mortality in non-Western (Gordon *et al.* 1963) and undoubtedly premodern societies. These events are often short in duration lasting 1–4 days, at which time the infant recovers or does not. Many of the bacterial and viral causes of this diarrhoea (especially cholera (*Vibrio cholera*), *Salmonella enteritidis, Vibrio parahaemolyticus, Escherichia coli, Campylobacter* sp., *Shigellosis* spp., *Clostridium* spp.) are rapid onset, short duration (2–5 days) of expression of the primary symptoms, and lacking a chronic infection and not necessarily accompanied by fevers (Connor and Gibson 1988). Although the diarrhoea can arise from a number of different causes, a common source of these transitory and acute infections seems to accompany the onset of weaning, with the introduction of novel foodstuffs and pathogens. In most societies, this begins early in the first year to as late as the end of the second year, accompanied by an increase in sickness and death. This schedule of illness mirrors the temporal patterning of PS in enamel. In contrast, surface defects are most prevalent after the second year, longer in duration, and do not demonstrate disruptions in enamel prism structure. Often, the two enamel irregularities will co-occur, especially at the period of onset of the surface defect. Many diseases begin with both a fever and diarrhoea or vomiting. Thus, their co-occurrence is not unexpected but it must be recognised that their coincidental appearance is not necessarily the product of a common physiological disruption.

Conclusion

A series of individuals represented by their mandibular canines were macroscopically and microscopically scored for presence and location of defects in enamel formation in a sample of native American mandibular canine teeth recovered from archaeological sites dated between AD 1 and AD 1704. The sample was divided into four groups (prehistoric preagricultural, prehistoric agricultural, historic non-mission and historic mission) and the frequency and location of enamel defects was compared between the different subsamples. The frequency of microdefects increased dramatically following the introduction and elaboration of the Spanish missions. This observation is consistent with a variety of other historical and hard tissue studies documenting the deterioration of post contact Native American populations. A chronology of canine enamel crown development was generated, allowing direct comparison of the distribution and timing of surface defects and PS. PS and surface defects have a different structural and temporal signature, suggesting that they are products of physiological disruptions with different courses, timings and durations. Surface defects and PS cannot be considered as different responses to a common underlying cause. A wide variety of external stressors, especially febrile diseases and dietary irregularities, have been implicated in the formation of surface defects. PS are here suggested to have formed as a response to acute dehydration (due to severe diarrhoea and vomiting) perhaps in association with the onset of weaning.

Acknowledgements

I thank Robert Hoppa and Charles FitzGerald for inviting me to participate in this volume. This research was funded by a NSF grant (SBR-9305391) and a Case Western Reserve University Research Initiation Grant.

References

Berti PR and Mahaney MC (1992) Quantification of the confidence intervals for linear enamel hypoplasia chronologies. *Journal of Paleopathology* **2**, 19–30.

Bromage TG and Dean MC (1985) Re-evaluation of the age at death of Plio-Pleistocene fossil hominids. *Nature* **317**, 981–983.

Condon KW (1981) Correspondence of developmental enamel defects between the mandibular canine and the first premolar. MA thesis, University of Arkansas.

Condon KW and Rose JC (1992) Intertooth and intratooth variability in the occurrence of developmental enamel defects. *Journal of Paleopathology* **2**, 61–77.

Connor DH and Gibson DW (1988) Infectious and parasitic diseases. In *Pathology*, ed. E. Rubin and J.L. Farber, pp. 326–451. Philadelphia: JB Lippincott Co.

Dean MC (1987) Growth layers and incremental markings in hard tissues; a review of the literature and some preliminary observations about enamel structure in *Paranthropus boisei*. *Journal of Human Evolution* **16**, 157–172.

Demirjian A and Levesque GY (1980) Sexual differences in dental development and prediction of emergence. *Journal of Dental Research* **59**, 1110–1122.

Deutsch D (1989) Structure and function of enamel gene products. *Anatomical Record* **224**, 189–210.

Doi Y, Eanes ED, Shimokawa H and Termine JD (1984) Inhibition of seeded growth of enamel apatite crystals by amelogenin and enamelin proteins in vitro. *Journal of Dental Research* **63**, 98–105.

FDI (Fédération Dentaire International) (1982) An epidemiological index of developmental defects of dental enamel (DDE) *International Dental Journal* **32**, 159–167.

FitzGerald CM (1998) Do enamel microstructures have regular time dependency? Conclusions from the literature and a large-scale study. *Journal of Human Evolution* **35**, 371–386.

Gohdo S (1982) Differential rates of enamel formation on human tooth surfaces deduced from the striae of Retzius. *Archives of Oral Biology* **27**, 289–296.

Goodman AH (1988) The chronology of enamel hypoplasias in an industrial population: a reappraisal of Sarnat and Shour (1941, 1942). *Human Biology* **60**, 781–791.

Goodman AH (1989) Dental enamel hypoplasias in prehistoric populations. *Advances in Dental Research* **3**, 265–271.

Goodman AH and Armelagos GJ (1985) Factors affecting the distribution of enamel hypoplasias within the human permanent dentition. *American Journal of Physical Anthropology* **68**, 479–493.

Goodman AH, Allen LH, Hernandez GP, Amador A, Arriola LV, Chavez A and Pelto GH (1987) Prevalence and age at development of enamel hypoplasias in Mexican children. *American Journal of Physical Anthropology* **72**, 7–19.

Goodman AH, Martin DI, Perry A, Martinez C, Chavez A and Dobney K (1989) The effect of nutritional supplementation of permanent tooth development and morphology. *American Journal of Physical Anthropology* **78**, 129–130.

Goodman AH, Martin DL, Klein CP, Peele MS, Cruse NA, McEwen LR, Saeed A and Robinson BM (1992) Cluster bands, Wilson bands, and pit patches: histological and enamel surface indicators of stress in the Black Mesa Anasazi population. *Journal of Paleopathology* **2**, 115–127.

Goodman AH and Rose JC (1990) Assessment of systemic physiological perturbations from dental enamel hypoplasias and associated histological structures. *Yearbook of Physical Anthropology* **33**, 59–110.

Gordon JE, Chitkara ID and Wyon JB (1963) Weanling diarrhea. *American Journal of Medical Science* **245**, 345.

Harris EF and McKee JH (1990) Tooth mineralization standards for blacks and whites from the middle southern United States. *Journal of Forensic Science* **35**, 859–872.

Hutchinson DL and Larsen CS (1990) Stress and lifeway change: the evidence from enamel hypoplasias. *Anthropological Papers of the American Museum of Natural History* **68**, 50–65.

Lanphear KM (1990) Frequency and distribution of enamel hypoplasias in a historic skeletal sample. *American Journal of Physical Anthropology* **81**, 35–43.

Larsen CS (1982) The anthropology of St. Catherines Island: 3. Prehistoric biological adaptation. *Anthropological Papers of the American Museum of Natural History* **57**(3).

Larsen CS (ed.) (1990) The archaeology of Mission Santa Catalina de Guale: 2. Biocultural interpretations of a population in transition. *Anthropological Papers of the American Museum of Natural History* **68**.

Larsen CS (1993) On the frontier of contact: mission bioarchaeology in La Florida. In *The Spanish Missions*, ed. B.G. McEwan, pp. 322–356. Gainesville, FL: University of Florida Press.

Larsen CS (ed.) (1999) *The Bioarchaeology of La Florida*. Gainesville, FL: University of Florida Press, in press.

Larsen CS and Hutchinson DL (1992) Dental evidence for physiological disruption: biocultural interpretations from the eastern Spanish Borderlands, USA. *Journal of Paleopathology* **2**, 151–169.

Logan WHG and Kronfeld R (1933) Development of human jaws and surrounding structures from birth to age of fifteen years. *Journal of the American Dental Association* **20**, 379–427.

Marks MK (1992) Developmental dental enamel defects: an SEM analysis. *Journal of Paleopathology* **2**, 79–90.

Marks MK, Rose JC and Davenport, Jr WD (1996) Technical note: thin section procedure for enamel histology. *American Journal of Physical Anthropology* **99**, 493–498.

Massler M, Schour I and Poncher HG (1941) Developmental pattern of the child as reflected in the calcification pattern of the teeth. *American Journal of Diseases in Children* **62**, 33–67.

Meindl RS and Swedlund AC (1977) Secular trends in mortality in the Connecticut valley, 1700–1850. *Human Biology* **49**, 389–414.

Nanci A, Hashimoto J, Zalzal S and Smith CE (1996) Transient accumulation of proteins at interrod and rod enamel growth sites. *Advances in Dental Research* **10**, 135–149.

Nolla CM (1960) The development of the permanent teeth. *Journal of Dentistry for Children* **27**, 254–266.

Osborn JW (1973) Variations in structure and development of enamel. *Oral Sciences Reviews* **3**, 3–83.

Risnes S (1986) Enamel apposition rate and the prism periodicity in human teeth. *Scandinavian Journal of Dental Research* **94**, 394–404.

Risnes S (1990) Structural characteristics of staircase-type Retzius lines in human dental enamel analysed by scanning electron microscopy. *Anatomical Record* **226**, 135–146.

Robinson C, Kirkham J, Stonehouse NJ and Shore RC (1989) Control of crystal growth during enamel maturation. *Connective Tissue Research* **22**, 139–145.

Robinson C, Brookes SJ, Kirkham J, Bonass WA and Shore RC (1996) Crystal growth in dental enamel: the role of amelogenins and albumin. *Advances in Dental Research* **10**, 173–180.

Rose JC (1977) Defective enamel histology of teeth from Illinois. *American Journal of Physical Anthropology* **46**, 439–446.

Rose JC (1979) Morphological variation of enamel prisms within abnormal striae of Retzius. *Human Biology* **51**, 139–151.

Rose JC, Armelagos GJ and Lallo JW (1978) Histological enamel indicators of childhood stress in prehistoric skeletal samples. *American Journal of Physical Anthropology* **49**, 511–516.

Rose JC, Condon KW and Goodman AH (1985) Diet and dentition: developmental disturbances. In *The Analysis of Prehistoric Diets*, ed. R.I. Gilbert and J.H. Mielke, pp. 281–305. New York: Academic Press.

Schour I and Poncher HG (1937) Rate of apposition of enamel and dentin, measured by the effect of acute fluorosis. *American Journal of Diseases of Children* **54**, 757–776.

Schultz PD and McHenry HM (1975) Age distribution of enamel hypoplasias in prehistoric California Indians. *Journal of Dental Research* **54**, 913.

Sciulli PW (1992) Estimating age of occurrence of dental defects in deciduous teeth. *Journal of Paleopathology* **2**, 31–39.

Shellis RP (1984) Variations in growth of the enamel crown in human teeth and a possible relationship between growth and enamel structure. *Archives of Oral Biology* **29**, 697–705.

Simmer JP and Fincham AG (1995) Molecular mechanisms of dental enamel formation. *Critical Reviews in Oral Biology and Medicine* **62**, 84–108.

Simmons S, Larsen CS and Russell KF (1989) Demographic interpretations frm ossuary remains during the late contact period in northern Spanish Florida. Paper presented at the Annual meeting of the American Association of Physical Anthropologists, San Diego, California.

Simpson SW (1999) Patterns of growth perturbation in La Florida: evidence from mandibular canine enamel structure. In *Bioarchaeology of La Florida*, ed. C.S. Larsen. Gainesville, FL: University of Florida Press, in press.

Simpson SW and Kunos CA (1998) A radiographic study of the development of the human mandibular dentition. *Journal of Human Evolution* **35**, 479–505.

Simpson SW, Hutchinson DL and Larsen CS (1990) Coping with stress: tooth size, dental defects, and age-at-death. *Anthropological Papers of the American Museum of Natural History* **68**, 66–77.

Skinner M and Goodman AH (1992) Anthropological uses of developmental defects of enamel. In *Skeletal Biology of Past Peoples*, ed. S.R. Saunders and M.A. Katzenberg, pp. 153–174. New York: Wiley-Liss.

Smith BH (1991) Standards of human tooth formation and dental age assessment. In *Advances in Dental Anthropology*, ed. M. Kelley and C.S. Larsen, pp. 143–168. New York: Wiley-Liss.

Suckling G (1989) Developmental defects of enamel – historical and present-day perspectives of their pathogenesis. *Advances in Dental Research* **3**, 87–94.

Suckling GW, Nelson DGA and Patel MJ (1989) Macroscopic and scanning electron microscopic appearance and hardness values of developmental defects in human permanent tooth enamel. *Advances in Dental Research* **3**, 219–233.

Swanson JH (1931) Age-incidence of lines of Retzius in the enamel of the permanent teeth. *Journal of the American Dental Association* **18**, 819–826.

Swärdstedt T (1966) *Odontological Aspects of a Medieval Population from the Province of Jamtland/Mid-Sweden*. Stockholm: Tiden Barnangen, AB.

Weber DF and Eisenmann DR (1971) Microscopy of the neonatal line in developing human enamel. *American Journal of Anatomy* **132**, 375–392.

Weber DF, Eisenmann DR and Glick PL (1974) Light and electron microscope

studies of Retzius lines in human cervical enamel. *American Journal of Anatomy* **141**, 91–104.
Whittaker DK and Richards D (1978) Scanning electron microscopy of the neonatal line in human enamel. *Archives of Oral Biology* **23**, 45–50.
Wilson DF and Shroff FR (1970) The nature of the striae of Retzius as seen with the optical microscope. *Australian Dental Journal* **15**, 162–171.
Wright LE (1990) Stresses of contact: a study of Wilson bands and enamel hypoplasias in the Maya of Lamanai. *American Journal of Human Biology* **2**, 25–35.

11 *Estimation of age at death from dental emergence and implications for studies of prehistoric somatic growth*

LYLE KONIGSBERG AND DARRYL HOLMAN

Introduction

Analysis of prehistoric somatic growth is a tool commonly applied in an effort to characterise past population health and physiological adaptation to the environment. Saunders and Hoppa (1993) have recently reviewed a number of the interpretive problems that can arise in attempting to study human growth in the past, and all of the chapters in this volume address at least some facets of the factors that continue to complicate analyses of prehistoric growth. We can identify at least three factors that complicate the analytical framework for studies of past human somatic growth, as follows:

(1) Growth is studied cross-sectionally rather than longitudinally.
(2) Mortality sampling across age may vary in intensity and nature.
(3) Ages at death are estimated rather than known.

The first problem, that prehistoric growth studies are necessarily cross-sectional in nature, has some well-documented effects on reconstruction of growth curves. Because of timing and rate differences, important periods of growth may be missed or mis-characterised when one is looking at a conglomeration of single observations on individuals (Tanner 1978; Sinclair 1985; Bogin 1988). There is simply less information available when growth curves are fitted from individual observations rather than from serial observations on individuals. While this is lamentable, there is nothing we can do about it. Tabloid newspapers notwithstanding, individuals only die once, and death is an unfortunate requirement for entering prehistoric growth studies.

The second problem is dealt with quite extensively in Saunders and Hoppa's article, so we do not revisit it here. We focus fairly narrowly on the third problem, because the fact that ages are estimated rather than known has some important implications for studies of past human somatic growth.

Estimation of ages from skeletal remains has long been an important technique in physical anthropology. Even so, it has only been recently that a number of statistical issues have been recognised and addressed in the ageing of skeletons. For example, Bocquet-Appel and Masset (1982) argued that traditional methods of ageing yield age-at-death distributions that mimic the reference sample on which the methods for ageing were developed. Konigsberg and Frankenberg (1992) examined this issue in more detail and proposed a method that yields a target age-at-death distribution that is independent of the reference population. However, their approach has not been met with unbridled enthusiasm (Bocquet-Appel and Masset 1996). A second shortcoming of traditional age estimation methods is that estimated ages are treated and used as if they are exact ages, when, in fact, age estimates should form a distribution of ages within individuals (Whittington 1991; Konigsberg and Frankenberg 1992; O'Connor 1995). The second issue is the primary focus of this chapter. We discuss methodological and statistical issues in developing standards for ageing, and using those standards to generate age-at-death distributions, which, in turn, are needed to estimate parameters of somatic growth from a sample of skeletons.

The methods developed in this chapter require that we begin with raw data on a reference sample, and develop the age standards from the beginning. We do this because nearly all methods of ageing in current use do not make proper use of the statistical nature of age estimates. Some methods treat the estimates as if age were found without error. Other methods produce mean ages and standard errors for a number of correlated traits, but provide no way of properly combining the individual ages into a single estimate of age and a standard error. As we discuss in some detail below, age estimation from one or more skeletal traits is a process of generating the *distribution* of possible chronological ages for a skeleton. Point estimates of age may be acceptable in some circumstances. Nevertheless, when trying to test statistical hypotheses, like differences in growth trajectories, one must use the entire distribution of age for all skeletons for proper inference. In other words, by throwing out the distributional information around the mean or median age, we gain a false sense of statistical power about statements based on that age.

Given a meaningful age standard, we generate an age-at-death distribution, which we apply to the study of somatic growth. The whole process can be separated into four tasks. First, we describe the use of multivariate probit estimation to produce statistical estimates for a set of (possibly correlated) markers in a reference population. The method begins with the raw data from a reference population with known ages. Given the markers, we generate the multivariate means and variances for the time to first appearance of the marker. Since different age markers probably show some degree of correlation, the full matrix of covariances among all pairs of markers is estimated as well. We demonstrate the method using a sample of living Bangladeshi children as our reference population, and deciduous tooth emergence in these children as the multivariate marker. This part of the study results in estimates of the mean eruption times for ten teeth, and the 55 variance–covariance terms for eruption timing.

Our second task is to develop the method to estimate a distribution of ages for a *single* target individual, given the particular configuration of markers for that individual, as well as the parameters from the reference population. For example, given a target skeleton in which a particular set of teeth had emerged, we want to generate a probability distribution for all possible ages for that individual. We develop a Monte Carlo method to generate this 'posterior distribution of age' for individuals.

Our third task is to develop methods for estimating an age-at-death distribution for a sample of individuals. We intend to publish a more detailed treatment elsewhere, and so we will skip many of the details here. Instead, we treat the problem somewhat differently from previously published approaches. We assume that the age-at-death distribution follows a particular parametric form. Specifically, mortality throughout childhood is assumed to follow a negative exponential distribution, corresponding to the first component of the Siler model of mortality with the a_1 term assumed equal to 1 (Gage 1989). Thus, the entire age-at-death distribution is determined by one parameter. We use a maximum likelihood method to estimate the age-at-death distribution from the individual distributions of age.

The final task of this chapter is to develop a method to estimate parameters for age-specific somatic growth from a target sample. Now we assume that some somatic growth parameter (stature, femur length etc.) has been measured on target individuals, in addition to the markers used for age estimation. Again, the methods described here differ from previous work in that we make use of the entire distribution of ages estimated for individuals. In effect, individual measurements of somatic growth are probabilistically weighted according to individuals' distributions of ages. By so doing, we can generate more realistic standard errors on growth parameters.

Ageing markers and methods

Degenerative changes – such as age-related remodelling of the pubic symphysis – are frequently used to age adult skeletons. With subadult skeletons, methods of ageing can be based on more clock-like developmental events. Many developmental events show a regular sequence of stages that are likely to be under strong genetic control. Even so, all developmental markers are not equal. Some exhibit more variability among individuals for a given age, and some markers are more susceptible to effects of undernutrition, diseases or other environmental conditions. Additionally, developmental markers may differ among populations, reflecting underlying genetic differences, so that population-specific standards may be required. The ideal set of developmental markers would not be susceptible to environmental conditions, would exhibit little variability among individuals of a given age within a population, and would not show differences among populations.

The ideal markers have not and probably will not be found, but tooth development has long been considered a set of markers that show the least variability against chronological age (Holtz 1959; Lewis and Garn 1960; Krogman 1968a,b; Demirjian 1986). In particular, studies of tooth formation (mineralisation or calcification), have yielded age estimation methods that have been usefully applied to numerous archaeological (Johnson 1969; Saunders *et al.* 1993) and medico-legal (Stewart 1973) situations.

Ageing by dental development is not completely free from problems. Perhaps the most important difficulty is that population-specific standards have not been developed. Indeed, almost all standards are based on serial radiographs taken from children of European ancestry (Smith 1992). The recognised dangers of X-rays pose serious limitations on future large-scale radiographic studies in many populations. Without comparative standards, it is difficult to assess the validity of these methods for non-European populations (Krogman 1968a,b). Finally, sex-specific standards may be necessary for accurate ageing by dental development (Smith 1991). Unfortunately anatomical methods for determining the sex of subadult skeletons are not very accurate (Saunders 1992).

Tooth emergence is another type of dental marker that has long been used to age children (e.g. studies reviewed in Haavikko 1970; Townsend and Hammel 1990). Methods of ageing based on deciduous emergence are limited to the range over which deciduous teeth emerge. The range is from before birth to just over 2 years in European populations, and later in some non-European populations where emergence of the last deciduous tooth can occur after age 3 (Holman and Jones 1998). It is commonly accepted that methods of ageing based on dental calcification produce more accurate values then those based on emergence. In part, this is because almost all previous methods of ageing children by deciduous tooth emergence have relied on counts of teeth in a child's mouth (e.g. papers in Jellife and Jellife 1973) giving a maximum of 11 discrete ages. Calcification methods rely on timing of a number of stages for multiple teeth, so that the possible number of discrete configurations of traits is much greater than 11. Nevertheless, times to any given stage of calcification may genuinely be less variable and more correlated to chronological time than times to emergence. Emergence is an easy trait to study, and age standards based on deciduous tooth emergence have been developed for many populations (Townsend and Hammel 1990).

A common feature of all methods of ageing from skeletal markers is that they cannot provide exact chronological ages. One reason for this is that there is variation among children in the age at which they enter a given developmental state. Even if this variation were to be eliminated, individual children, upon entering a given state, remain in that state for some period of time. This second point implies that age estimation from a marker that is *perfectly* correlated with chronological age (i.e. the 'ideal' marker) will still yield some distribution of ages (unless, of course, there are infinitely many stages for the marker).

For this chapter, we focus on deciduous tooth emergence as the developmental marker of age, although the method can be applied to other developmental events as well. The practical reason for doing so is that we have the raw tooth emergence data required to fully develop standards from the start. We acknowledge that ages based on tooth development (and the same methods) would probably be better than ages based on emergence, but we simply do not have the raw data. The 'fully statistical' method that we adopt requires that all traits be considered simultaneously to generate the entire matrix of means, variance and covariance terms among the marker traits be estimated. These are rarely found in the published literature.

An apparent difficulty of our approach is that the number of covariance terms grows exponentially with each new trait, so that the size of the reference sample must, likewise, increase exponentially with each trait. Even so, the traditional approach, by not incorporating the covariances between traits, provides no way of combining individual ages, such that a statistically valid distribution around those ages can be estimated.

Before we get down to the 'nuts and bolts' of this chapter, we examine some of the characteristics of deciduous tooth emergence as a marker of age when somatic growth is the trait of interest. Specifically, we want to know whether standards developed on living children of known age can be applied to skeletal populations. Additionally, we examine whether deciduous tooth emergence is population specific, whether it is affected by undernutrition, disease and other environmental factors, and whether it is a sexually dimorphic trait.

Ginvival vs. alveolar emergence

Emergence can refer to either alveolar emergence, when the tooth first breaks through the alveolar bone, or gingival emergence (also called clinical emergence), when the tooth first pierces the gum (Garn *et al.* 1957). All current deciduous tooth standards of ageing are based on gingival emergence in living reference populations. Clearly, alveolar emergence should be used, if at all possible, with skeletal samples. Can we easily convert from one to another? Garn and colleagues (1957) suggested that the two types of emergence are not the same in that each may yield a different order of emergence among permanent teeth. Even so, Garn and co-workers were interested primarily in the emergence *sequence* of the permanent second premolar and second molar for purposes of phylogenetic analysis. The emergence sequence of these two teeth was of interest precisely because of the contentious order of emergence among species. We note that Haavikko's (1970) cross-sectional study of 1162 children living in Helsinki did not replicate the Garn *et al.* (1957) findings. In that study, upper and lower P2 and M2 showed the same mean alveolar and gingival sequences for both boys and girls, but for two other pairs of teeth (C^1 and P^2 and M_1 and I_1) girls showed a reversal from alveolar to gingival emergence.

For purposes of ageing skeletons, the sequence is unimportant because two teeth with a very similar distribution of emergence times will provide the same information about age regardless of the specific order in which they emerge. What is important in ageing is the absolute time to alveolar emergence, which probably occurs some short time before gingival emergence.

Ideally, we would like to be able to take gingival emergence standards from living populations and apply them to alveolar emergence in skeletal populations. One strategy would be to subtract some small difference from gingival emergence reflecting either the average difference for all teeth or, better yet, a tooth-specific difference. The rate of intraosseous eruption (1–10 μm/day; Parfitt 1984) is substantially slower than the rate of eruption after the tooth has broken the alveolar crest (75 μm/day; Proffit *et al.* 1991). The acceleration in eruption after alveolar emergence suggests that differences among individuals in gum thickness and condition will make little difference to the resulting estimates of age.

Haavikko (1970) documented tooth-specific differences between alveolar and gingival emergence for the permanent teeth, but we know of no research that has documented these same differences for the deciduous dentition. In principle, this could be done from serial radiographs of children following the methods of Haavikko (1970). In the absence of this information, we do not make any correction, so that skeletal ages based on the method described in this chapter will be biased toward later ages.

Sex differences

The male deciduous dentition is slightly more developed than the female dentition by the end of the first trimester (Burdi *et al.* 1970). A number of studies have found that deciduous emergence shows slight sexual dimorphism in which boys lead girls in emergence of the anterior dentition. Some time after age 2 years, the girls 'catch up' with the boys (Demirjian 1986). Rarely do studies find sex-specific differences between boys and girls in the posterior dentition (Holman and Jones 1991). These small differences between boys and girls in deciduous tooth emergence seem unlikely to create serious biases in age estimates.

Tooth emergence, malnutrition and illness

Many studies have attempted to look at the relationship between somatic growth, nutrition, disease and tooth emergence. In 1973, Jellife and Jellife brought together a series of symposium papers that examined whether deciduous tooth emergence could be relied upon to estimate the ages of living children in studies of growth and nutrition (Jellife and Jellife 1973). They summarised the collection of 14 papers and their own literature review with the following conclusions (Jellife and Jellife 1973):

(1) Emergence times are similar across ethnic groups among healthy children.
(2) Slight to moderate undernutrition does not delay deciduous emergence.
(3) Severe undernutrition does delay dental emergence, but the effect on the dentition is less severe than on weight and height.
(4) Most childhood illnesses do not delay dental eruption.

The more recent literature has largely confirmed the conclusions that only severe forms of undernutrition delay the deciduous teeth (Billewicz and McGregor 1975; el Lozy *et al.* 1975; Khan *et al.* 1981; Rao 1985; Alvarez *et al.* 1990; Kodali *et al.* 1993), and that illness does not delay the deciduous teeth (Khan *et al.* 1981). Saleemi and colleagues (1993) found slight differences between Pakastani children from high socioeconomic areas and slums, and Kodali and co-workers (1993) found that abnormally high levels of fluoride delays emergence. Overall, deciduous tooth emergence is seen to be well buffered against environmental influences. A consistent finding from some longitudinal studies is that in children of higher birth weight, teeth emerge faster (Billewicz *et al.* 1973; Delgado *et al.* 1975; Khan *et al.* 1981). Apparently this is because many heavier infants are born at later gestational ages. For the purposes of studying somatic growth in skeletons, any bias introduced by shortened and lengthened gestation will have no effect on estimates.

The possibility of genetic differences in the timing of emergence among populations is more of a concern. Saleemi *et al.* (1993) suggested that genetic differences are responsible for slow emergence among Pakastani children. Likewise, Holman and Jones (1998) found large differences among four populations in the timing of deciduous tooth emergence, but it is unclear whether the differences are environmental (particularly nutritional) rather than genetic. If so, the differences among populations appear to be large enough to warrant population-specific standards of ageing.

Estimating parameters of ageing from a reference sample

In this section, we estimate the means and matrix of variance–covariance terms for the times to tooth emergence for a reference sample. We assume that tooth emergence follows a normal distribution, so the problem reduces to estimating the parameters for a multivariate normal distribution, given observations of tooth emergence in a reference sample.

For our reference sample, we use anthropometric and tooth emergence data collected as part of the Meheran Growth and Development Study (Khan *et al.* 1979) conducted between 1974 and 1977 in the rural village of Meheran, Bangladesh. Dental records of 397 children were available from the study. Clinical examinations were conducted monthly for the first year of the study and quarterly thereafter. Most of the children were recruited into the study at birth, and all children were less than 1 year of age at the beginning of the study. Exact dates of birth were recorded for all children.

A description of the dental data can be found elsewhere (Khan and Curlin 1978; Khan *et al.* 1981; Holman and Jones 1998).

Anthropometric and emergence data were collected from periodic examinations of the children by a trained paramedic. Teeth were checked by fingertip palpitation of the gum by the paramedic during scheduled visits or when a sick child visited a health clinic (ICDDR 1990). From the longitudinal records, two dates are recorded for each tooth: the child's age at the last visit prior to emergence, and the child's age at the first visit after emergence. These ages form an interval within which each tooth emerged. When a tooth was recorded as having emerged before the child was enrolled in the study, birth was used as the first age of the interval. Children for whom a particular tooth had not emerged by the last visit had the closing age of the interval set to infinity. These observations are called 'interval censored' in statistical parlance.

If the children had had oral exams given every day, then we could simply calculate the vector of means (**m**) and the variance–covariance matrix (**V**) among eruption times, taking the day of emergence as the observations. But, because the observations of emergence are interval censored, this task is made more difficult. We circumvent this problem by using a relatively new method of Monte Carlo integration. Throughout this chapter we make extensive use of Monte Carlo estimation procedures, and in particular the Markov Chain Monte Carlo (MCMC) or Gibbs Sampler procedure. MCMC has been extensively described by Smith and Roberts (1993) and in a volume edited by Gilks *et al.* (1996).

We will not belabour the mechanics behind our implementation of the multivariate model here. Chib and Greenberg (1998) describe a multivariate probit analysis that is fitted using the Gibbs Sampler, while Bock and Gibbons (1996) describe a multivariate probit fitted by traditional maximisation of the likelihood. The latter application removes the assumption of conditional independence previously used by Kolakowski and Bock (1981). We use a simpler MCMC approach, which is a stochastic Expectation–Maximisation (or SEM) algorithm described by Diebolt and Ip (1996).

Estimation of the parameters **m** and **V** using this approach is quite intuitive, so we will describe the SEM for our particular application and ignore the mathematics as much as possible. The underlying idea of the method is repeatedly to simulate exact observations that fall within the intervals of the real observations, use the simulated observations to estimate the parameters **m** and **V**, and repeat the process. If we repeat this process many times, the values of **m** and **V** will 'settle' on their maximum likelihood estimates.

Table 11.1. *Means and variance–covariance matrix for eruption times of deciduous teeth (months)*

I_1	I_2	C_x	M_1	M_2	I^1	I^2	C^x	M^1	M^2
9.560									
9.556	20.113								
9.333	15.082	25.685							
6.040	8.445	10.370	14.669						
6.582	10.561	13.405	8.569	28.946					
8.120	11.448	11.326	6.900	9.765	12.722				
9.049	12.511	11.601	6.519	11.094	10.579	16.535			
7.285	12.312	17.684	8.641	12.578	9.157	10.844	17.464		
5.244	6.522	9.327	7.625	10.159	6.320	6.535	8.088	9.476	
8.673	11.764	17.586	10.963	28.376	11.866	12.604	15.853	9.631	35.886
Means									
10.507	16.360	22.491	18.010	27.143	11.809	13.928	21.086	16.004	27.098

Super- and subscripts denote upper and lower jaws, respectively.

Since we know the ages between which each tooth of an individual have emerged, we can simulate exact emergence time by drawing times from a doubly truncated normal distribution. Here the lower truncation point corresponds to $\mathbf{t_u}$ (the last age at which teeth were observed to be unemerged) and the upper truncation point is $\mathbf{t_e}$ (the first age at which teeth were observed to be emerged). The normal distribution is found from a current estimate of **m** and **V**, where the eruption time for the specific tooth is predicted by conditioning (i.e. regressing) on the eruption times for the other nine teeth. Any simulated eruption time comes by taking a stochastic element around the expected value (from the regression), and so this constitutes an 'E-step' (for 'expectation'). Once all the eruption times have been simulated for all the teeth in the 397 individuals, it is a simple matter to treat the simulation as real data and calculate a new vector of means and a variance–covariance matrix. These calculations lead to a maximum likelihood estimate of the parameters, and hence constitute the 'M' step of the EM. From this 'M' step we can return to simulation in the 'E' step.

For our application we started with the mean eruption times from the Holman and Jones (1998) study, and a variance–covariance matrix that was diagonal, with the variances from Holman and Jones on the diagonal. We then ran 200 iterations of the SEM, and averaged the means and variance–covariance matrices from the last 150 iterations. Table 11.1 contains the estimates from this analysis. Our analysis differs from Chib and Greenberg's multivariate probit analysis in that ours does not include draws from the vector of means or from the variance–covariance matrix itself. As such, we do not have full alternative conditioning such as would be found in a true Gibbs Sampler. Bocquet-Appel and Masset (1996) previously criticised an approach to age estimation used by Konigsberg and Frankenberg (1992), citing the fact that they disregarded a variance term due to the finite size of the reference sample. Our use of the SEM rather than the Gibbs Sampler could be criticised on the same grounds, but we feel that the large size of our reference sample militates against Bocquet-Appel and Masset's criticism, and that the additional complications of the Gibbs Sampler are therefore unnecessary.

Finding the posterior density of age for a single individual

Our second task is to find the distribution of ages for a skeleton given that skeleton's particular configuration of teeth. Death occurs at a single age, so all we can tell from the dentition is that each tooth has emerged or has not emerged. From the means and variance–covariance matrix for emergence times that we found in the last section, it is possible to find the distribution of possible ages for the skeleton conditional on their eruption status. Following traditional Bayesian terminology (Box and Tiao 1992; Gelman *et al.* 1995), we refer to this as finding the 'posterior density' of age, since we are finding the age distribution conditional on (i.e. 'posterior to') the eruption status. Numerical integration can be used to find this distribution (Schervish 1984; Drezner 1992), but integration of a 10–fold multivariate normal is a fairly onerous task. We instead use another Monte Carlo technique to find the posterior distribution of age.

If we knew the actual age for an individual (which, of course, we do not), then we could simulate the eruption ages for each of their teeth from a truncated multivariate normal distribution. Here we would use parameters $\mathbf{m}$ and $\mathbf{V}$ for the multivariate normal that we found in the previous section (Table 11.1), and truncate the distribution so that teeth that have already erupted must have done so prior to the known age, while those yet to erupt must erupt after the known age. If, on the other hand, we knew the eruption times for all of the teeth, but we did not know the actual age of the individual, then we could simulate their age by selecting a random uniform deviate between the highest age for a tooth that had erupted and the lowest age for a tooth yet to erupt. We use a uniform distribution in this instance because there is no further information on age beyond what we can tell from the teeth. In the next section we consider the case where there is additional information. If we start with a guess at the individual's age, then

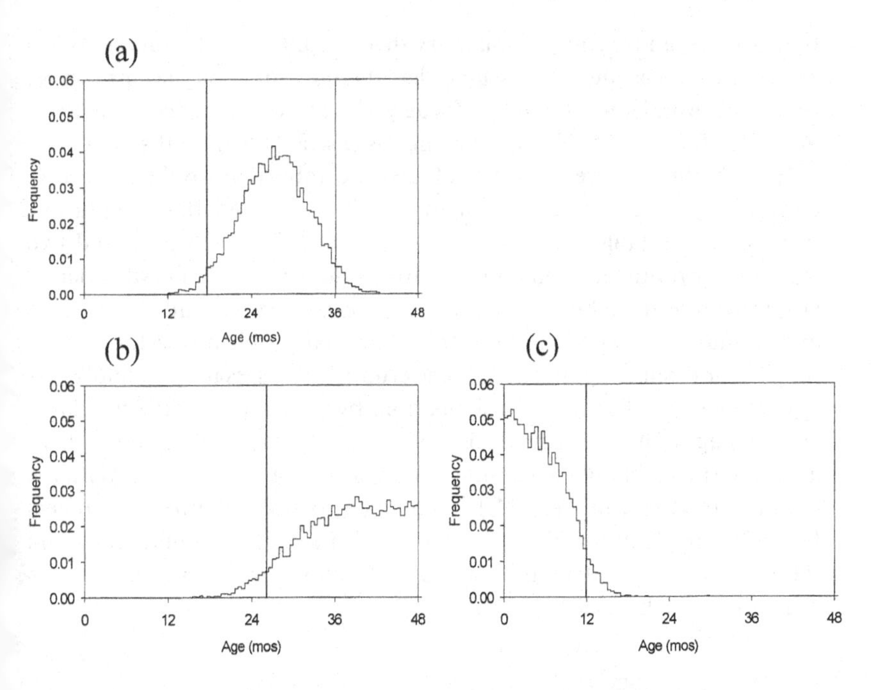

Fig. 11.1. Examples of Monte Carlo posterior densities of age. (a) Case where all mandibular teeth except the second molar are erupted, (b) case where all mandibular teeth are erupted, and (c) case where no mandibular teeth are erupted. The histograms were drawn using 10 000 replications of simulation. The vertical lines represent the 95% highest posterior density. mos, months.

we can cycle between simulation of eruption times and of age, and store the simulated ages to build the posterior distribution of age.

Figure 11.1 shows a few examples of such simulations, where we have used 10 000 rounds of simulation. In the first example we assume that all teeth save the second mandibular molar have erupted. From the simulations we can select the 250th and 9750th values as representing the central 95% highest posterior density (HPD). In this case the 95% HPD runs from 17.55 to 36.08 months. This is fairly remarkable, since we would have to state the individual's age as falling within a year and a half span. We could presumably refine this estimate by using dental formation standards, but we currently lack the reference information for such an analysis. In the second and third examples we consider the cases where all teeth are erupted and where no teeth are erupted. When all teeth are erupted it is impossible to give an upper bound on age using the dentition itself, and this consequently affects where the lower 95% boundary (the 500th value) is found. In this example, we have assumed that nobody in the sample is older than 4 years, and the lower boundary then occurs at 26.16 months. When all teeth are unerupted we assume that the individual has been born, so there is a natural lower boundary for age. This allows us to specify an upper boundary for the 95% HPD, which in this case is 11.94 months.

It is tempting to give some type of 'easy use' table that would allow us to convert eruption statuses into ages. We have resisted this temptation, because such a table would be utterly unwieldy. With ten teeth and two statuses per tooth (erupted and unerupted) there are 1024 possible dental eruption patterns. While many of these patterns are very unlikely to occur, and so may be skipped, we have the additional problem that for archaeological or forensic applications some eruption statuses may be unobservable. We can handle this in the simulations by not truncating the simulated eruption ages for the missing observations, but it is simply impossible to anticipate in a table all the possible combinations of missing data. With the parameters we give in Table 11.1 it should be possible for other researchers to apply either numerical integration or Monte Carlo simulation to find the posterior age distribution. We are currently developing software to accomplish the later task.

Finding the posterior density of an age at death distribution

In palaeodemographic work it is not the individual ages at death that are of interest, but instead the age-at-death structure of a mortality sample. Konigsberg and Frankenberg (1992) recognised this problem and suggested that hazards analysis (Gage 1988; Wood *et al.* 1992) could be used to represent age distributions with a small set of parameters, while at the same time accounting for uncertainty in individual age estimates. Bocquet-Appel and Masset (1996) also argued for using reduced parameterisations, but in their case they simply calculate the first moment (mean) of the age-at-death distribution. In this section we give a simple implementation, again using Monte Carlo simulation, that allows hazard models to be fitted when age at death is estimated, rather than known. Obviously, in studies of prehistoric ontogenetic growth it is necessary to have unbiased age estimates, and also to have some idea of the precision (or uncertainty) of the estimates.

In the previous section we described simulation of age at death for an individual case. For this context, we simulated age from a uniform distribution between the highest age for an erupted tooth and the lowest age for a tooth yet to erupt. If we have multiple individuals, then we can simulate using a current guess at the probability density function for age at death. To specify this distribution we use an exponential hazard, which is a very

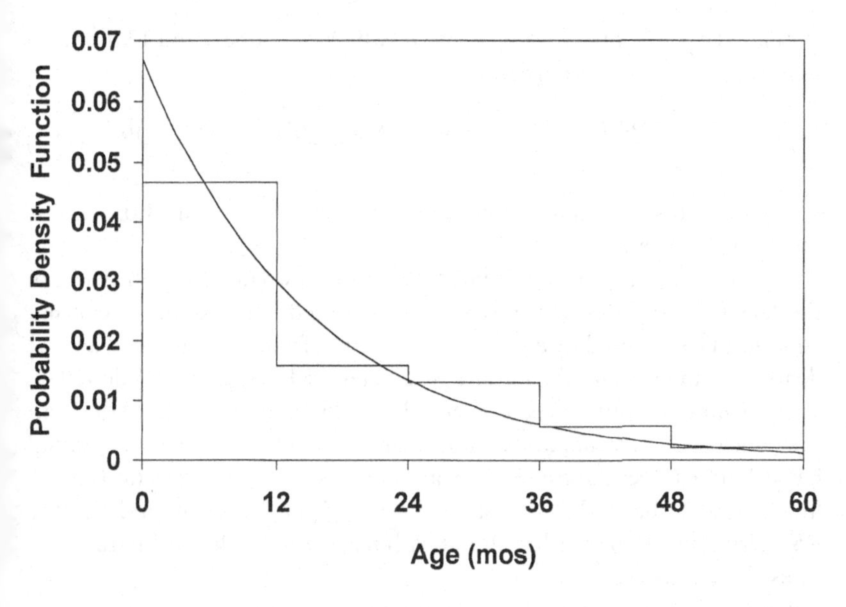

Fig. 11.2. Negative exponential probability density function fit to age-specific (by annum) frequencies of death from the Matlab region (Muhuri 1996). The smooth curve is the negative exponential, while the step function is the observed frequency within a year. mos, months.

common hazard function used to represent mortality in the first few years of life. As an example of fitting an exponential hazard model by traditional likelihood analysis, Fig. 11.2 shows a histogram of proportions of deaths by age up to 5 years for the Matlab region of Bangladesh. These data were taken from Muhiri's (1996) tabulation by yearly intervals for deaths of 2596 children. Also shown in this figure is the exponential probability density function for age at death fit using version 1.3.8 of MLE, written by one of the authors (D.J.H.).

We can also fit the exponential hazard model by Monte Carlo simulation, in this particular instance using the Gibbs Sampler. For any individual death, our observations are interval censored. For example, there were 1451 deaths between birth and 1 year, but we cannot tell for any individual the exact age at death. However, we can simulate 'exact' ages at death for each individual. It would obviously be inappropriate to simulate the ages from uniform distributions (as we did in the previous section). Instead, we should simulate the ages from a current guess at the probability density function described by the single parameter exponential distribution. If t_l and t_h are, respectively, the lower and higher ages for censoring (i.e. the simulated age must be between t_l and t_h, which might be 0 and 12 months), then we can simulate an age (t_i) as:

$$t_i = \frac{-\log\{u \times [\exp(-\lambda t_h) - \exp(-\lambda t_l)] + \exp(-\lambda t_l)\}}{\lambda}, \quad (11.1)$$

where u is a pseudorandom uniform deviate between 0.0 and 1.0, and λ is the hazard parameter.

Equation 11.1 appears problematic, because we do not know the value of the hazard parameter. If we assume some initial value, then we can use equation 11.1 to simulate ages at death. In turn, from the simulated ages at death we can simulate the hazard parameter itself. Gilks and Wild (1992) showed how adaptive rejection sampling can be used to simulate from log-concave density functions. The likelihood function for the exponential hazard model does have a log-concave curve, so adaptive rejection sampling can easily be applied. To sample we need the log-likelihood and the first derivative of the log-likelihood with respect to the hazard parameter. These functions are:

$$\ln LK(\lambda) = N \times \ln(\lambda) - \lambda \sum_{i=1}^{N} t_i$$

$$\frac{\partial \ln LK(\lambda)}{\partial \lambda} = \frac{N}{\lambda} - \sum_{i=1}^{N} t_i, \quad (11.2)$$

where t_i are the ages at death, N is the number of individuals, and λ is the exponential hazard parameter. If we cycle between simulating ages at death (conditional on the hazard parameter), and then simulating the hazard parameter (conditional on the ages at death), then the simulated hazard parameters will converge on the true value. We show this in Fig. 11.3, which plots a histogram of simulated values from the Matlab example and the normal distribution for the parameter estimated using the program MLE. To draw the histogram we did 200 simulation runs as a 'burn-in' or 'dememorisation', and then run 5000 simulations taking every fifth value (for a total of 1000 simulations). To summarise, we start with a value for λ, then simulate exact ages at death using equation 11.1 (here t_l and t_h are the yearly censoring). Then we treat these exact ages as if they were real known ages, and use equation 11.2 with the adaptive rejection sampler to draw a new value for λ. As is clear from Fig. 11.3, the agreement between the Gibbs Sampler and the maximum likelihood estimator is high.

To find the posterior density for the exponential hazard parameter in a palaeodemographic setting we need to combine the simulation we have just described above with the simulation described in the previous section.

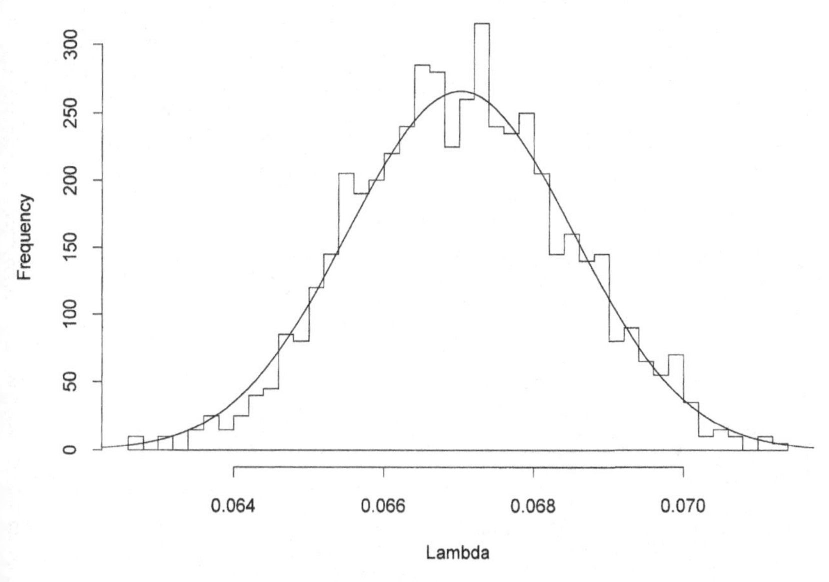

Fig. 11.3. Comparison of the maximum likelihood estimate (MLE) and the Gibbs Sampler estimate for the negative exponential hazard in an interval censored case (data from Muhuri 1996). The smooth curve is the posterior density for the hazard from MLE, while the histogram is from the Gibbs Sampler. Note the good agreement between the two methods of estimation.

This Gibbs Sampler can be encapsulated as follows:

(1) Initialise the simulation by assigning a value for λ and ages at death for all individuals.
(2) Simulate the ages at eruption for each individual from truncated multivariate normal distributions (as described above).
(3) Simulate an age at death for each individual using equation 11.1, so that the age is between t_l and t_u (from eruption ages).
(4) Simulate a new value of λ from equation 11.2, store this value, and return to step 2.

As an example, we have taken a portion of the sample of 397 Bangladeshi children described earlier, and used cross-sectional observations on dental eruption status to determine age at death. We used records on 218 individuals from the total sample. We limited ourselves to only about 55% of the original sample, because we needed to select the ages such that the individuals formed a plausible 'death' sample, in this case following an exponential hazard. Given the mixed longitudinal nature of the sample, we could not use the entire data set in order to construct a new sample that even remotely approximated normal mortality. We also limited the sample to

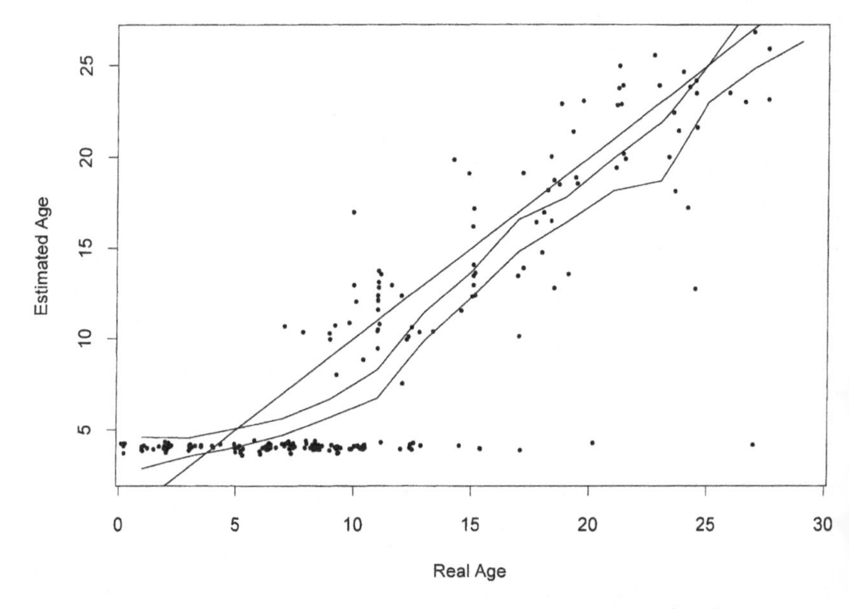

Fig. 11.4. Comparison of real ages with estimated ages. The straight line represents the case of unbiased estimation, while the two other lines represent the 90% bootstrap confidence interval from a cubic B-spline fit.

only those individuals who had one or more unemerged teeth from the left maxilla and mandible. This reduced the actual age range, so that no individual older than about 30.5 months was included. This limitation on the sample was necessary because an individual with all ten left teeth erupted has no upper bound on their posterior age distribution.

Figure 11.4 shows a plot of the estimated 'ages at death' against the real ages. To estimate age at death we ran an initial 200 simulation runs, and then averaged the age for each individual across the next 1000 simulations. The plot shows that there are some individuals whose ages are substantially under estimated. This occurs because of delayed eruption for some children. As a result, the true average age of 11.09 months is underestimated at 9.48 months. Although a 1.5 month difference may seem trivial, the difference is highly significant ($p < 0.0001$) on a two-tailed paired t-test. Nevertheless, the correlation between real and estimated age is fairly high ($r = 0.85$). The linear model relating real and estimated ages is, however, a bit misleading. This is shown clearly in Fig. 11.4, where we have plotted the 90% confidence interval from 199 bootstraps using a cubic B-spline with four degrees of freedom. The method for constructing this confidence interval is given by Davison and Hinkley (1997). One potential

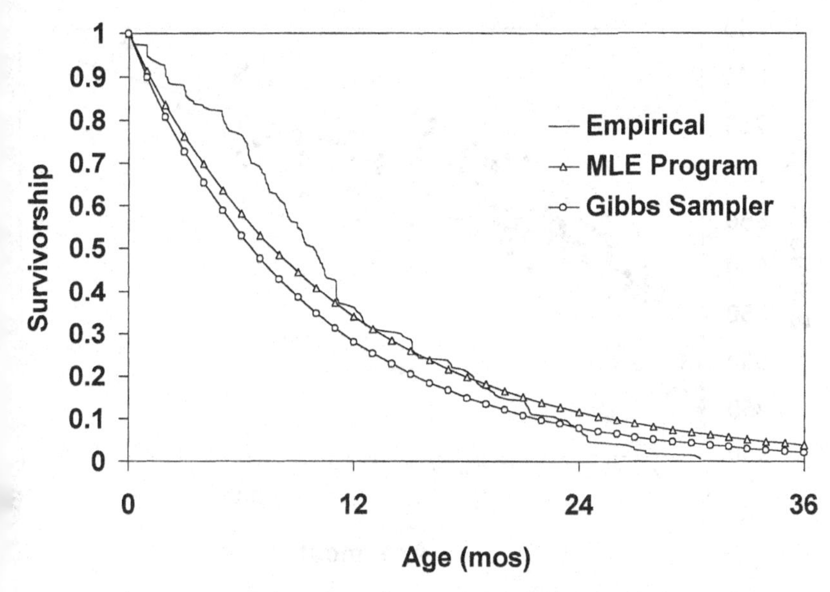

Fig. 11.5. Comparison of actual survivorship (labelled 'empirical'), the MLE of survivorship in an exponential hazard model, and the Gibbs Sampler estimate. Note that the known ages were used to construct the MLE of survivorship, while ages estimated from dental emergence were used in the Gibbs Sampler. mos, months.

problem in age estimation here is that we were not very successful in forming a sample of individuals with an exponential hazard. This is shown in Fig. 11.5, which compares the empirical survival function, the survival function from an exponential hazard fit to the known ages, and the survival function from an exponential hazard estimated from the Gibbs Sampler (where ages are treated as estimated, rather than known).

Finding the posterior density of somatic growth parameters

As we mentioned in the introduction, previous studies of prehistoric somatic growth have treated point estimates of ages as if these estimates were known with complete certainty. We are now in a position to include the uncertainty of age estimates in the estimation of growth parameters. For the following example we will focus on statural growth, though obviously our comments would apply just as well to long bone growth.

To model stature growth we use a very simple logarithmic model described in Count (1943). Here stature (s) is regressed on age (t) and the

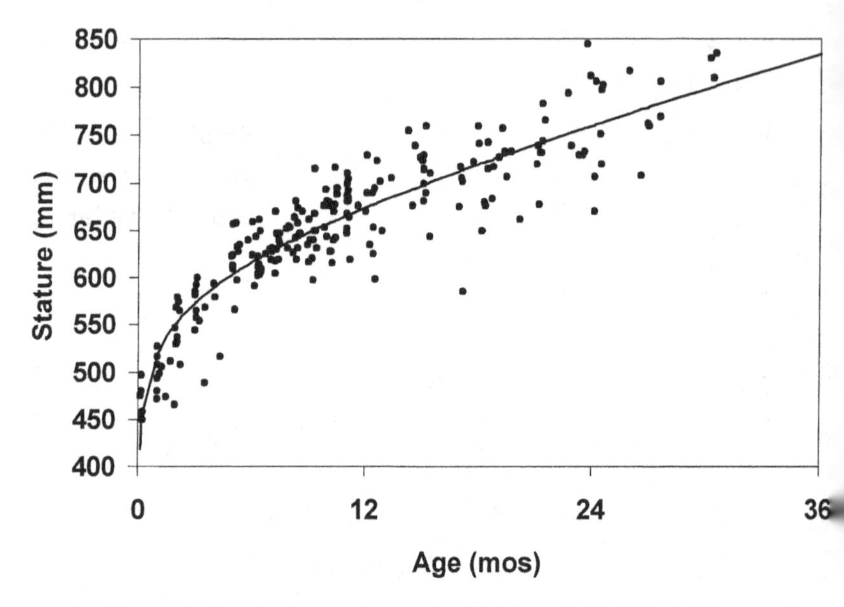

Fig. 11.6. Plot of stature against age for 218 Bangladeshi children. The regression line shown is $s_i = \beta_0 + \beta_1 \times t + \beta_2 \times \ln(t)$ where s is stature and t is age.

natural log of age, so that there are three parameters to be estimated (an intercept and two slopes). This model can be written as:

$$s_i = \beta_0 + \beta_1 \times t + \beta_2 \times \ln(t), \tag{11.3}$$

where t is the age in months. We can first fit this model using actual known ages (in this case for the 218 individuals used in the previous section). Figure 11.6 shows a plot of this line and the 218 cases. We take a fairly complicated approach to fitting the model because it will serve as an introduction to the estimation procedure we use when we assume that ages are estimated rather than known. In this method we use Monte Carlo simulation to draw numerous times from the posterior distributions of these three parameters as well as from the conditional variance of stature. This method is described in detail by Gelman *et al.* (1995).

It is relatively simple to adapt the Monte Carlo method given by Gelman *et al.* (1995) to incorporate uncertainty in age estimates. Essentially, we merely need to combine the simulations described on pp. 271–276 with simulation from Gelman *et al.* (1995). This gives the following algorithm:

(1) Assume initial values for λ (the exponential hazard parameter), for each individual's age, and for eruption times of all teeth.

(2) Simulate exact ages at eruption for each tooth within each individual. These ages are simulated by conditioning on all other teeth's eruption times, and by truncating so that teeth that have yet to erupt will erupt at greater than the current age, while teeth that have erupted did so earlier than the current age.
(3) For each individual, simulate their age from a truncated exponential distribution (equation 11.1).
(4) Treat the simulated ages for individuals as if these were real observed data, and then sample from the posterior distribution for regression parameters of stature on age (again, as in Gelman *et al.* 1995).
(5) Sample a new exponential hazard using adaptive rejection sampling and equation 11.2.
(6) Return to step 2.

Figure 11.7 compares the parameters (β_0, β_1, β_2, and σ^2) from the case where age is known to that where age is unknown but estimated. The parameters are drawn as histograms and kernel density estimates (Silverman 1986). For the case where age is known, the histograms represent 1000 simulations. For the case where age is estimated, we used 200 iterations of simulation as a 'burn-in' and then every fifth iterate of 5000 iterations. The initial cycles are run to 'dememorise' (i.e. 'forget' the starting values), while sampling every fifth iteration is done so that the simulations will not be serially correlated. There are a few important points to be made from Fig. 11.7. First, all parameters are more efficiently estimated in the case where age is known. This is reflected in the more concentrated distributions. Further, the error variance is obviously higher when age is unknown. Taken together, these observations mean that our estimates of average growth curves must have larger standard errors when age is estimated rather than known. This is exactly as we would suspect, as the loss of power entailed in using estimated rather than known ages should result in broader confidence intervals around summaries of growth. Taken to its logical conclusion, this also means that tests for growth differences between samples will be too liberal if we do not account for uncertainty in age estimates. Once we make the adjustment for unknown ages, there is necessarily a loss of power, so that we will require even larger samples to detect significant differences between samples.

Another point apparent from Fig. 11.7 is that the regression parameter for log age when age is unknown is not significantly different from zero (i.e. 0.0 is well within the 95% HPD). This is clear when we use the information represented in Fig. 11.7 to draw the 95% confidence intervals around average stature for age. Figure 11.8 shows a comparison of the 95%

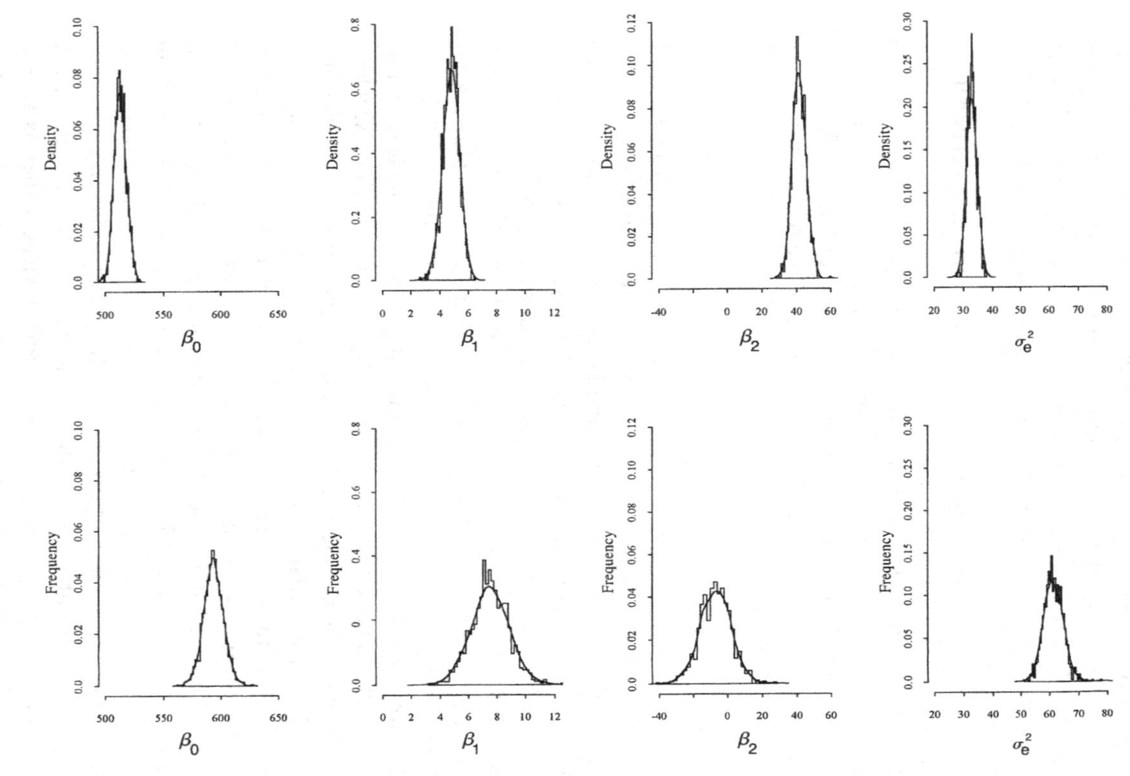

Fig. 11.7. Histograms and kernel densities for regression parameters of stature on age. The top row shows these posterior densities for the case where age is known (i.e. using the actual ages), while the bottom row shows the case where age is estimated from dental eruption.

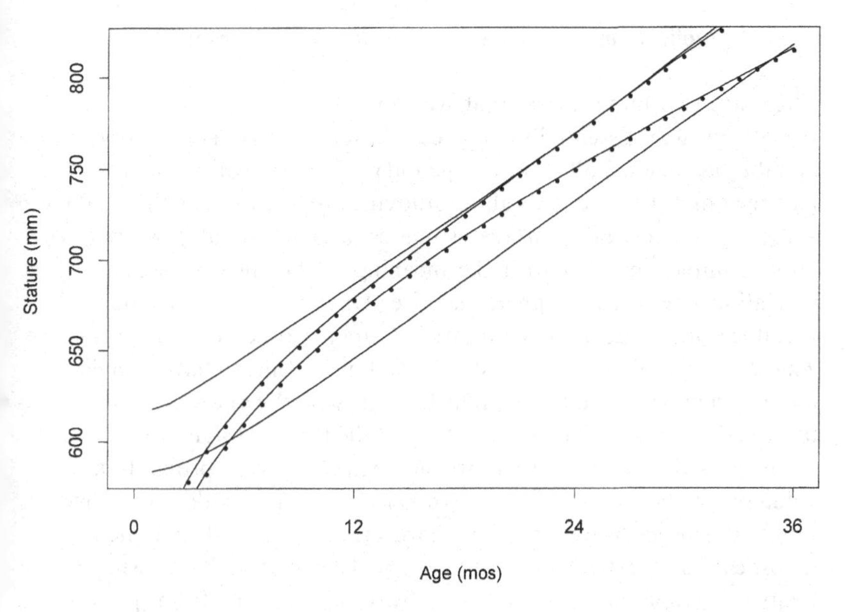

Fig. 11.8. Comparison of 95% confidence intervals (CIs) for the regression of stature on age when age is known (shown with dots) vs. when age is estimated (shown without dots). The dots shown on the lines are from standard regression theory CIs, while the lines are from Monte Carlo simulation. The case where age is estimated was fitted via the Gibbs Sampler. mos, months.

confidence intervals when age is known versus when age is estimated. To draw the confidence intervals when age is known we use the variance–covariance matrix among simulated parameters (from the 1000 replicates), the average parameter values, and the t-distribution with 215 degrees of freedom. For comparison, we also show the ordinary intervals that would be calculated with standard regression formulae. The intervals when age is unknown (but estimated) are drawn using the average and variance–covariance matrix from the Gibbs Sampler. There is both good and bad news to take away from Fig. 11.8. The good news is that, with the exception of growth within the first 6 months, the unknown age case gives a fairly unbiased estimate of actual (cross-sectional) growth. This is shown by the fact that the unknown-age growth confidence intervals enclose the actual cross-sectional curve. Only in the early months is there a departure, where, as we previously mentioned, the exponential nature of the growth curve is missed. The bad news is that when age is estimated rather than known, the confidence interval for cross-sectional growth is (necessarily) larger.

Implications for studies of prehistoric somatic growth

There are two implications that we can make for studies of prehistoric somatic growth. Essentially, these can be summarised as the 'good news' and the 'bad news'. Following an age-old practice, we will dispatch with the bad news first. The bad news about studying prehistoric growth is that if we account for uncertainty in age estimates (and we should), we will need larger samples in order to make meaningful statistical statements. This revelation comes as no great surprise. Uncertainty in an independent variable leads to greater uncertainty in estimates of regression parameters. This is a very well-known statistical fact, but it seems to have been largely ignored in previous studies of prehistoric growth. We know of no nicer way to state this, so we will be blunt. Many of the reported significant differences in growth curves between archaeological samples probably are not actually significant. We have shown here how it is possible to adjust the estimation procedure so that standard errors of growth parameters are realistically stated. Bad news aside, the good news is that it is possible to get relatively unbiased estimates of growth parameters from prehistoric samples, provided a reasonable estimation scheme is in place.

Acknowledgements

We thank Bob Jones, George Milner, Kathleen O'Connor, and Jim Wood for valuable discussions and insights. For assistance with the Bangladesh data, we thank the International Centre for Diarrhoeal Disease Research, Bangladesh, and a fellowship from the American Institute of Bangladesh Studies (D.J.H.). This research was supported in part by NICHD grant F32–HD07994–02.

References

Alvarez JO, Eguren JC, Caceda J and Navia JM (1990) The effect of nutritional status on the age distribution of dental caries in the primary teeth. *Journal of Dental Research* **69**, 1564–1566.

Billewicz WZ and McGregor IA (1975) Eruption of permanent teeth in West African (Gambian) children in relation to age, sex and physique. *Annals of Human Biology* **2**, 117–128.

Billewicz WZ, Thomson AM, Baber FM and Field CE (1973) The development of primary teeth in Chinese (Hong Kong) children. *Human Biology* **45**, 229–241.

Bock RD and Gibbons RD (1996) High-dimensional multivariate probit analysis.

Biometrics **52**, 1183–1194.

Bocquet-Appel J-P and Masset C (1982) Farewell to paleodemography. *Journal of Human Evolution* **11**, 321–333.

Bocquet-Appel JP and Masset C (1996) Paleodemography: expectancy and false hope. *American Journal of Physical Anthropology* **99**, 571–583.

Bogin B (1988) *Patterns of Human Growth.* Cambridge: Cambridge University Press.

Box GEP and Tiao GC (1992) *Bayesian Inference in Statistical Analysis.* Wiley Classics Library edn. New York: John Wiley and Sons.

Burdi AR, Garn SM and Miller RL (1970) Developmental advancement of the male dentition in the first trimester. *Journal of Dental Research* **49**, 889.

Chib S and Greenberg E (1998) Analysis of multivariate probit models. *Biometrika* **85**, 347–361.

Count EW (1943) Growth patterns of the human physique: an approach to kinetic anthropometry. *Human Biology* **15**, 1–32.

Davison AC and Hinkley DV (1997) *Bootstrap Methods and their Application.* New York: Cambridge University.

Delgado H, Habicht J-P, Yarbrough C, Lechtig A, Martorell R, Malina RM and Klein RE (1975) Nutritional status and the timing of deciduous tooth eruption. *American Journal of Clinical Nutrition* **28**, 216–224.

Demirjian A (1986) Dentition. In *Human Growth*, ed. F. Falkner and J.M. Tanner, pp. 269–298. New York: Plenum Press.

Diebolt J and Ip EHS (1996) Stochastic EM: method and application. In *Bayesian Data Analysis*, ed. W.R. Gilks, S. Richardson and D.J. Spiegelhalter, pp. 259–273. New York: Chapman & Hall.

Drezner Z (1992) Computation of the multivariate normal integral. *ACM Transactions on Mathematical Software* **18**, 470–480.

el Lozy M, Reed RB, Kerr GR, Boutourline E, Tesi G, Ghamry MT and Stare FJ (1975) Nutritional correlates of child development in Southern Tunisia. IV. The relation of deciduous eruption to somatic development. *Growth* **39**, 209–221.

Gage TB (1988) Mathematical hazard models of mortality: an alternative to model life tables. *American Journal of Physical Anthropology* **76**, 429–441.

Gage TB (1989) Bio-mathematical approaches to the study of human variation and mortality. *Yearbook of Physical Anthropology* **32**, 185–214.

Garn SM, Koski K and Lewis AB (1957) Problems in determining the tooth eruption sequence in fossil and modern man. *American Journal of Physical Anthropology* **15**, 313–331.

Gelman A, Carlin JB, Stern HS and Rubin DB (1995) *Bayesian Data Analysis.* London: Chapman & Hall.

Gilks WR and Wild P (1992) Adaptive rejection sampling for Gibbs sampling. *Applied Statistics* **41**, 337–348.

Gilks WR, Richardson S and Spiegelhalter DJ (eds.) (1996) *Markov Chain Monte Carlo in Practice.* New York: Chapman & Hall.

Haavikko K (1970) The formation and the alveolar and clinical eruption of the permanent teeth: an orthopantomographic study. *Proceedings of the Finnish*

Dental Society **66**, 101–165.

Holman DJ and Jones RE (1991) Longitudinal analysis of deciduous tooth emergence in Indonesian children. I. Life table methodology. *American Journal of Human Biology* **3**, 389–403.

Holman DJ and Jones RE (1998) Longitudinal analysis of deciduous tooth emergence. II. Parametric survival analysis in Bangladeshi, Guatemalan, Japanese and Javanese children. *American Journal of Physical Anthropology* **105**, 209–230.

Holtz R (1959) The relation of dental calcification to chronological and skeletal age. *Transactions of the European Orthodontic Society* **35**, 140–149.

ICDDR B (1990) *Meheran Growth and Development Study Data Set Description and Technical Documentation.* V1.10 1990. Dhaka, Bangladesh: International Centre for Diarrhoeal Disease Research, Bangladesh.

Jellife EFP and Jellife DB (1973) Deciduous dental eruption, nutrition and age assessment. *Journal of Tropical Pediatrics and Environmental Child Health* **19**, 193–248.

Johnson FE (1969) Approaches to the study of developmental variability in human skeletal populations. *American Journal of Physical Anthropology* **31**, 335–341.

Khan M and Curlin GT (1978) *Development of Milk Teeth in Rural Meheran Children of Bangladesh.* Working Paper no. 8, Cholera Research Laboratory, Dacca, Bangladesh.

Khan M, Curlin GT and Chakraborty J (1979) Growth and development studies: rural Meheran, Comilla. *Bangladesh Medical Journal* **7**, 74–90.

Khan MJ, Chakraborty J and Paul SR (1981) Teeth as index to health and age in Bangladeshi children. *Nutrition Reports International* **24**, 963–971.

Kodali VRR, Krishnamachari KAVR and Gowinathstry J (1993) Eruption of deciduous teeth: influence of undernutrition and environmental fluoride. *Ecology of Food and Nutrition* **30**, 89–97.

Kolakowski D, and Bock RD (1981) A multivariate generalization of probit analysis. *Biometrics* **37**, 541–551.

Konigsberg LW and Frankenberg SR (1992) Estimation of age structure in anthropological demography. *American Journal of Physical Anthropology* **89**, 235–256.

Krogman WM (1968a) Biological timing and the dento-facial complex. Part I. *Journal of Dentistry for Children* **35**, 175–185.

Krogman WM (1968b) Biological timing and the dento-facial complex. Parts II and III. *Journal of Dentistry for Children* **35**, 328–381.

Lewis AB and Garn SM (1960) The relationship between tooth formation and other maturational factors. *Angle Orthodontist* **30**, 70–77.

Muhuri PK (1996) Estimating seasonality effects on child mortality in Matlab, Bangladesh. *Demography* **33**, 98–110.

O'Connor KA (1995) The age pattern of mortality: a micro-analysis of Tipu and a meta-analysis of twenty-nine paleodemographic samples. PhD dissertation, University of Albany.

Parfitt AM (1984) The cellular basis of bone remodeling: the quantum concept reexamined in light of recent advances in the cell biology of bone. *Calcified*

Tissue International **36**, S37–S45.

Proffit WR, Prewitt JR, Baik HS and Lee CF (1991) Video microscope observations of human premolar eruption. *Journal of Dental Research* **70**, 15–18.

Rao KV (1985) Nutritional status, deciduous dental eruption and age assessment. In *Dental Anthropology, Applications and Methods*, ed. V.R. Reddy, pp. 3–11. New Delhi: Inter-India Publications.

Saleemi M, Jalil F, Karlberg J and Hagg U (1993) Early child health in Lahore, Pakistan: XVIII. Primary teeth emergence. *Acta Paediatrica Supplement* **390**, 159–167.

Saunders SR (1992) Subadult skeletons and growth related studies. In *Skeletal Biology of Past Peoples: Research Methods*, ed. S.R. Saunders and M.A. Katzenberg, pp. 1–20. New York: Wiley-Liss.

Saunders SR and Hoppa RD (1993) Growth deficit in survivors and non-survivors: biological mortality bias in subadult skeletal samples. *Yearbook of Physical Anthropology* **36**, 127–151.

Saunders S, DeVito C, Herring A, Southern R and Hoppa R (1993) Accuracy tests of tooth formation age estimations for human skeletal remains. *American Journal of Physical Anthropology* **92**, 173–188.

Schervish MJ (1984) Multivariate normal probabilities with error bound. *Applied Statistics* **33**, 81–94.

Silverman BW (1986) *Density Estimation for Statistics and Data Analysis.* New York: Chapman & Hall.

Sinclair D (1985) *Human Growth After Birth.* 4th edn. New York: Oxford University Press.

Smith AFM and Roberts GO (1993) Bayesian computation via the Gibbs sampler and related Markov chain Monte Carlo methods. *Journal of the Royal Statistical Society* B **55**, 3–23.

Smith BH (1991) Standards of tooth formation and dental age assessment. In *Advances in Dental Anthropology*, ed. M. Kelley and C.S. Larsen, pp. 143–168. New York: Wiley-Liss.

Smith BH (1992) Life history and the evolution of human maturation. *Evolutionary Anthropology* **1**, 134–142.

Stewart TD (1973) *Essentials of Forensic Anthropology.* Springfield, IL: Charles C. Thomas.

Tanner JM (1978) *Foetus into Man: Physical Growth from Conception to Maturity.* Cambridge, MA: Harvard University Press.

Townsend N and Hammel EA (1990) Age estimation from the number of teeth erupted in young children: an aid to demographic surveys. *Demography* **27**, 165–174.

Whittington SL (1991) Detection of significant demographic differences between subpopulations of prehispanic Maya from Copan, Honduras, by survival analysis. *American Journal of Physical Anthropology* **85**, 167–184.

Wood JW, Holman DJ, Weiss KM, Buchanan AV and LeFor B (1992) Hazards models for human population biology. *Yearbook of Physical Anthropology* **35**, 43–87.

12 *Linear and appositional long bone growth in earlier human populations*: *a case study from Mediaeval England*

SIMON A. MAYS

Introduction

In the human skeleton, long bones attain their adult forms via a combination of longitudinal and appositional growth (Tanner 1989: 31–35). Increase in length occurs via endochondral growth. The epiphyseal cartilage or growth plate, which separates the epiphysis from the metaphysis, grows away from the shaft centre, and the cartilage is replaced by bone on the metaphyseal side of the plate so that the shaft increases in length. Growth in bone width takes place as a result of the apposition of new bone on existing cortical walls without the intervention of cartilage.

The gross features of appositional growth in human long bones have been characterised using radiographic studies of living subjects. This work has indicated that bone is added to the outer surface of the cortex beneath the periosteum throughout the growth period. During infancy and childhood, bone is simultaneously resorbed from the endosteal surface, so that the medullary cavity increases in width, although endosteal resorbtion generally occurs at a slower rate than periosteal apposition (Garn 1970). In adolescence, cortical bone is added at both the periosteal and endosteal surfaces (Frisancho *et al.* 1970). The upshot of all this is that, as well as showing increase in overall bone width, the thickness of the cortical walls of a long bone increases throughout the growth period (Garn 1970). Slight apposition of bone continues beneath the periosteum throughout adult life (Lazenby 1990). However, during middle and later adulthood, bone begins to be resorbed once more from the endosteal surface, so that from middle age there is a net loss of cortical thickness (Garn 1970).

Since the pioneering work of Johnston (1962), there have been many studies of longitudinal growth in earlier human populations using skeletal remains (for references, see Saunders *et al.* 1993). By comparison, appositional bone growth in earlier populations has been somewhat neglected. Longitudinal growth is widely accepted as a good indicator of child health and nutrition, poor growth being viewed as an indicator of unfavourable conditions. There is also abundant evidence (e.g. Adams and Berridge 1969; Himes *et al.* 1975) for a link between poor nutrition and deficient appositional growth in the long bones of immature individuals. Although it is influenced by other factors, such as activity patterns and changes in hormonal balance, cortical thickness is potentially a very useful, if rather overlooked, stress indicator in earlier populations.

This chapter seeks to show that, although the study of longitudinal bone growth alone may yield valuable insights into growth patterns and child health and nutritional status in earlier populations, examination of appositional bone growth also has an important role to play in these respects. In particular, it is suggested that if longitudinal and appositional growth are analysed in conjunction, additional insights may emerge over and above those which might be obtained were either aspect studied in isolation. A case study using skeletal material excavated from the Mediaeval site at Wharram Percy, England, is used to support these contentions.

Growth at Wharram Percy

Wharram Percy is a deserted Mediaeval village situated in the Yorkshire Wolds, an area of chalk uplands in northern England (Fig. 12.1). The site was subject to a long-running research excavation (Beresford and Hurst 1990). An important focus of the fieldwork was the church and churchyard, one of the aims being to obtain a substantial series of burials in order to shed light on a Mediaeval rural population. This aim was realised, as the excavations have resulted in the recovery of 687 articulated skeletons. The burials, which date primarily from the 10th–16th century, are mainly of ordinary peasants who lived in Wharram Percy itself or elsewhere in the parish, according to the archaeological and historical evidence. Nearly half (327 burials) of the assemblage is composed of individuals who were aged less than about 18 years when they died. This large number of immature skeletons, together with the excellent bone preservation which characterises the site, means that the assemblage presents a major opportunity to investigate growth patterns in Mediaeval children.

In order to investigate growth in skeletal remains, an accurate and independent estimate of age at death in individual skeletons is needed against which to plot bone dimensions. In the present study dental calcification was used to determine age. Individual age was estimated by comparison of dentitions with the chronology of Schour and Massler (1941).

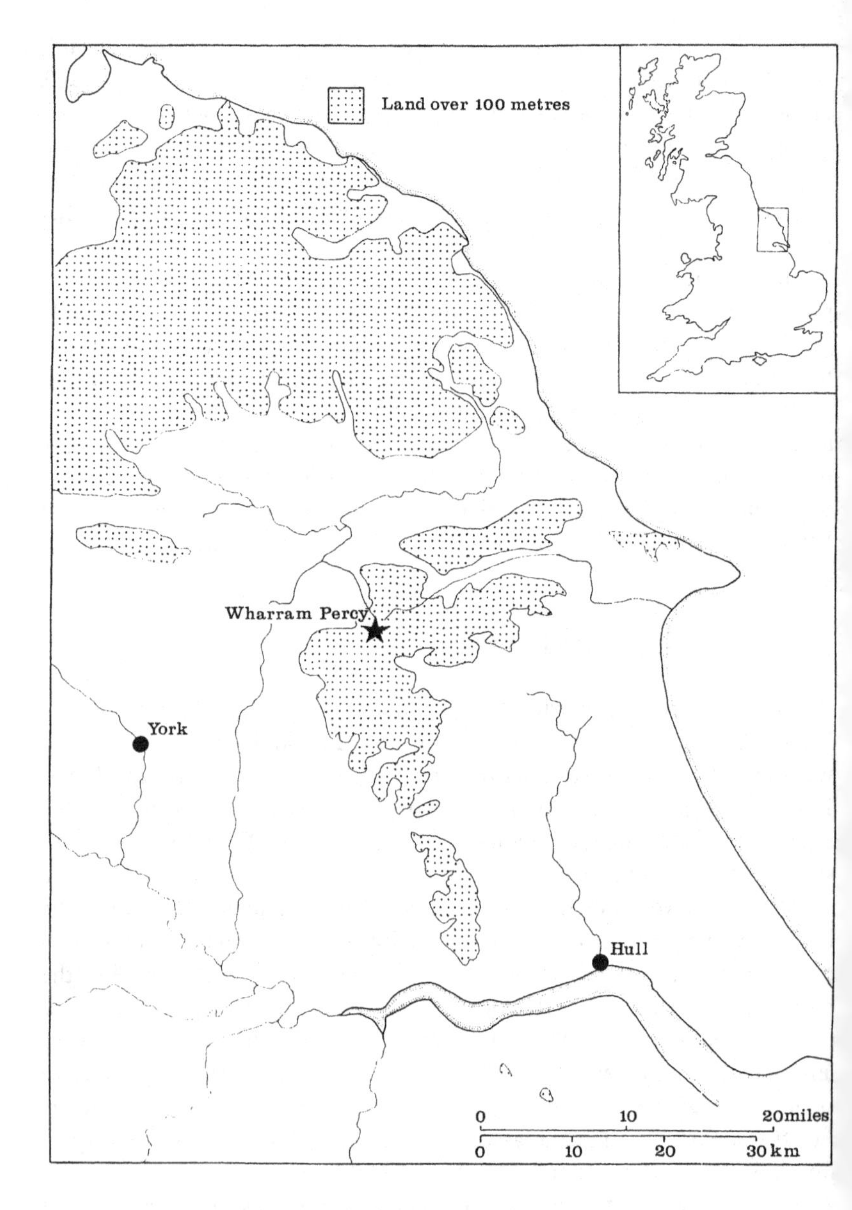

Fig. 12.1. The location of Wharram Percy.

Longitudinal bone growth

Longitudinal growth in the femur was studied by plotting diaphyseal length against dental age. In an attempt to place them in context, the Wharram Percy results were compared with others from populations of northern European origin, both ancient and modern. Availability of suitable archaeological comparisons was constrained by a number of factors. Firstly, a large number of juveniles are needed, with a good representation of different age groups, in order to establish a skeletal growth profile. During many periods in the past, immature individuals are scarce in cemetery material. A major reason for this is probably differential burial practices. For example, relatively few immature remains are known from the prehistoric periods in Britain – presumably they were often afforded mortuary treatment that has left no trace in the archaeological record. Even if all age groups are accorded cemetery burial, there may be relatively few older children and adolescents as mortality in these age groups is generally fairly low; this accentuates the need for large assemblages. In addition, only that subgroup of immature individuals who can be aged using dental calcification, and who also have the long bone of interest intact for measurement, can be used for the preparation of skeletal growth profiles. This can cut sample size markedly, even where soil conditions are conducive to good bone preservation. A final consideration is that not all publications of skeletal growth profiles present their findings in such a way as to be suitable for comparison with the present data. In light of the above, four assemblages of populations of European origin, dating from the Romano-British period to the 19th century were selected (Table 12.1).

The classic study of Maresh (1955), on mid 20th century US White children, was used as a recent comparandum. He gives femur lengths obtained from radiographs of living subjects. These are not precisely comparable with measurements from dry bones as there will have been a degree of radiographic enlargement. The enlargement factors estimated for Maresh's study by Feldesman (1992) have been used to estimate true femur length for Maresh's subjects. The femur diaphyseal data for Wharram Percy are listed in Table 12.2. They, and those for the ancient and modern comparanda are plotted in Fig. 12.2.

Looking first at the comparison between the Wharram Percy data and those from the recent subjects studied by Maresh (1955), the difference in femur length is very marked, for example the diaphyseal length for a 14 year old from Wharram Percy approximates to that of a recent 10 year old. One could argue that mortality bias may be playing a part in producing this deficit in femur length in the archaeological group. In our archaeologi-

Table 12.1. *Archaeological* comparanda *for femoral longitudinal growth*

Site	Date	*N*	Reference
Poundbury, England	AD 3rd–4th century	55	S. Mays (unpublished data)
Mikulčice, Czech Republic	AD 9th century	249	Stloukal and Hanáková (1978)
Ensay, Scotland	AD 1500–1850	118	Miles and Bulman (1994)
Belleville, Canada	AD 1821–1874	156	Saunders *et al.* (1993)

N, sample size.

Table 12.2. *Diaphyseal lengths for femurs from juvenile skeletons from Wharram Percy, Yorkshire*

Age	*N*	Mean	SD
1	10	125.5	10.9
2	10	151.9	10.3
3	11	173.1	12.8
4	11	186.0	12.1
5	6	195.5	9.9
6	21	214.6	12.8
7	8	241.9	23.9
8	4	262.3	16.8
9	6	272.7	14.1
10	9	285.0	26.4
11	6	300.3	24.0
12	2	306.0	12.7
14	5	346.0	12.0
17	8	376.9	22.6

Notes for Tables 12.2–12.4: Age is mid point of age category in years (e.g. the 1 year age class includes individuals whose estimated ages fall into the band 6 months to 1.49 years). Diaphyseal lengths and cortical thicknesses are in millimetres. SD, standard deviation; *N*, sample size.

cal studies we are clearly looking at growth in those who died in childhood, and it might be suggested that this may not be representative of growth patterns in healthier children who survived into adult life. However, a recent review of this problem (Saunders and Hoppa 1993) concluded that the biasing effect on skeletal studies is probably minor. An additional consideration is that age in the Wharram Percy children was estimated using dental development, whereas for Maresh's subjects age was known exactly. Dental development is relatively little affected by extrinsic factors (Lewis and Garn 1960) and in general has been shown to be a highly reliable technique for estimating age in immature individuals (Smith 1991).

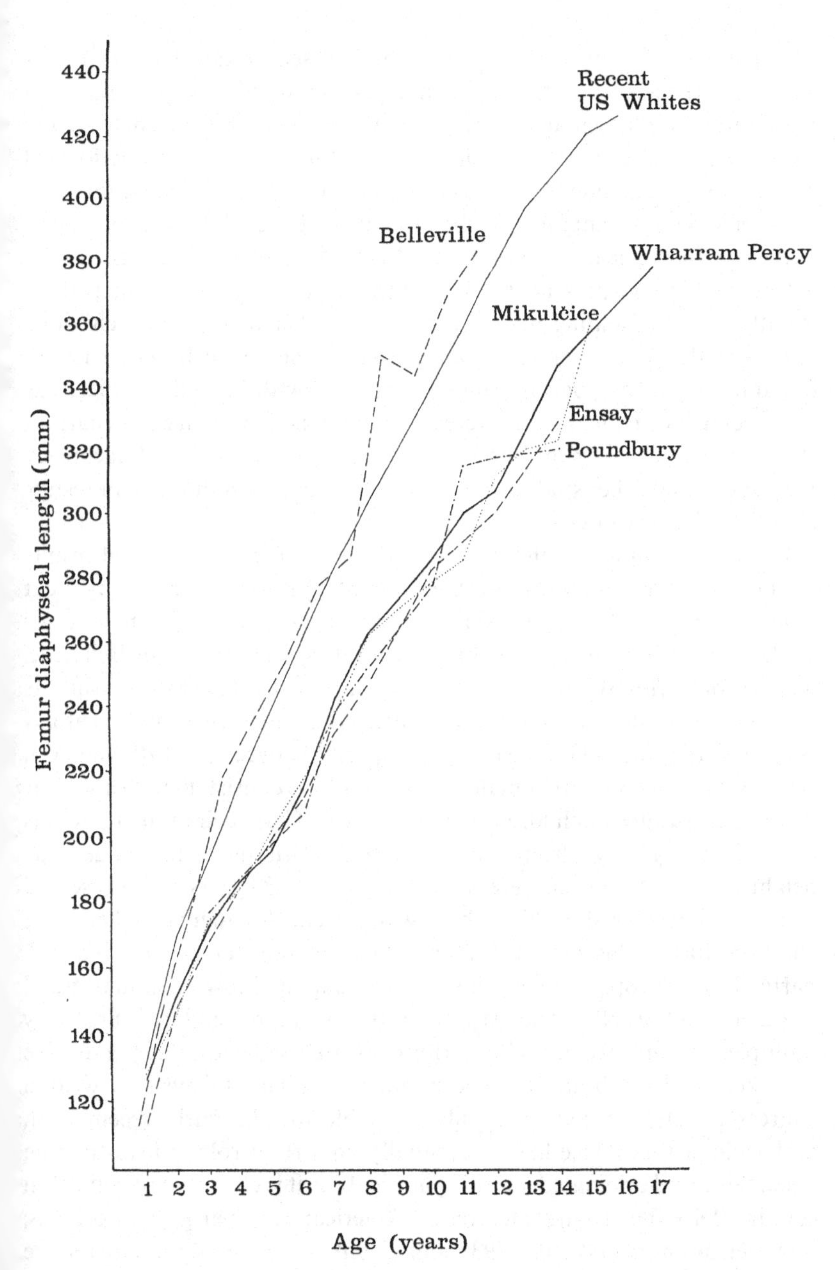

Fig. 12.2. Femur diaphyseal length plotted against dental age for Wharram Percy, and for some other archaeological assemblages taken from the literature (for data sources, see the text). Data for recent US children (Maresh 1955) have also been included. Maresh's data have been adjusted for radiographic enlargement, and for the older age groups diaphyseal length has been estimated from total bone length (see the text).

This has been demonstrated directly for archaeological material. Liversidge (1994) tested a variety of dental ageing standards on a series of immature skeletal remains of known age at death recovered from the 18th–19th century crypt at Spitalfields, London. All standards performed well, although all tended to underage slightly. For example the Schour and Massler standards employed in the present work were found to underestimate age by a mean of 0.11 years. If the Spitalfields findings can be extrapolated to the present case, then it would seem that we are likely to be slightly underestimating age at death in the Wharram Percy children. However, this factor cannot be implicated in the deficit in femur length found in the archaeological material compared with Maresh's subjects, as the effect will be to lead to an overestimation of height for age at Wharram Percy. In any event, given the high accuracy found for dental ageing by Liversage's and other studies, the effect of this factor on the current results is likely to be very minor.

Despite spanning about 1500 years, most of the archaeological results tend to closely resemble one another, providing no evidence for any great difference in growth profiles over many centuries. The exception, in terms of the archaeological series, is the 19th century material from Belleville, which groups closely with the recent data, suggesting that stressors such as malnutrition or disease may not have affected the Belleville children greatly (Saunders *et al.* 1993). The marked disparity between the 19th and 20th century data and the earlier material in Fig. 12.2 is consistent with a secular increase in stature in children in recent times. That such an increase has occurred is well known from written sources. In Britain, the first large-scale height surveys of children were carried out about 150 years ago. These and subsequent studies show that there was a trend for increased height in children during this period. Other data from the industrialised world, particularly Europe, North America and Japan, show a similar trend (Tanner 1989; 156ff.). The causes of this secular trend are probably multiple, but improvements in nutritional status and a lessening of disease are likely to have been of prime importance (*ibid.*). Given that written sources on child growth are only available for the fairly recent past, archaeological evidence has a potentially important role in investigating when this trend originated. Looking at Fig. 12.2, it is clear that the mid 19th century Belleville group match recent Americans in their growth profiles; however, as Saunders *et al.* (1993) pointed out, there is no reason to assume that the growth profiles of the Belleville children are representative of those of other contemporaneous White groups, such as 19th century British children. In order to investigate the origin of the secular trend for increased height for age in Britain, we need large, well-dated assemblages of post-

Mediaeval juvenile skeletal remains. At the time of writing, none is available, although this situation may change in the future. In the meantime, another way of approaching this problem is to compare the archaeological growth data we do have with 19th and 20th century studies of living British subjects. In order to do this we need to convert bone lengths into estimated statures for juveniles.

Feldesman (1992) gathered data on the relationship between femur length and standing height in immature individuals using published radiographic studies of living subjects (US children of European ancestry) aged 8–18 years. He found that, although there were differences in femur: stature ratio according to the age and sex of the child, within sex and age classes femur length bore a close relationship to standing height, with little variation in femur: stature ratio between different studies. I used his femur: stature ratio data to estimate stature in the Wharram Percy children. An illustration of the method is as follows. Feldesman (1992: Table 5) gives the femur: stature ratio for 10 year olds as 0.2691 for females and 0.2675 for males, producing a mid-sex mean of 0.2683. Feldesman's figures refer to total femur length, with epiphyses. In his study, Maresh (1955) gives both diaphyseal length and total femur length with epiphyses for some age groups. These data suggest that diaphyseal length makes up about 91% of total femur length in the immature individual. For some of the Wharram Percy femora, proximal and distal epiphyses were recovered and were sufficiently well preserved that they could be accurately refitted to their diaphyses. Although numbers were small, measuring these femurs consistently indicated that the diaphyseal length formed 91% of total length, in agreement with Maresh. For Wharram Percy 10 year olds, mean femur diaphyseal length is 28.5 cm, so the estimated mean total length is 31.3 cm. Using Feldesman's femur: stature ratio, the estimated mean stature of 10 year olds at Wharram Percy is 117 cm. Working through the different age groups between 8 and 17 years old in this fashion, using Feldesman's femur: stature ratios, enabled a growth profile using estimated stature rather than femur length to be constructed. For comparison I used some data on recent and 19th century British children. The former are from Tanner *et al.* (1966). These are broadly indicative of modern subjects, although there are indications that height for age in the age range of interest has increased by about 1 cm since Tanner and co-workers' data were gathered (Freeman *et al.* 1995). The 19th century data are from the survey of British children conducted in 1833, by Samual Stanway and J.W. Cowell, who measured children in Manchester and Stockport. Their data, for children employed in factories, are reproduced by Tanner (1981). The results are shown in Fig. 12.3.

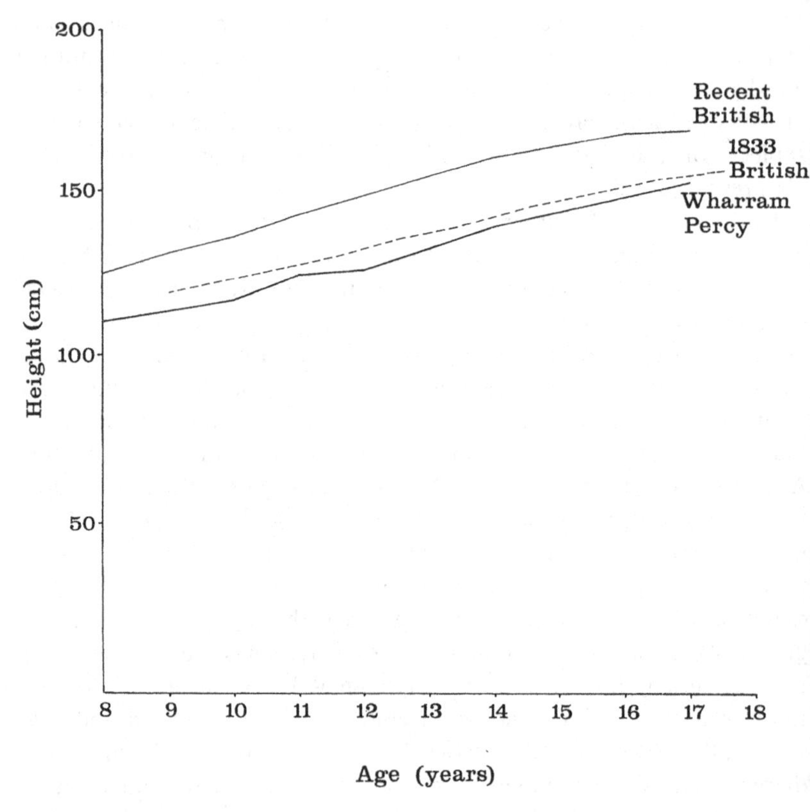

Fig. 12.3. Estimated stature vs. dental age for the Wharram Percy children. Stature figures taken from recent and 19th century living subjects of known age are included for comparative purposes (for data sources, see the text). For the 19th century figures, adjustment has been made to take account of footwear (Tanner 1981: note 7.5).

In addition to the caveats discussed above, which apply generally for comparisons between archaeological growth data and data taken from living subjects, there is also the consideration in the present instance that we are comparing children whose heights were measured exactly with those whose statures have been estimated from long bone lengths. An assumption is that femur: stature ratio was similar at Wharram Percy to that in the reference groups upon which Feldesman based his analysis. Although the potential problems should be kept in mind, some tentative comparisons between the Wharram Percy and the other data can be drawn. As with the comparison of femur lengths, the height deficit compared with recent children is very striking. For example, at age 10 the estimated mean stature of the Wharram Percy children is 20 cm less than the modern figure of 137 cm. To express it another way, the data suggest that the height of a typical 14 year old at Wharram Percy would only have been about the same as that of a modern 10 year old. This is in agreement with the comparison of femur lengths with the Maresh data above, and indirectly lends support to the idea that our construction of stature from femur lengths at Wharram Percy is reliable.

The difference between the Wharram Percy and the 19th century data in Fig. 12.3 is much less pronounced. Although the growth profiles lie quite close together, it is perhaps worth noting that in every age class, the mean height of Wharram Percy children is lower than that for the 1833 subjects.

Growth in children is strongly dependent upon disease and nutrition (Eveleth and Tanner 1990), so height is often used as a proxy for health when making comparisons at the population level (Tanner 1989; 163). In this light, the present results suggest, not surprisingly, that the Mediaeval children from Wharram Percy suffered a greater disease load and had poorer nutrition than recent subjects. That the height in the Wharram Percy children is not greater than that of the 19th century subjects suggests that child health among the rural Mediaeval population at Wharram Percy was no better than among the urban poor of the Industrial Revolution; given that for every age class, mean height for the Wharram Percy children was somewhat less, it may even have been worse. That the Wharram children are perhaps shorter than those in the 1833 sample might also be taken to indicate that the secular trend for increase in height for age in British children originated well before the first child height surveys were undertaken in the 19th century. However, it would be premature to arrive at firm conclusions; further studies on well-dated post-Mediaeval juvenile skeletal remains would be needed to substantiate this.

In instances of long-term suboptimal nutrition, growth is retarded, so that children are short for their ages. However, under such circumstances the growth period is extended so that final adult stature may be unaffected, or at least affected rather less than height for age in children, although it will be reached at a later age (Eveleth and Tanner 1990). As might be anticipated from this, concomitant with the secular increase in height for age in children there has also been a trend toward earlier completion of growth. In modern populations in the developed world, growth in height is virtually complete by about 18 years (Tanner 1989: Fig. 4). Records show that in the past, final adult stature was not attained until somewhat later. For example, Morant (1950) studied growth data gathered in 19th and early 20th century England. Growth was prolonged, particularly during the earlier periods for which data were available, and among the lower social classes. For example in the period 1857–1883, final adult stature was not attained until about 29 years of age in the lowest social class. The age of attainment was lower among the more privileged, but still later than in recent populations – for example, it was about 21 years in the 'professional' classes.

Aspects of the Wharram Percy data provide evidence that growth may have been prolonged in this group compared with modern British subjects. The estimated mean height for 10 year olds at Wharram Percy is about 117 cm. Mean adult stature, derived from long bone lengths using the regression equations of Trotter and Gleser (1952, 1958) is 169 cm for males and 158 cm for females, giving a mid-sex mean of 163.5 cm. Thus height at 10 years is about 71% of final adult height. This compares with a figure of 81% for recent 10 year olds (Tanner *et al.* 1966). Consistent with this, the difference in stature between Mediaeval and recent adults is less than that during childhood. For example, at 10 years the estimated height of children from Wharram Percy was about 85% of that reported by Tanner *et al.* (1966) for recent 10 year olds, but the Mediaeval adult height was about 97% of that quoted by Tanner *et al.* (1966). These observations are consistent with the hypothesis of prolonged growth in the Wharram Percy group.

Appositional bone growth

At Wharram Percy, appositional bone growth was monitored at the femur midshaft using radiogrammetry. Total bone width (T) and the medullary width (M) were measured from anterio-posterior radiographs (for technical details see Mays 1995). From these, cortical thickness ($T - M$) and the cortical index ($100 \times (T - M)/T$) were calculated. Cortical index is cortical thickness expressed as a percentage of bone width; in other words it is a measure of cortical thickness standardised for bone size. The two-dimensional analogues of cortical thickness and cortical index, cortical area and percentage cortical area, were also calculated (from T and M, using the formulae of Garn (1970: 12)). Growth profiles were examined by plotting results against estimated age. The patterning in cortical area and percentage cortical area with respect to age was found to be similar to that in their linear analogues, so these data are not reproduced here.

Virtama and Helelä (1969) have published data for cortical thickness and cortical index at the femur midshaft for a modern northern European population (Finns). Their data are used for comparison with the Wharram Percy results. Owing to possible differences in the degree of radiographic enlargement, cortical thickness data derived from archaeological and living subjects are not, strictly speaking, precisely comparable, although

Table 12.3. *Cortical thickness for femurs from juvenile skeletons from Wharram Percy, Yorkshire*

Age	*N*	Mean	SD
Perinatal	36	3.6	0.7
1	9	3.9	0.8
2	9	4.5	0.9
3	11	5.2	1.2
4	11	5.1	0.8
5	5	5.1	0.5
6	21	5.8	1.0
7	8	7.3	1.0
8	4	8.4	1.3
9	6	7.9	1.5
10	7	8.9	1.6
11	6	9.4	0.8
12	2	9.3	1.4
14	5	10.5	1.5
17	7	12.0	2.0

See notes for Table 12.2.

comparisons of general patterns should still be broadly valid. Effects of differential radiographic enlargement on cortical index are likely to be negligible. The results for cortical thickness and cortical index at Wharram Percy are summarised in Tables 12.3 and 12.4. These data are plotted in Fig. 12.4 and 12.5. The data for bone width and medullary width from Wharram Percy are also shown (Fig. 12.6).

As was found for bone length, the Wharram Percy children are markedly deficient in cortical thickness and cortical index compared with recent subjects, although both ancient and modern groups show the same general pattern of a slow increase in both parameters with age. For cortical thickness, the Wharram Percy children lag behind their modern counterparts to a greater degree than was the case for bone lengths. The difference is about 2 years, even at age 3, and this rises to about 7 years by age 17. The rise in cortical index with age at Wharram Percy from about 1 year onward is rather less consistent, although still statistically significant (Mays 1995). For cortical index there is no overlap in the mean values between ancient and modern children – between the ages of 1 and 17 the highest mean value for any of the Wharram Percy age groups is lower than the lowest mean for any age group in the modern data. As for diaphyseal length, the cortical thickness and cortical index results are consistent with idea that the Wharram Percy people suffered poorer health during the growth period

Table 12.4. *Cortical index for femurs from juvenile skeletons from Wharram Percy, Yorkshire*

Age	*N*	Mean	SD
Perinatal	36	58.0	13.7
1	9	37.5	7.0
2	9	38.1	7.1
3	11	40.4	8.3
4	11	38.6	3.4
5	5	35.3	1.7
6	21	41.1	7.4
7	8	46.0	6.9
8	4	50.6	7.7
9	6	47.1	4.8
10	7	48.4	7.0
11	6	50.8	3.5
12	2	47.7	6.0
14	5	48.5	4.8
17	7	50.6	6.0

See notes for Table 12.2.

than recent children. That the deficiency in cortical bone is greater than that in diaphyseal length may indicate that cortical bone apposition was more affected by the stresses experienced during the growth period than was longitudinal bone growth, and hence that the former may be a more sensitive stress indicator in this group.

There is already a marked difference in cortical index between the Mediaeval and the modern group at 1 year of age. It is unclear whether this arose during the first year of postnatal life or whether it was present from birth, as Virtama and Helelä did not perform measurements on individuals under 12 months old. There is a very marked drop in cortical index, although not in cortical thickness, during the first year of life of the Wharram Percy infants. Plotting cortical index vs. gestational age, estimated from long bone lengths (Scheuer *et al.* 1980), for the perinatal material from Wharram Percy (Fig. 12.7) indicates that this fall in cortical index during the early postnatal period is a continuation of a process that was occurring in later foetal life. Indeed, Fig. 12.7 suggests that most of the reduction from the mean perinatal value to that at 1 year occurs during later foetal life and the first few weeks after birth. Garn noted that cortical thickness was static or even fell slightly during the first year of life in living subjects, due to the fact that at this age the rate of subperiosteal deposition tends to fall faster than the rate of endosteal resorbtion (Garn 1970: 46–47).

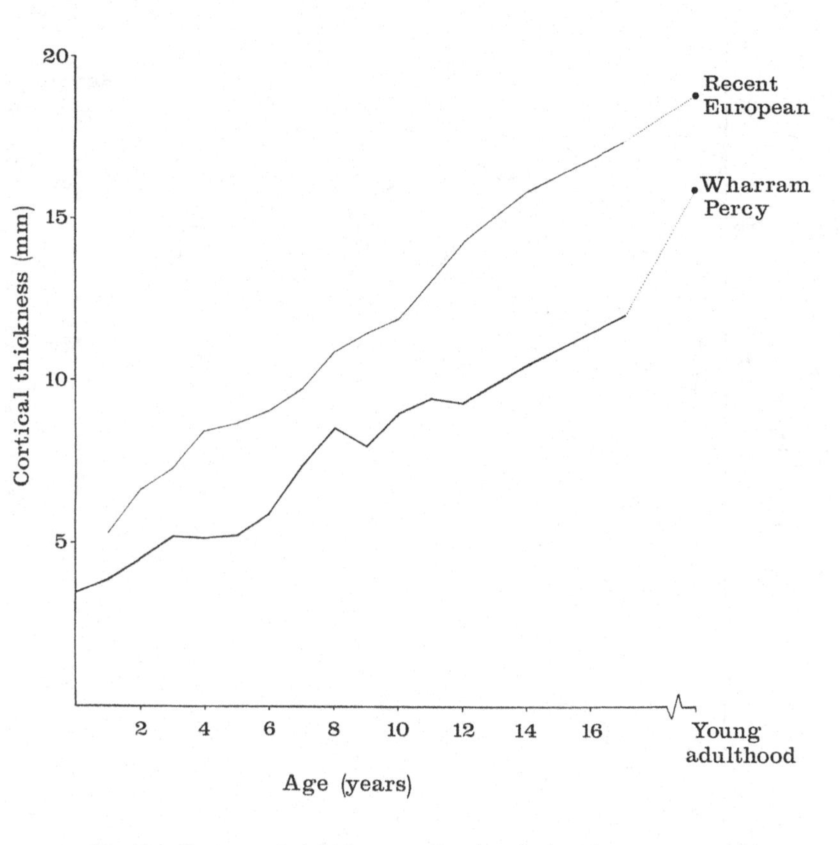

Fig. 12.4. Femur cortical thickness vs. dental age in the Wharram Percy children. Recent data (Virtama and Helelä 1969) have been included for comparison. In Figs. 12.4–12.6, 'young adulthood' denotes about 18–30 years of age.

Looking at archaeological material from Sudan, Hummert (1983) found :hat there was a fall in percentage cortical area in the first year of life. Cook 1979) observed a fall in cortical index in the year following birth in two orehistoric North American assemblages. It would seem likely that the drop in cortical index observed during the first year at Wharram Percy is, at least in part, a general feature of human growth, rather than of itself ndicating any special problems among the Wharram Percy infants.

Huss-Ashmore *et al.* (1982) also investigated cortical bone growth using orehistoric remains from Sudan. They found that cortical thickness failed :o increase during childhood, suggesting severe nutritional problems for :hat group. Although the cortical thickness for the Wharram bones is much ower than for recent children, there is, unlike the Sudanese example, an ncrease in thickness with age. It is possible that this reflects lesser stress

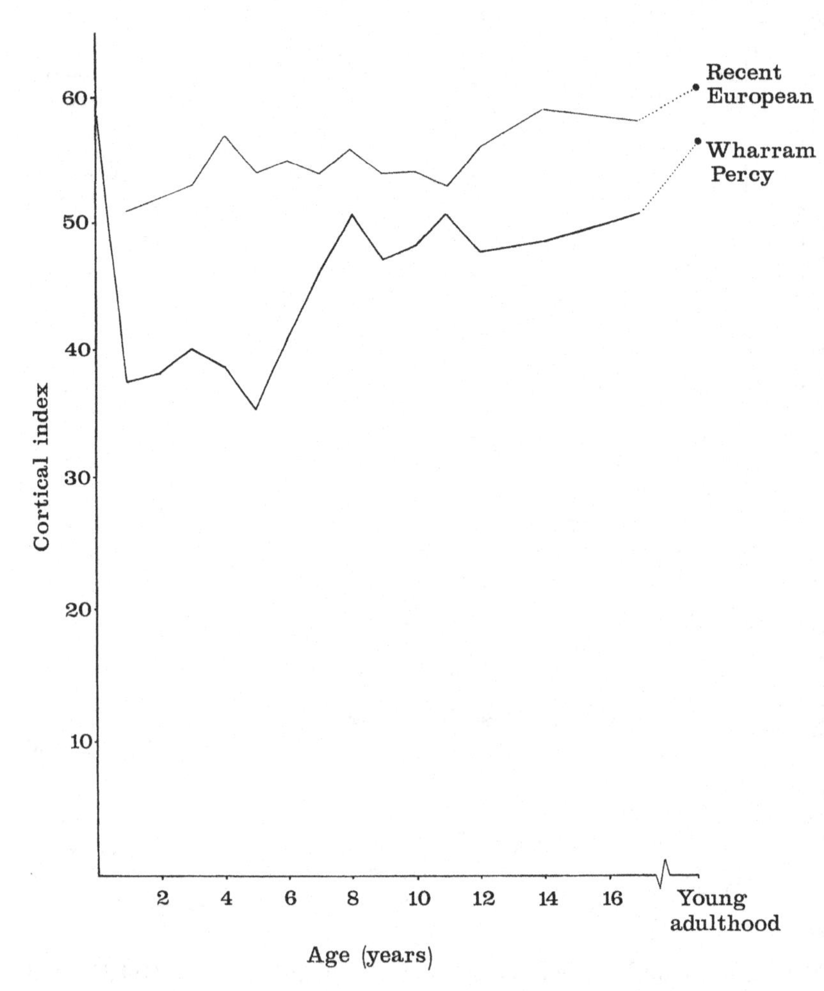

Fig. 12.5. Femur cortical index vs. dental age in the Wharram Percy children. Recent data (Virtama and Helelä 1969) have been included for comparison.

among the Wharram people, due to their less harsh environment compared with the prehistoric Sudanese.

Psychologists have long known that at about the age of 5–7 years a shift in cognitive abilities occurs, and at this time a child starts to become more adventurous and independent (Bogin 1997). This pattern is cross-cultural (Rogoff *et al.* 1975). At Wharram Percy, cortical index, and to a lesser extent cortical thickness, appears to show a sharper increase in the age range 5–8 years than in younger years. It is tempting to explain this is terms of increasing physical activity and muscular strength sufficient to stimulate

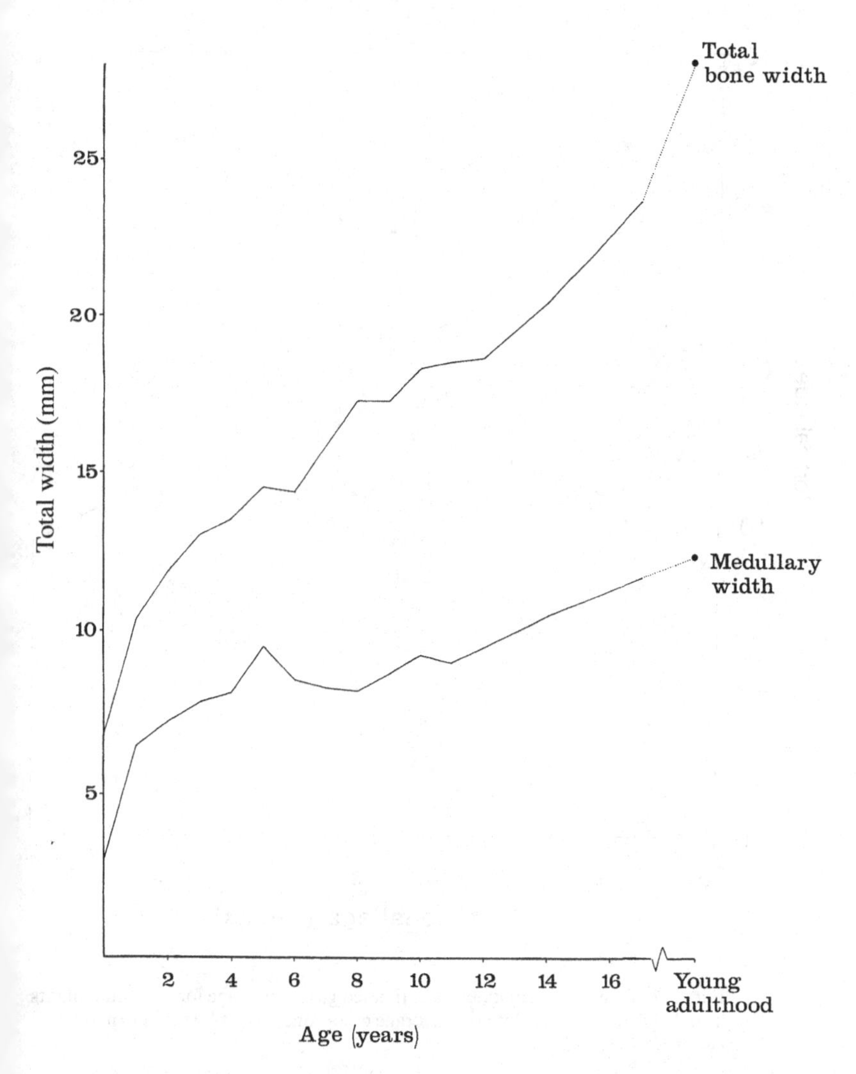

Fig. 12.6. Total bone width and medullary width vs. dental age for the Wharram Percy children.

increased appositional bone growth at this age. However, the evidence is insufficient to allow this to be offered as anything more than a tentative suggestion.

Frisancho *et al.* (1970), studying growth in living subjects, found that medullary area decreased from age 12 in girls and age 16 in boys, due to apposition of bone at the endosteal surface. There was no evidence for this in the Wharram Percy juveniles. Here medullary cavity width continued to

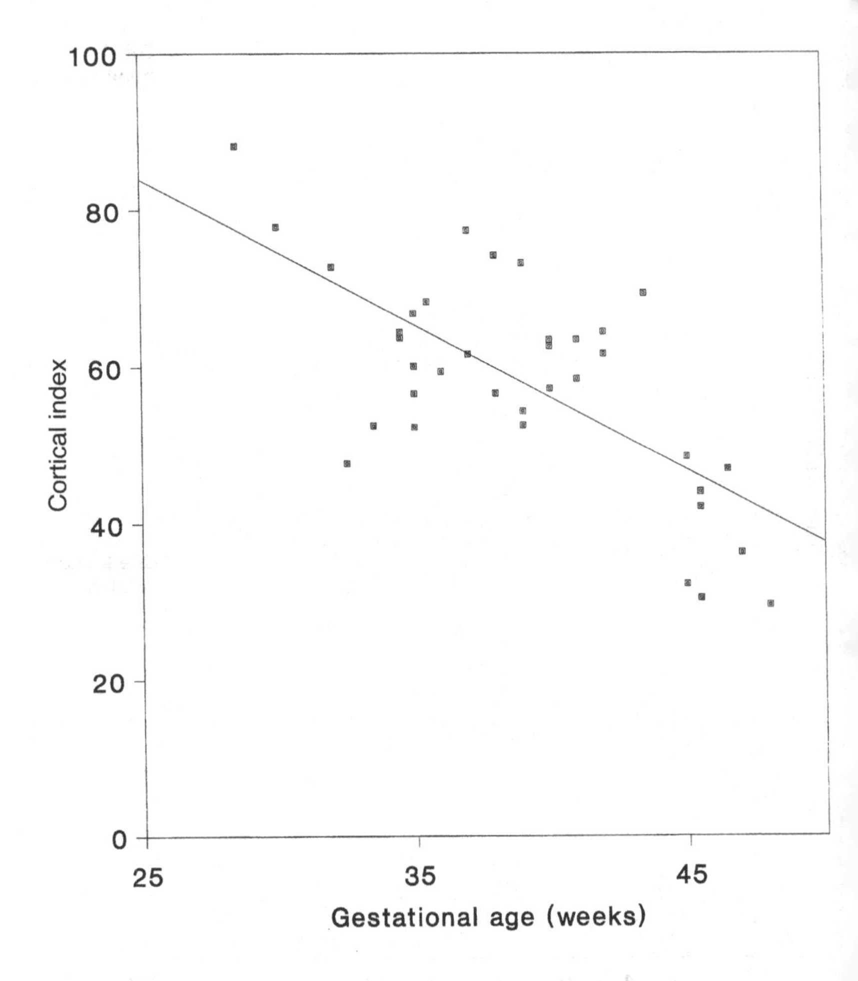

Fig. 12.7. Femur cortical index vs. estimated gestational age for perinatal infants from Wharram Percy. Note that average gestational age at birth is normally about 39–41 weeks.

increase throughout the growth period (Fig. 12.6). In contrast to recent populations, there is no good evidence of endosteal apposition at any age at Wharram Percy, although due to differences in the timing of the onset of endosteal apposition in males and females, it may be more difficult to detect this effect in mixed sex samples.

In both the Mediaeval and the recent data, cortical thickness and cortical index each continue to increase after longitudinal bone growth has ceased, peak values not being attained until early adult life. For both cortical thickness and cortical index, the difference between adult values in the Mediaeval and modern samples is less than that during childhood (Figs. 12.4 and 12.5). Just as for final adult height, this finding may be viewed as consistent with the inference that in the Mediaeval group peak cortical thickness was attained somewhat later than it is in modern subjects.

Growth interruption and recovery

Harris lines record interruptions in longitudinal bone growth, which may occur due to non-specific stress (*sensu* Cox *et al.* 1983) in the immature individual. Complete arrest of longitudinal growth, followed by its resumption is needed for a Harris line to form, mere slowing of growth is not sufficient. Harris lines are sometimes thought of as indicators of acute rather than chronic stress, but this is not so. For a line to form, complete recovery from an episode of disease or poor nutrition is not needed, merely sufficient resources to permit resumption of endochondral bone growth, even at a slowed rate. In fact, chronically ill or malnourished children are likely to show many growth arrest lines (Park 1964: 637).

In a study of Harris lines in the femurs of individuals from Wharram Percy (Mays 1995), it was found that femur length in children who showed Harris lines was similar to that in those who did not show lines. However, those with Harris lines were deficient in cortical index. A similar study on juveniles from Romano-British Poundbury (Mays 1985) revealed a rather different pattern: those with Harris lines were not deficient in either bone length or cortical bone.

Following a period of growth arrest, growth will, if sufficient resources are available, restart at an accelerated rate, quickly returning the individual to his or her original growth trajectory (Prader *et al.* 1963; Tanner 1986). Alternatively, growth may restart at a normal or slowed rate if insufficient resources are available following the stress episode that caused the growth arrest to sustain a spurt of catch-up growth. That the Wharram Percy children who showed Harris lines did not show reduced femur diaphyseal lengths indicates that, following the stress episodes that led to line formation, recovery was sufficient to permit a spell of catch-up growth. This was likewise the pattern at Poundbury, and these individuals also seem to have caught up any deficiency of cortical bone which may have resulted. However, at Wharram Percy, those with Harris lines showed reduced cortical bone compared with those without, indicating that catch-up growth had not occurred in cortical bone apposition. It would appear that, at Wharram Percy, in those subject to stress during the growing years, longitudinal bone growth was maintained at the expense of cortical bone apposition.

This supports the suggestion, above, that cortical apposition may be a more sensitive indicator of stress in childhood at Wharram Percy than long bone length.

Consistent with the present results, Huss-Ashmore *et al.* (1982), investigating Mediaeval Sudanese skeletal remains, felt that longitudinal bone growth was being maintained at the expense of cortical thickness, as did Hummert (1983) also studying material from the Sudan. Garn *et al.*'s (1964) results on living subjects are in accord with this. They found that, during recovery in Guatemalan children hospitalised for protein-calorie malnutrition, catch-up in longitudinal bone growth occurred at the expense of cortical bone deposition. However, others have found a different pattern. Barr *et al.* (1972), studying infant malnutrition due to coeliac disease, found that during recovery cortical thickness reached normal levels sooner than did height. Briers *et al.* (1975) studied Ugandan children, with and without histories of malnutrition. Those who had been malnourished were short for their ages but were not deficient in cortical bone. Why the different studies should have produced these apparently inconsistent findings is unclear.

Unlike at Poundbury, it was clear that among those from Wharram Percy suffering arrest in endochondral bone growth, as indicated by Harris lines, insufficient resources were available to sustain catch-up growth in cortical bone, implying that the stress episodes which led to the arrest of endochondral bone growth were more severe or more chronic in nature in the Mediaeval than in the Romano-British sample. Alternatively, the insults that caused the Harris lines may have been similar in the two groups, but their living conditions may have been such that the Wharram children would have had feebler powers of recovery. In either case, it would appear to imply that chronic poor health or nutrition was more frequent at Wharram Percy than at Poundbury.

The soils at Wharram Percy are thin and prone to nutrient exhaustion (Hayfield 1988: 25–26). Wharram Percy's more northerly location and greater elevation (it lies some 150 m above sea level) mean that it has a harsher climate than the part of southern England in which Poundbury is situated. Crop failures, with associated food shortages, might be expected to have been more common at Wharram Percy. Mediaeval peasants occupied a lower stratum of what was a highly stratified society, and hence would be expected to be vulnerable to dietary stress. Indeed, documentary sources indicate that dietary shortages were common among them (e.g. Gies and Gies 1990: 96–98). By contrast, the Poundbury people were a sophisticated, heavily Romanised group (Molleson 1992), who were doubtless prosperous compared with the inhabitants of Wharram Percy. Therefore the differences in stress experience between the two groups, indicated by the bone data, are understandable in terms of the harsher environment and lower socioeconomic status of the Wharram Percy people.

Conclusions

It is to be hoped that appositional bone growth in earlier populations will start to receive more attention among osteoarchaeologists. For example, it would be interesting to determine, by examination of archaeological assemblages, the nature of the secular trend for increasing cortical thickness for age, and whether it parallels that in height for age. The value of measures of cortical bone as stress indicators in immature individuals needs to be more widely appreciated. Indeed, at Wharram Percy it appeared that appositional long bone growth was a more sensitive indicator of stress than was longitudinal growth, although it is unclear whether this is generally so.

In the present work it was noted that archaeological assemblages from Roman times to the early modern era show very similar patterns of longitudinal bone growth. This would at first sight appear a somewhat surprising result, given that the samples span more than 1500 years, and range from Romano-British townsfolk, to Mediaeval peasants, to post-Mediaeval Scottish islanders. However, this should not be taken to indicate that all necessarily had similar childhood nutrition and disease experiences. This is illustrated by the comparison between Wharram Percy and Poundbury – although longitudinal growth profiles for these sites were similar, differences in growth experience emerged once appositional growth was considered alongside longitudinal growth. Perhaps, in future, measurements of cortical bone will start to be used routinely in conjunction with longitudinal bone growth when skeletal growth in earlier populations is investigated.

The potential contribution of archaeological evidence to the study of the history of human growth has generally been rather overlooked by researchers in this area. There are parallels here with the role of palaeopathology in the history of medicine. For a long time, the value of studying ancient bones was not recognised by medical historians, but this situation has changed recently, as the potential of this sort of evidence has been more widely appreciated. It is to be hoped that, in analogous fashion, archaeological studies of growth will begin to impact upon those who have hitherto relied solely on documentary evidence. Reliable documentary sources on growth only cover the very recent past; the great majority of the history of human growth will be written from archaeological evidence.

Acknowledgements

Thanks are due to Sebastian Payne for his comments on an earlier draft of this chapter.

References

Adams P and Berridge FR (1969) Effects of Kwashiorkor on cortical and trabecular bone. *Archives of Disease in Childhood* **44**, 705–709.

Barr DGD, Shmerling DH and Prader A (1972) Catch-up growth in malnutrition, studied in celiac disease after institution of a gluten-free diet. *Paediatric Research* **6**, 521–527.

Beresford M and Hurst J (1990) *Wharram Percy Deserted Mediaeval Village*. London: Batsford/English Heritage.

Bogin B (1997) Evolutionary hypotheses for human childhood. *Yearbook of Physical Anthropology* **40**, 63–89.

Briers PJ, Hoorweg J and Stanfield JP (1975) The long-term effects of protein-energy malnutrition in early childhood on bone age, bone cortical thickness and height. *Acta Paediatrica Scandinavica* **64**, 853–858.

Cook DC (1979) Subsistence base and health in prehistoric Illinois Valley: evidence from the human skeleton. *Medical Anthropology* **4**, 109–124.

Cox T, Cox S and Thirlaway M (1983) The psychological and physiological response to stress. In *Physiological Correlates of Human Behaviour*, vol. 1, ed. A. Gale and J.A. Edwards, pp. 255–276. London: Academic Press.

Eveleth PB and Tanner JM (1990) *Worldwide Variation in Human Growth*, 2nd edn. Cambridge: Cambridge University Press.

Feldesman MR (1992) Femur/stature ratio and estimates of stature in children. *American Journal of Physical Anthropology* **87**, 447–459.

Freeman JV, Cole TJ, Chinn S, Jones PRM, White EM and Preece MA (1995) Cross-sectional stature and weight reference curves for the UK, 1990. *Archives of Disease in Childhood*, **73**, 17–24.

Frisancho AR, Garn SM and Ascoli W (1970) Subperiosteal and endosteal bone apposition during adolescence. *Human Biology* **42**, 639–664.

Garn SM (1970) *The Earlier Gain and Later Loss of Cortical Bone* Springfield, IL: Charles C. Thomas.

Garn SM, Rohmann CG, Viteri F and Guzman MA (1964) Compact bone density in protein-calorie malnutrition. *Science* **144**, 1444–1445.

Gies F and Gies J (1990) *Life in a Mediaeval Village*. London: Harper and Row.

Hayfield C (1988) Cowlam Deserted Village: a case study of post-Mediaeval village desertion. *Post-Mediaeval Archaeology* **22**, 21–109.

Himes JH, Matorell R, Habicht J-P, Yarbrough C, Malina RM and Klein RE (1975) Patterns of cortical bone growth in moderately malnourished preschool children. *Human Biology* **47**, 337–350.

Hummert JR (1983) Cortical bone growth and dietary stress among subadults from

Nubia's Batn El Hajar. *American Journal of Physical Anthropology* **62**, 167–176.

Huss-Ashmore R, Goodman AH and Armelagos GJ (1982) Nutritional inference from palaeopathology. *Advances in Archaeological Method and Theory* **5**, 395–474.

Johnston FE (1962) Growth of the long bones of infants and young children at Indian Knoll. *American Journal of Physical Anthropology* **20**, 249–254.

Lazenby RA (1990) Continuing periosteal apposition. I. Documentation, hypotheses and interpretation. *American Journal of Physical Anthropology* **82**, 451–472.

Lewis AB and Garn SM (1960) The relationship between tooth formation and other maturation factors. *Angle Orthodontist* **30**, 70–77.

Liversidge HM (1994) Accuracy of age estimation from developing teeth of a population of known age (0 to 5.4 years). *International Journal of Osteoarchaeology* **4**, 37–45.

Maresh MM (1955) Linear growth of the long bones of the extremities from infancy through adolescence. *American Journal of Diseases in Children* **89**, 725–742.

Mays SA (1985) The relationship between Harris line formation and bone growth and development. *Journal of Archaeological Science* **12**, 207–220.

Mays S (1995) The relationship between Harris lines and other aspects of skeletal development in adults and juveniles. *Journal of Archaeological Science* **22**, 511–520.

Miles AEW and Bulman JS (1994) Growth curves of immature bones from a Scottish island population of 16th to mid-19th century: limb bone diaphyses and some bones of the hand and foot. *International Journal of Osteoarchaeology* **4**, 121–136.

Molleson TI (1992) The anthropological evidence for change through Romanisation of the Poundbury population. *Anthropologischer Anzeiger* **50**, 179–189.

Morant GM (1950) Secular changes in the heights of British people. *Proceedings of the Royal Society of London* **137B**, 443–452.

Park EA (1964) Imprinting of nutritional disturbances on the growing bone. *Paediatrics* **29**, *Supplement*, 815–862.

Prader A, Tanner JM and von Harnack GA (1963) Catch-up growth following illness or starvation. *Journal of Paediatrics* **62**, 646–659.

Rogoff B, Sellers MJ, Pirotta S, Fox N and White SH (1975) Age of assignment of roles and responsibilities in children: a cross cultural survey. *Human Development* **18**, 353–369.

Saunders SR and Hoppa RD (1993) Growth deficit in survivors and non-survivors: biological mortality bias in subadult skeletal samples. *Yearbook of Physical Anthropology* **36**, 127–151.

Saunders SR, Hoppa RD and Southern R (1993) Diaphyseal growth in a nineteenth century skeletal sample of subadults from St Thomas' Church, Belleville, Ontario. *International Journal of Osteoarchaeology* **3**, 265–281.

Scheuer JL, Musgrave JH and Evans SP (1980) Estimation of late foetal and perinatal age from limb bone lengths by linear and logarithmic regression. *Annals of Human Biology* **7**, 257–265.

Schour I and Massler M (1941) The development of the human dentition. *Journal of the American Dental Association* **28**, 1153–1160.

Smith BH (1991) Standards of human tooth formation and dental age assessment. In *Advances in Dental Anthropology*, ed. M.A. Kelley and C.S. Larsen, pp. 143–168. Chichester: Wiley-Liss.

Stloukal M and Hanáková H (1978) Die Länge der Längsknochen Altslawischer Bevölkerungen – Unter besonderer Burücksichtigung von Wachstumfragen. *Homo* **29**, 53–69.

Tanner JM (1981) *A History of the Study of Human Growth.* Cambridge: Cambridge University Press.

Tanner JM (1986) Growth as a target-seeking function. In *Human Growth. A Comprehensive Treatise* vol. 1, *Developmental Biology and Prenatal Growth*, ed. F. Falkner and J.M. Tanner, pp. 167–179. London: Plenum Press.

Tanner JM (1989) *Foetus Into Man*, 2nd edn. Ware: Castlemead.

Tanner JM, Whitehouse RH and Takaishi M (1966) Standards from birth to maturity for height, weight, height velocity and weight velocity; British children. *Archives of Disease in Childhood* **41**, 454–471; 613–635.

Trotter M and Gleser GC (1952) Estimation of stature from long-bones of American Whites and Negroes. *American Journal of Physical Anthropology* **10**, 463–514.

Trotter M and Gleser GC (1958) A re-evaluation of estimation of stature based on measurements of stature taken during life and long-bones after death. *American Journal of Physical Anthropology* **16**, 79–123.

Virtama P and Helelä T (1969) Radiographic measurements of cortical bone. Variations in a normal population between 1 and 90 years of age. *Acta Radiologica, Supplementum* **293**.

Index

Italic page numbers denote illustrations

Adaptive immunity 161, 166, 170, 173
Adolescence 54, 91–3, 99, 132, 290
Age estimation
  dental development 1, 3–5, 10, 13–15, 65, 89, 95, 199, 267, 293
  dental emergence 3, 67, 267–70
  dental microstructure 3, 18–19, 120–5, 252
  epiphyseal union 89
  pubic symphysis 89, 267
Agriculturists 162, 211, 245, 248, 258
Alaskans (Inupiaq Eskimos) 128-57
Aleutians 128–57
Allometry 43–7, 57, 128, 132–3, 137, 139, 148–56
  allometric equation 33
Altriciality 90, 93
Altun Ha 212, 224, 227, 229
Ancient DNA 3, 190–1, 198, 201
Archaic, Late 163, 174
Arikara 164, 175, 200, 236

Belleville, Ontario 164, 192, 199, 296
Bias, mortality 14, 70, 191, 193–4, 293
Birth weight 165, 170, 211, 271
Body mass 33–58, 88–9
Body size 33, 43, 51, 54–7, 66, 88, 91, 103
Breastfeeding 168, 170–3, 193, 201
Breastmilk 165, 168, 170–3, 191, 195

Caries 70, 162, 175, 242
Cartesian transformation grid 144, 150, *151*
Cause of death 189, 197–9, 202
Childhood period 8, 89–93, 98, 102
Chronological age 18, 65, 67, 71, 84, 102, 210, 213, 267–8
Circadian rhythm 113–14
Cranial capacity 17, 55–6, 92–3
Cranial morphology 17, 128, 132
Cribra orbitalia 162, 175

Demographic transition 183
Dental development *see* Age estimation
Dentine 10, 18, 194
  calcospheritic lines 114, 118, 120
  circumpulpal dentine 111, 118
  odontoblasts 111–12, 246
  peritubular dentine 112
  predentine 111–12, 117–19
Dickson Mounds 8, 163, 174, 216
Diet 54, 99, 192–3, 196, 201, 242–3, 247, 259, 308
  and growth 162–3
Disease 186, 197–9, 243, 258–9, 269–70, 299, 307–8
  diarrhoea 165, 167, 169, 171, *172*, 173, 175–6, 258
  and growth164
  gastrointestinal 195, 197–9
  and nutrition 165–8, *167*, 296
  respiratory infection 169, 171, *173*, 175, 195, 197–8

Enamel 9, 18, 186, 193–4, 242
  ameloblasts 210, 220, 243, 246, 252
  buried (cuspal) enamel 220–4
  cross-striations *11*
  crown formation 123, 215, 220, 230, *231*, 232–6, 242, 246, 251–5
  perikymata 120, 212, 242, 254–5
  striae of Retzius 120–1, 212, 220, 235–6, 241, 243, 245, 252–5
  Tomes process 118, 120, 257
  Wilson bands 243, *253*, *257*
Encephalisation 71
Endochondral ossification 53, 129, 290, 307–8

Endosteal apposition 306
Endosteal resorption 290, 303
Ensay, Isle of 164, *294*
Epiphyseal growth plate 52, 57, 290
Euclidean distance matrix analysis 140, 142

Fat, increase 35; *see also* Body Mass

Gambia 169–70, 173
Geometric morphometrics 128, 139, 141, 156
Gibbs Sampler 272, 274, 277, 279, 281, *281*, 285
Gompertz equation 48, 68, *79*
Growth
  brain 9, 51, 54–5, 89–90, 94
  dentition *see* Teeth
  hypertrophic 55
Growth hormone 52–4, 57, 170
Growth interruption 307–8 *see also* Hypoplasia, Harris lines

Hamann–Todd 212, 221, 223, 227, 231–2, 234
Harris lines 18, 176, 307–8
Hazards analysis 276
Height *see* Stature
Heterochrony 18
Hominids
  Australopithecines 88–9, 91
  Homo erectus 12, 17, 32, 56, 88, 90, 92–3, 98, 102–3
    KNM-WT 15000 12, 32, 92–93
  Homo habilis 84, 88, 92, 103
  Homo sapiens 2, 34, 93
  Neandertal 88–103
  Paranthropus robustus 123
Hominoids
  Gorilla 32–58, 66–85
  Pan 32–58, 66–85
  Proconsul 18, 121–3
Horticulturalists 163, 193
Hyaline cartilage 129
Hunter–gatherers 170
  Late Archaic 163
Hypermorphosis 45, 150, 152
Hypoplasia 17–18, 175, 192, 210–36, 241–59

Immunocompetence 165–6
Infancy period 54, 56, 79, 90–3, 102, 171, 183–202, 290
Infection, *see* Disease
Inupiaq Eskimos 128–57

Juvenile period 56, 102

La Florida 242–3, *250*
Late Woodland 162–3, 174
Libben 163
Life history events 9, 32, 51, 54–6, 71, 85, 91, 94

Macaques 115
Malnutrition 161–77, 184, 270–1, 296, 308
Maya 212, 221, 224
Meheran Growth and Development Study 271
Menarche, age of onset 69, 91
Miocene 121
Mississippian 162–3, 174
Monte Carlo estimation 266, 272, 274, *275*, 276, 282
Mortality 12, 17, 168, 174, 183, 241, 266, 276, 279, 293
  infant 16, 183–90, 195, 197, 199–202
  neonatal 187–9, 195
  perinatal 186, 188
  postneonatal 187–8, 195
  seasonality 195–7
Muscle mass, increase *see* Body mass

Ontogeny 1, 9, 11, 19, 33–5, 43–7, 51, 53–8, 88, 129, 156
Ontogenetic scaling 43

Pleiotropic effects 55–7
Porotic hyperostosis 175
Posterior density of age 274–81
Poundbury *294*, *295*, 307–9
Principal components analysis (PCA) 142–3, *146*, 147, *148*, 152, *153*, *154*
Probit analysis 266, 272–3
Procrustes analysis 141–3, 147
Progenesis 45
Puberty 91, 132, 190

Reduced Major Axis (RMA) analyses 44–45, *44*

Santa Maria ossuary *244*, 245–6, *247*
Secondary sex ratio (SSR) 189–91
Sex, determination of 2–3, 189–91
Sexual dimorphism 3, 88, 132, 190, 270
Sexual maturity 54, 91
Siler model of mortality 266
Spheno-occipital synchondrosis 130–1, 138
Spitalfields 5, 66, 70, 296
Stable isotope analysis 192–5
St Bride's Church 4, 66
St Thomas' Church 4–5, 175, 192, 199, *294*, *295*, 296
Stature
  as a measure of somatic maturity 32
  as an indicator of health changes 242
  differences, influence on SGP 7, 8
  estimation using femur: stature ratio 95, 96, 296–300, *298*
  of *Homo erectus* 92
  modelling growth of 281–5, *282*, *284*, *285*
  of Neandertals 94–100
  secular increase in 296
Swartkrans 123

Teeth
  CEJ (cement–enamel junction) 121, *215*, 215, 216, *219*, 219, 222, 223
  crown height *215*, 215, 216, 218, *219*, 223, 226–9, 232, 235, 245–6, 248, 250, 252, *253*
  deciduous dentition 3, 67–8, 70, 122, 134, 137, 219, 235, 266–71
  DEJ/EDJ (enamel dentine junction) 243, 245–7, *247*, 253, 255
  extension rate 121, 122
  neonatal line 186, 194, 235
  permanent dentition 3–5, 32, 43, 48–50, 67–71, 77–8, 83, 123, 212, 220, 230, 234, 269, 270
  root length 121
  tooth size 83, 212
  see also Enamel, Dentine, Age estimation
Trace element analysis 192–5
Turkana children, growth in 99

Weaning 17, 162, 177, 201, 234–5, 258
  and disease 174, 189,
  and growth 165, 170,
  timing of 90, 169, 191–3, 212,

Weight *see* Body mass
Wharram Percy 176, 291–309